Lecture Notes in Mathematics

Edited by A. Dold, Heidelberg and B. Eckmann, Zürich

391

John W. Gray

University of Illinois at Urbana-Champaign, Urbana, Il/USA

Formal Category Theory: Adjointness for 2-Categories

Springer-Verlag

Berlin · Heidelberg · New York 1974

AMS Subject Classifications (1970): Primary: 18D05, 18D25
Secondary: 18A25, 18A40

ISBN 3-540-06830-9 Springer-Verlag Berlin · Heidelberg · New York
ISBN 0-387-06830-9 Springer-Verlag New York · Heidelberg · Berlin

Offsetdruck: Julius Beltz, Hemsbach/Bergstr.

Contents

Introduction

The purpose of category theory is to try to describe certain general aspects of the structure of mathematics. Since category theory is also part of mathematics, this categorical type of description should apply to it as well as to other parts of mathematics. When I first conducted a seminar on this subject during the Bowdoin Summer Session on Category Theory in 1969, Saunders Mac Lane suggested the name "Formal Category Theory" for this study. The basic idea is that the category of small categories, Cat, is a 2-category with properties and that one should attempt to identify those properties that enable one to do the "structural parts of category theory". The results of this present study suggest the following analogy with homological algebra: Cat corresponds to the category of abelian groups; the categories, Cat χ, of category objects in a category χ with pullbacks correspond to categories of modules; and representable 2-categories correspond to abelian categories.

There has been a considerable amount of work on various aspects of this study. Much of the work of Ehresmann is directly concerned with, or at least, relevant to it. In particular, he has studied Cat χ under the name of "p-structured categories", see, e.g., [12], [13], [14], [15]. Similarly, Benabou's work, both published, e.g., [BC], [3], and unpublished, [4] is relevant. Maranda's paper, "Formal Categories" [36] contains in its first part, a discussion of adjoint squares very similar to I,§6. However, by a formal category, he means a category over a monoidal category in the sense of [E-K]. The papers of Bunge [7] and Palmquist [37], are in the same spirit, as is recent unpublished work of Street. In general, much of the work in closed categories needs only to be rephrased to apply to 2-categories. Some of this is briefly done in I,§6. However, the main thrust of this work does not seem to have been noticed before; namely, that complete, representable 2-categories are quasi-complete (III,§4) and that this is what is needed to do category theory (I, §1 and IV).

This volume is the first part of a projected three part work and its sections are numbered I,1 through I,7. In it are to be found general background material and the various relevant notions of adjointness. In I,§1 the notations for categories are fixed and the various aspects of the category $\underline{2}$, which are responsible for the properties of Cat we wish to abstract, are discussed, leading up to a precise statement (in I,1.9 through I,1.13) of those "structural parts of category theory" which are amenable to the sort of treatment presented here.

If one tries to formulate standard results about categories as global statements about the 2-category Cat in order to try to discuss them in other 2-categories, then it frequently happens that this involves "functors" that may not really be functors but only pseudo-functors and which look like adjoints, except that there are extra natural transformations expressing various kinds of global compatibility. Thus, it is not sufficient to deal solely with 2-categories and 2-functors. There are, in fact, four types of structures which arise in trying to discuss these situations:

a) 2-categories. (I,2.1 - I,2.3)

b) The category of 2-categories with the closed structure given by the 2-category $\mathrm{Fun}(\mathfrak{A},\mathfrak{B})$ of quasi-natural transformations, denoted by $2\text{-Cat}_{\otimes}$. (I,2.4 and I,4)

c) Bicategories. (I,3)

d) The partial structure for bicategories corresponding to b). (I,4.21)

Frequently, the most difficult thing is to recognize which pattern a given adjointness situation fits. One reason for this is that, for instance, quasi-natural transformations appear as 1-cells in the 2-category $\mathrm{Fun}(\mathfrak{A},\mathfrak{B})$ and as 2-cells in the situation described in b). Besides these structures, there are others which enter in describing their properties.

e) 2-comma categories. (I,2.5)

f) 3-categories and 3-comma categories. (I,2.6; I,2.7)

g) Double and triple categories. (I,2.8)

Because of this profusion of structures, I,2 and I,3 should be regarded as encyclopedias to be refered to as needed. However, scattered through them are a number of examples which might be helpful to the reader. In particular, in I,3.4 (3) the important bicategory $\mathrm{Bim}(\mathrm{Spans}\ \chi)$ is described. Proofs of assertions for which no specific reference is given (either later in this paper or elsewhere) can be found in the works of Benabou, Ehresmann, Gray, and Lawvere cited in the bibliography.

The material proper to this work begins in I,§4 with the construction of the non-symmetrical monoidal closed category built on the category of 2-categories, one of whose internal homs is given by Fun($\mathfrak{A}$,$\mathfrak{B}$) , and the detailed examination of the relations between the two internal homs. Two constructions are given for the tensor product of 2-categories, an explicit one in terms of cells and relations (I,4.9), and, in an appendix (I,4.23), one using fibred categories, inspired by a more general assertion of Benabou, [4]. In I,§5, this material is used to derive the algebraic properties of 2-comma categories which are central to the discussion of quasi-adjoints.

In I,§6, the properties of adjoint 1-cells in a 2-category are discussed. Using the language of adjoint squares and Kan extensions, one easily derives the expected formal properties. This is to be contrasted with the situation in I,§7 on quasi-adjoints where these formal properties fail dramatically, and thereby provide a number of new phenomena. Many examples are discussed at the end of this section. A list of them will be found at the beginning of I,7.

Much of the content of these last four sections was first presented in a much cruder fashion at the Bowdoin Summer Session mentioned above, and later, approximately in the form given here, in a series of lectures at the Forschungsinstitut für Mathematik of the ETH in Zürich.

Part II will be devoted to the study of the categories Cat χ (II,§1) and the categories of χ-valued functors on a category object in χ (II,§2). These really make up the bicategory Bim(Spans χ) and we provide a Yoneda embedding into this bicategory. In II,§3, we study the relations of these notions to fibrations in order to have the results - via the embedding theorem below - for strongly representable 2-categories. Finally, in II,§4, we treat the important special case of category objects in the category of triples.

Part III will be concerned with representable 2-categories (III,§1). The term was suggested by Jon Beck since they are characterized by the property that the 2-cells are "representable" by 1-cells. In III,§2, we give a classification theorem in terms of category objects in the category of triples. These 2-categories were discussed from this point of view at Bowdoin. From the present point of view, a representable 2-category is a 2-category with a suitable property rather than a category with additional structure. In III,§3, strongly representable 2-categories are introduced. They admit "strict" embeddings into 2-categories of the form Cat χ . In III,§4, completeness theorems are given for representable 2-categories. Under ordinary completeness hypotheses, these share with Cat the property of admitting all cartesian quasi-limits, while corepresentable, cocomplete 2-categories are cartesian quasi-cocomplete. Thus, in particular, using the results of Street, these 2-categories admit constructions of Kleisli categories and Eilenberg-Moore categories for triples (cf. I,7). III,§5 is devoted to many examples of representable and corepresentable 2-categories as well as examples of weaker types of structures. We mention in particular that Cat χ and $\mathfrak{D}$-Cat (for a closed monoidal $\mathfrak{D}$ with strong pullbacks) are strongly representable. In III,§6, the algebraic structure of comma objects necessary for category theory is discussed.

Part IV will study the structural theorems of I, in the context of representable 2-categories, in so far as this is possible.

In an Appendix to Part IV, we shall comment briefly on the implications of this for foundations, the idea being that, since the notion of a representable 2-category is elementary as is the material in Chapters III and IV, much of category theory can be carried out in the context of the elementary theory of representable 2-categories. This serves to identify those parts of things like the adjoint functor theorem and Kan extensions (which have concerned logicians) that are purely syntactical and hence completely independent of set-theoretical considerations.

This lengthy work would never have reached its present form of a reasonable complete and self-contained treatment of the main results without the long term support of the National Science Foundation and the Forschungsinstitut für Mathematik of the ETH, Zurich, as well as the impetus provided by the Bowdoin Summer Session mentioned above, for all of which I express my appreciation. I would also like to thank the Battelle Institute, Geneva, for making possible the early appearance of Part I.

Part I: Adjointness for 2-categories

I,1. Categories. In this work, Sets will denote the category of small sets in some fixed universe (or alternatively, a model of the theory of sets in the sense of Lawvere-Tierney [22].) All categories are assumed to have small hom sets. A category is small if its set of morphisms is small. The category of small categories will be denoted by Cat . When necessary, we shall write $_{\ell}$Cat for (large) categories belonging to some universe big enough to contain all the relevant constructions. Cat has many properties that we shall make frequent use of but which will not be incorporated in our discussion of representable 2-categories in Chapter III.

I,1.1. Cat is cartesian closed; i.e., exponentiation via functor categories $\underline{B}^{\underline{A}}$ describes a closed category structure (see [E-K]) such that for all $\underline{A}$,

$$\underline{A} \times - \longrightarrow\!\!\!\!\!\dashv (-)^{\underline{A}}$$

(The sign " $\longrightarrow\!\!\!\!\!\dashv$ " means "is left adjoint to".)

I,1.2. The underlying set functor $|(-)|$: Cat → Sets is part of a string of adjunctions. Let D : Sets → Cat be the inclusion of Sets as discrete (i.e., only identity morphisms) categories. Let π_o : Cat → Sets be the "set of (path) components" functor, where a path is a string of morphisms in either direction; e.g.

$$\cdot \rightarrow \cdot \leftarrow \cdot \; \ldots \; \cdot \rightarrow \cdot \leftarrow$$

Let $G : \text{Sets} \to \text{Cat}$ be the "trivial connected groupoid" functor; i.e., if X is a set then $G(X)$ is the groupoid (= category with every morphism an isomorphism) with objects X and such that there is exactly one morphism from x to y for all x and y in X. Then

$$\pi_o \longrightarrow D \longrightarrow |(-)| \longrightarrow G .$$

I,1.3. Cat is complete and cocomplete. Furthermore, limits and colimits are Cat-limits, in the sense of closed categories; i.e.,

$$\underline{B}^{\varinjlim \underline{A}_i} = \varprojlim(\underline{B}^{\underline{A}_i}) \quad ; \quad (\varprojlim \underline{B}_i)^{\underline{A}} = \varprojlim(\underline{B}_i^{\underline{A}})$$

This follows immediately from the cartesian closed structure of Cat . Limits are given by the usual subobjects of products. The structure of colimits is more complicated. Coproducts are disjoint unions, while a coequalizer

$$\underline{A} \underset{G}{\overset{F}{\rightrightarrows}} \underline{B} \xrightarrow{P} \underline{Q}$$

is described as follows: $|(-)|$ has a right adjoint so $|\underline{Q}|$ is the coequalizer of $|F|$ and $|G|$ in Sets. The hom set $\underline{Q}(Q,Q')$ is a coequalizer of the coproduct of all finite products of the form

$$\underline{B}(B_1,B_1') \times \underline{B}(B_2,B_2') \times \ldots \times \underline{B}(B_k,B_k')$$

where $P(B_1) = Q$, $P(B_i') = P(B_{i+1})$, $P(B_k') = Q'$. If $Q = Q'$, then an addition coproduct with $\underline{1}$ is taken. The two maps whose coequalizer is formed are constructed by inserting the maps

$$\underline{A}(A,A') \begin{array}{l} \xrightarrow{F_{A,A'}} \underline{B}(FA,FA') \\ \xrightarrow[G_{A,A'}]{} \underline{B}(GA,GA') \end{array}$$

$$\underline{B}(B,B')\times\underline{B}(B',B'') \begin{array}{l} \xrightarrow{id} \underline{B}(B,B')\ \underline{B}(B',B'') \\ \xrightarrow[o]{} \underline{B}(B,B'') \end{array}$$

$$1 \begin{array}{l} \xrightarrow{\{1_B\}} \underline{B}(B,B) \\ \xrightarrow{} 1 \end{array}$$

into all possible positions in all possible products in the coproduct. For an explicit formula, see Wolff [40]. Alternatively, the morphisms can be described as in Lawvere [29] as equivalence classes of admissable strings of morphisms from $\underline{B}$

$$f_1,\dots,f_n$$

where a string is admissable if the domain of f_i is equivalent to the codomain of f_{i+1} , and two strings are equivalent if they are made so by the smallest equivalence relation compatible with composition such that f',f and $f'f$ are equivalent whenever $f'f$ defined in $\underline{B}$, and $F(g)$ and $G(q)$ are equivalent for all morphisms $g \in \underline{A}$.

I,1.4. The properties of Cat which will be extended to representable 2-categories all depend ultimately on the category $\underline{2}$. Here, $\underline{1}$ denotes the category with a single identity morphism; $\underline{2}$ denotes the category that looks like

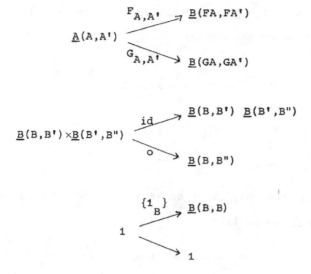

$0 \to 1$, with $\partial_i : \underline{1} \to \underline{2}$ the functors given by $\partial_i(\underline{1}) = i$, $i = 0,1$; $\underline{3}$ denotes the pushout

$$\begin{array}{ccc} \underline{1} & \xrightarrow{\partial_1} & \underline{2} \\ {\scriptstyle \partial_0}\downarrow & & \downarrow{\scriptstyle \alpha} \\ \underline{2} & \xrightarrow{\beta} & \underline{3} \end{array}$$

and $\gamma : \underline{2} \to \underline{3}$ the only other non-trivial functor from $\underline{2}$ to $\underline{3}$ (i.e., $\underline{3}$ looks like

$$\begin{array}{ccc} 0 & \xrightarrow{\alpha} & 1 \\ & {\scriptstyle \gamma}\searrow & \downarrow{\scriptstyle \beta} \\ & & 2 \end{array} \quad);$$

and $\underline{4}$ denotes the pushout

$$\begin{array}{ccc} \underline{2} & \xrightarrow{\beta} & \underline{3} \\ {\scriptstyle \alpha}\downarrow & & \downarrow{\scriptstyle \mu} \\ \underline{3} & \xrightarrow[\nu]{} & \underline{4} \end{array}$$

The higher ordinals can be defined similarly, or by induction as the pushout of

$$\underline{n} \xleftarrow{\partial_0} \underline{1} \xrightarrow{\partial_1} \underline{2} .$$

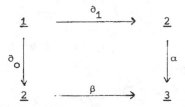

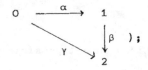

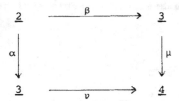

$\{\underline{4}\}$ will denote the category which looks like the full subcategory of Cat determined by $\underline{1},\underline{2},\underline{3}$ and $\underline{4}$ (cf. Lawvere [CCFM]). Alternatively, it is the category where objects are the ordinals 1,2,3 and 4 and whose morphisms are all order preserving maps between them. As such it is a full subcategory of the category $\underline{\Delta}$ of all finite ordinals (including 0) and all order preserving functions.

I,1.5. The entire theory presented in Part III of this work is a reflection of the relations between the categories $\underline{1},\underline{2},\underline{3},\underline{4}$, and $\underline{2}\times\underline{2}$. Of these, $\underline{1},\underline{2},\underline{3}$, and $\underline{4}$ determine the elementary structure of categories, in the sense that, for instance, Cat is isomorphic to the category of limit preserving functors from $\{\underline{4}\}^{op}$ to Sets (cf. [CCFM].) It is the thesis of the study of representable 2-categories that, by including $\underline{2}\times\underline{2}$ as well as exponentiation by these five categories, one recaptures much of category theory First of all, the non-trivial relations between $\underline{1},\underline{2}$ and $\underline{2}\times\underline{2}$ can be summarized by the statement that the functor

$$- \times \underline{2} : \text{Cat} \to \text{Cat}$$

is a cotriple (see I,7) with comultiplication the diagonal $\Delta : \underline{2} \to \underline{2}\times\underline{2}$ and counit the constant functor $\tau : \underline{2} \to \underline{1}$. By adjointness, $(-)^{\underline{2}}$ is a triple with multiplication $(-)^{\Delta}$ and unit $(-)^{\tau}$.

Now, small categories are the objects of another category, Cat_t , whose morphisms are natural transformations. These can be identified either with functors $\underline{A}\times\underline{2} \to \underline{B}$ or with functors $\underline{A} \to \underline{B}^{\underline{2}}$; i.e., as coKleisli morphisms for the cotriple or as Kleisli morphisms for the triple. Composition is given by Kleisli composition, so, for instance,

Cat_t is isomorphic to the Kleisli category of the triple $((-)^{\underline{2}}, (-)^{\Delta}, (-)^{\tau})$ in Cat . A similar representation theorem for the total category of a representable 2-category is given in III, §2. The "set of morphisms" functor

$$|(-)^{\underline{2}}| : \text{Cat} \to \text{Sets}$$

is also defined on Cat_t , a morphism $f : \underline{2} \to \underline{A}$ being taken by a natural transformation $\varphi : \underline{A}\times\underline{2} \to \underline{B}$ to the morphism

$$\varphi f = \varphi(f) : \underline{2} \xrightarrow{\Delta} \underline{2}\times\underline{2} \xrightarrow{f\times 2} \underline{A}\times\underline{2} \longrightarrow \underline{B}$$

Associativity of Kleisli composition shows that this is a functor. As such, it is right adjoint to $D : \text{Sets} \to \text{Cat}_t$; i.e.,

$$D \dashv |(-)^{\underline{2}}| \quad (\text{on } \text{Cat}_t).$$

I,1.6. The next thing to observe is that the structure of the functors between $\underline{1}, \underline{2}, \underline{3}$ and $\underline{4}$ implies that

$$\underline{1} \underset{\tau}{\overset{\partial_0, \partial_1}{\rightleftarrows}} \underline{2} \xrightarrow{\gamma} \underline{3}$$

constitutes a cocategory object in Cat (see II, §1) and hence $(-)^{\underline{2}}$ is a category object in the category of endofunctors on Cat ; i.e., there are a pullback diagram and functors for any $\underline{C} \in \text{Cat}$

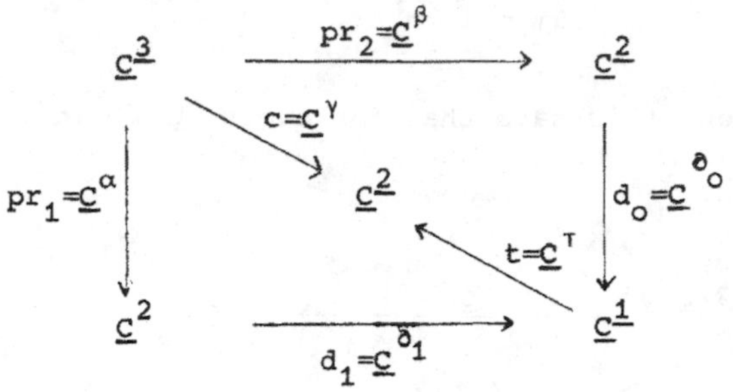

satisfying the equations for a category object.

I,1.7. It turns out that in the things treated in Parts III and IV - fibrations, the Yoneda lemma, adjointness, etc., - the crucial structure is provided by the functors and natural transformations between $\underline{3}$ and $\underline{2}\times\underline{2}$. Explicitely, there are functors u and ℓ taking $\underline{3}$ to the indicated triangles in $\underline{2}\times\underline{2}$:

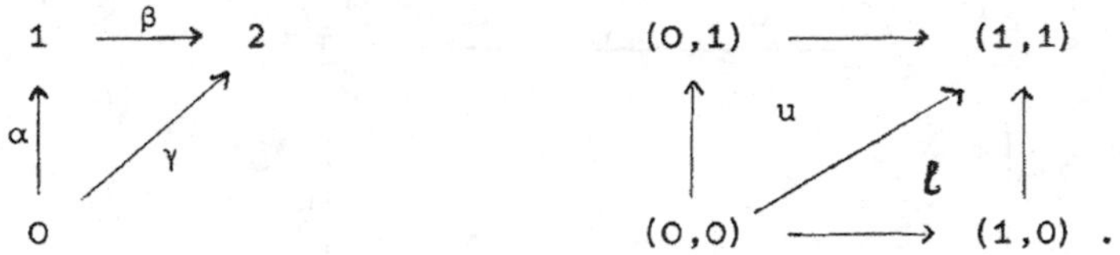

Or, treating $\alpha : \underline{2} \to \underline{3}$ as the first injection in the pushout and denoting maps into limits and out of colimits by $\{\ldots\}$, one can write

$$u = \{\{\partial_o, \underline{2}\} , \{\underline{2}, \bar{\partial}_1\}\}$$
$$\ell = \{\{\underline{2}, \bar{\partial}_o\} , \{\bar{\partial}_1, \underline{2}\}\}$$

where $\bar{\partial}_i = \partial_i\tau : \underline{2} \to \underline{2}$. These are the only non-trivial (i.e., not factoring through $\underline{2}$) functors from $\underline{3}$ to $\underline{2}\times\underline{2}$ and they satisfy the equations

$$u\gamma = \ell\gamma = \Delta \ .$$

By exponentiation, this says that for every $\underline{C} \in$ Cat , the diagram

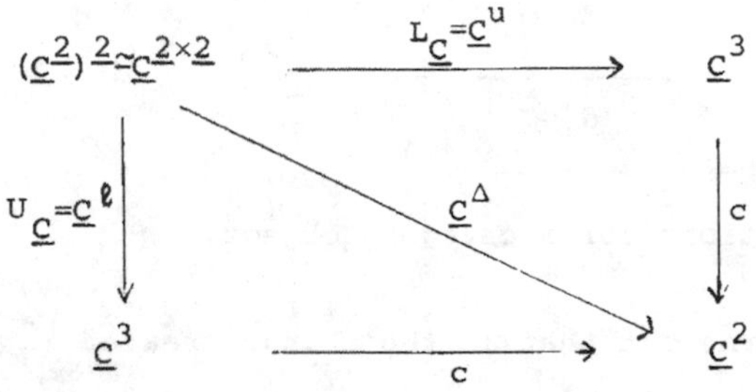

commutes. We shall see in II, §4 that this is equivalent to $(-)^{\underline{2}}$ being a category object in the category of triples on Cat .

I,1.8. It is of central importance that the square

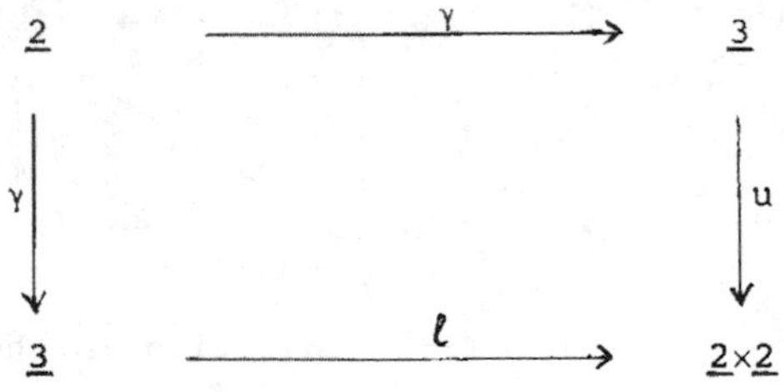

is a pushout (and hence for any $\underline{C}$, the preceeding square is a pullback). This allows one to describe the non-trivial functors from $\underline{2}\times\underline{2}$ to $\underline{3}$. There are five such. The "folding functor", fd = $\{\underline{3},\underline{3}\}$ and four others, described as follows (here u is treated as the first injection):

$\check{u} = \{\underline{3}, \gamma\{\underline{2}, \partial_1\}\}$

$\hat{u} = \{\underline{3}, \gamma\{\partial_o, \underline{2}\}\}$

$\check{\ell} = \{\gamma\{\underline{2}, \partial_1\}, \underline{3}\}$

$\hat{\ell} = \{\gamma\{\partial_o, \underline{2}\}, \underline{3}\}$

In terms of pictures, $\check{u}$ is the identity on u and takes $(1,0)$ to 2, while $\hat{u}$ is the identity on u and takes $(1,0)$ to 0. By definition

$$\check{u}u = \hat{u}u = \check{\ell}\ell = \hat{\ell}\ell = \underline{3}$$

and it is easily verified that there are adjunctions

$$\begin{array}{c} \check{u} \longrightarrow u \longrightarrow \hat{u} \\ \check{\ell} \longrightarrow \ell \longrightarrow \hat{\ell} \end{array}$$

and that $\{\ell\hat{\ell}, u\check{u}\}(\gamma\times\underline{2}) = id : \underline{2}\times\underline{2} \to \underline{2}\times\underline{2}$.

Formally, the required adjunction natural transformations can be represented as functors from $\underline{2}\times\underline{2}\times\underline{2}$ to $\underline{2}\times\underline{2}$. Any such φ is of the form $\{p_1\varphi, p_2\varphi\}$ where $p_i : \underline{2}\times\underline{2} \to \underline{2}$ is the i'th projection. Since $-\times\underline{2}$ has a right adjoint, it preserves pushouts and hence $p_i\varphi = \{a_{io}, a_{i1}\}$ where $a_{ij} : \underline{3}\times\underline{2} \to \underline{2}$. Similarly

$$a_{ij} = \{b_{ijo}, b_{ij1}\}$$

where $b_{ijk} : \underline{2}\times\underline{2} \to \underline{2}$, so

$$\varphi = \{\{\{b_{100}, b_{101}\}, \{b_{110}, b_{111}\}\}, \{\{b_{200}, b_{201}\}, \{b_{210}, b_{211}\}\}\}$$

Now functors from $\underline{2}\times\underline{2}$ to $\underline{2}$ are the same as positive Boolean functions of two variables. There are six such; the two projections $p_i : \underline{2}\times\underline{2} \to \underline{2}$ where $i = 1,2$, the two constant functors

$$\text{'true'} : \underline{2}\times\underline{2} \to \underline{1} \xrightarrow{\partial_o} \underline{2}$$
$$\text{'false'} : \underline{2}\times\underline{2} \to \underline{1} \xrightarrow{\partial_1} \underline{2},$$

and the two lattice operations, described as pairs of functors from $\underline{3} \to \underline{2}$ by

$$'\text{and}' = \wedge = \{\{\bar{\partial}_o,\underline{2}\},\{\bar{\partial}_o,\underline{2}\}\}$$

$$'\text{or}' = \vee = \{\{\underline{2},\bar{\partial}_1\},\{\underline{2},\bar{\partial}_1\}\}$$

If we abbreviate the first and second projections of $\underline{2}\times\underline{2}\times\underline{2}$ onto $\underline{2}$ by q_1 and q_2 ; i.e.,

$$q_1 = \{\{p_1,'\text{true}'\},\{'\text{false}',p_1\}\}$$

$$q_2 = \{\{'\text{false}',p_1\},\{p_1,'\text{true}'\}\}$$

then the four adjunction natural transformations are given by

$$(\psi_1 : u\hat{u} \to \underline{2}\times\underline{2}) = \{\{\{'\text{false}',p_1\},\{\wedge,\vee\}\},q_2\}$$

$$(\theta_1 : \underline{2}\times\underline{2} \to u\check{u}) = \{q_1,\{\{p_1,'\text{true}'\},\{\wedge,\vee\}\}\}$$

$$(\psi_2 : \ell\hat{\ell} \to \underline{2}\times\underline{2}) = \{q_1,\{\{\wedge,\vee\},\{'\text{false}',p_1\}\}\}$$

$$(\theta_2 : \underline{2}\times\underline{2} \to \ell\check{\ell}) = \{\{\{\wedge,\vee\},\{p_1,'\text{true}'\}\},q_2\}$$

(Here $\underline{2}\times\underline{2}$ means the identity functor on $\underline{2}\times\underline{2}$.)

All of this structure transports itself throughout Cat by exponentiation. So, refering to the diagram at the end of I,1.7, there are functors

$$\check{U}_{\underline{C}} = \underline{C}^{\hat{\ell}} \ , \ \hat{U}_{\underline{C}} = \underline{C}^{\check{\ell}} \ , \ \check{L}_{\underline{C}} = \underline{C}^{\hat{u}} \ , \ \hat{L}_{\underline{C}} = \underline{C}^{\check{u}}$$

satisfying

$$U_{\underline{C}}\check{U}_{\underline{C}} = U_{\underline{C}}\hat{U}_{\underline{C}} = L_{\underline{C}}\check{L}_{\underline{C}} = L_{\underline{C}}\hat{L}_{\underline{C}} = \underline{C}^{\underline{3}}$$

with adjunctions

$$\check{U}_{\underline{C}} \dashv U_{\underline{C}} \dashv \hat{U}_{\underline{C}}$$

$$\check{L}_{\underline{C}} \dashv L_{\underline{C}} \dashv \hat{L}_{\underline{C}}$$

and a commutative diagram

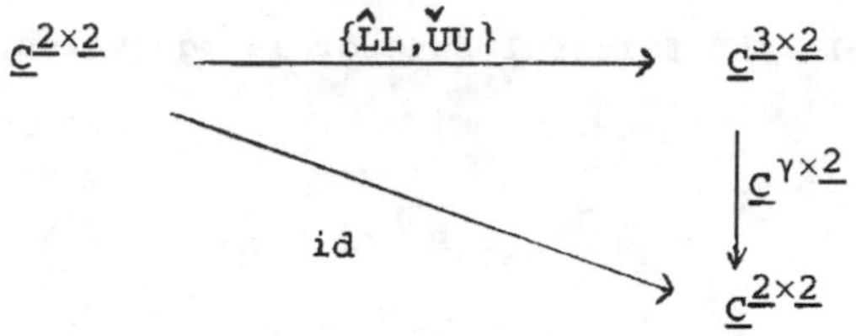

This structure gives rise to the properties of comma categories, where, if $F_i : \underline{A}_i \to \underline{B}$, $i = 1,2$, then

$$(F_1,F_2) = \varprojlim(\underline{A}_1 \xrightarrow{F_1} \underline{B} \xleftarrow{\underline{B}^{\partial_0}} \underline{B}^{\underline{2}} \xrightarrow{\underline{B}^{\partial_1}} \underline{B} \xleftarrow{F_2} \underline{A}_2) .$$

We always treat (F_1,F_2) as a category over $\underline{A}_1\times\underline{A}_2$ via the two projections of the limit. A number of properties of this construction will be found in [CCS] §2. They are discussed in a different form in Part III. We are interested in this work in the following kinds of results.

I,1.9. (Yoneda) If F and $G : \underline{A} \to \underline{B}$, then there is a 1-1 correspondence between natural transformations $\varphi : F \to G$ and functors

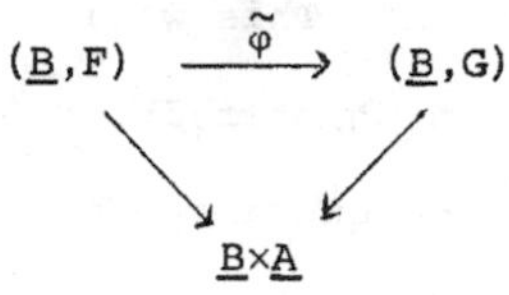

resp.,

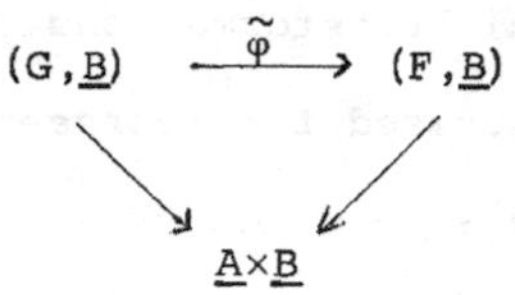

Actually, this determines two full and faithful embeddings

$$\underline{B}^{\underline{A}} \to (\text{Cat},\underline{B}\times\underline{A})$$
$$\underline{B}^{\underline{A}} \to (\text{Cat},\underline{A}\times\underline{B})^{op}$$

This is called a Yoneda lemma because the category $(\underline{B},F) \to \underline{B}\times\underline{A}$ is the bifibration given by the "basic construction" of [CCS], §5 applied to the functor $\underline{B}(-,F(-)) : \underline{B}^{op} \to$ Sets, so the above full and faithful functor is equivalent to the functor

$$\underline{B}^{\underline{A}} \to \text{Sets}^{\underline{B}^{op} \times \underline{A}}$$

taking $F : \underline{A} \to \underline{B}$ to $\underline{B}(-,F(-))$. Specializing to $\underline{A} = \underline{1}$, yields the ordinary Yoneda lemma. This construction will be discussed for category objects in an arbitrary category χ instead of Sets in II, §2 and for representable 2-categories in Part IV.

I,1.10. (Adjointness) If $F : \underline{A} \to \underline{B}$ and $U : \underline{B} \to \underline{A}$, then there is a 1-1 correspondence between natural transformations $\eta : \underline{A} \to UF$ (resp., $\varepsilon : FU \to \underline{B}$) and functors

$$\begin{array}{ccc} (F,\underline{B}) & \xrightarrow{\bar{\eta}} & (\underline{A},U) \\ & \searrow \quad \swarrow & \\ & \underline{A}\times\underline{B} & \end{array} \qquad \text{resp.,} \qquad \begin{array}{ccc} (\underline{A},U) & \xrightarrow{\bar{\varepsilon}} & (F,\underline{B}) \\ & \searrow \quad \swarrow & \\ & \underline{A}\times\underline{B} & \end{array}$$

Furthermore, ε and η satisfy the equations for adjunction natural transformations if and only if $\bar{\varepsilon} = \bar{\eta}^{-1}$. This will be discussed for representable 2-categories in Chapter IV. The ways in which this correspondence fails for quasi-adjunctions between 2-categories are important for expressing the global forms of many constructions in category theory. This is discussed in I, §7 .

I,1.11. (Fibrations) $P : \underline{E} \to \underline{B}$ is called a fibration if there exists a functor $L : (\underline{B},P) \to \underline{E}^{\underline{2}}$ which is right adjoint right inverse to $S = (P^{\underline{2}},\underline{E}^{\partial_1}) : \underline{E}^{\underline{2}} \to (\underline{B},P)$. (See [CCS] and [FCC].) A split normal fibration is a fibration with a choice of L which satisfies certain equations. (See II, §3). The category of split normal fibrations and cleavage (i.e., L) preserving functors is isomorphic to $\mathrm{Cat}^{\underline{B}^{op}}$ and the forgetful functor into $(\mathrm{Cat},\underline{B})$ has a left adjoint, the fibration associated to $F : \underline{A} \to \underline{B}$ being $(\underline{B},F) \to \underline{B}$. Furthermore, the adjunction morphism (in $(\mathrm{Cat},\underline{B})$)

$$\begin{array}{ccc} \underline{A} & \underset{\overset{\dashleftarrow}{P}}{\xrightarrow{Q_F}} & (\underline{B},F) \\ & {\scriptstyle F}\searrow \quad \swarrow{\scriptstyle P_F} & \\ & \underline{B} & \end{array} \qquad , \; P_F Q_F = F$$

has a left-adjoint left inverse $P \longrightarrow Q_F$ which is just the projection of $(\underline{B},F)$ to $\underline{A}$.

I,1.12. (Adjoint Functor Theorem) $F : \underline{A} \to \underline{B}$ has a left adjoint if and only if P_F has a left adjoint right inverse.

In Cat this leads to the usual adjoint functor theorem by the following steps:

i) Use the dual of Prop. 4.4 in [FCC] to show P_F has such an adjoint if and only if each fibre $(\ulcorner B \urcorner,F)$ has an initial object.

ii) An initial object in a category is an inverse limit of the identity functor of the category.

iii) Hence F has a left adjoint if and only if for every $B \in \underline{B}$, $\varprojlim((\ulcorner B \urcorner,F) \to \underline{A})$ exists and is preserved by F. (J. Beck has called this the basic adjoint functor theorem.)

iv) Add a solution set condition saying that the categories $(\ulcorner B \urcorner,F)$ have small initial subcategories, and assume F preserves small limits.

I,1.13. (Kan extensions) Let $F : \underline{A} \to \underline{B}$ be fixed. Given $\underline{X}$, a functor

$$E^F : \underline{X}^{\underline{A}} \to \underline{X}^{\underline{B}}$$

having $\underline{X}^F$ as left adjoint, is called the left Kan extension along F . It can be constructed in the following way.

i) Let $[{}^{op}Cat,\underline{X}]_o$ be the category whose objects are functors $K : \underline{A} \to \underline{X}$ and whose morphisms are pairs $\langle G,t\rangle$ in diagrams

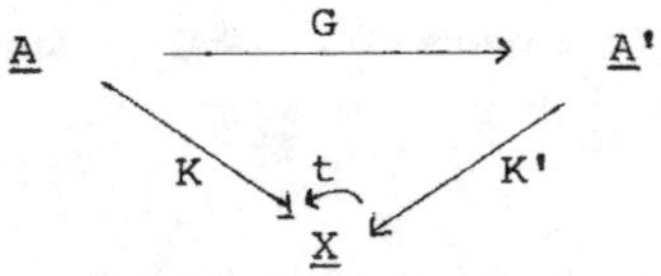

where $t : K'G \to K$ is a natural transformation. Composition is $\langle G',t'\rangle\langle G,t\rangle = \langle G'G,t\cdot t'G\rangle$. (See [CCS] and I, §2 for 2-comma categories $[F_1,F_2]$.) Then $\varinjlim : [{}^{op}Cat,\underline{X}]_o \to \underline{X}^{op}$ and is left adjoint to the "name" functor taking X to $\ulcorner X \urcorner : \underline{1} \to \underline{X}$ and $f : X \to X'$ to $\langle \underline{1},f\rangle$.

ii) If $P : \underline{E} \to \underline{B}$ is a split normal fibration then there is a functor $\tilde{P} : \underline{B} \to [{}^{op}Cat,\underline{E}]_o$ given by

$\tilde{P}(B) = (J_B : \underline{E}_B \hookrightarrow \underline{E})$, where $\underline{E}_B = P^{-1}(B)$

$\tilde{P}(f)$: 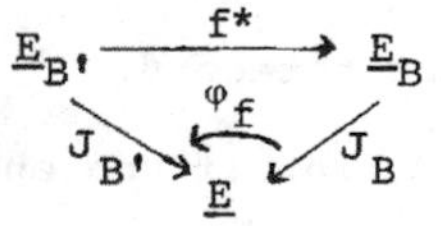 $f : B \to B'$

Here $(\varphi_f)_{E'} = L(f,E') : J_B f^*(E') \to J_{B'}(E')$.

Given $H : \underline{E} \to \underline{X}$, then $E^P(H)$ is "integration along the fibre" ; i.e., it is the composition

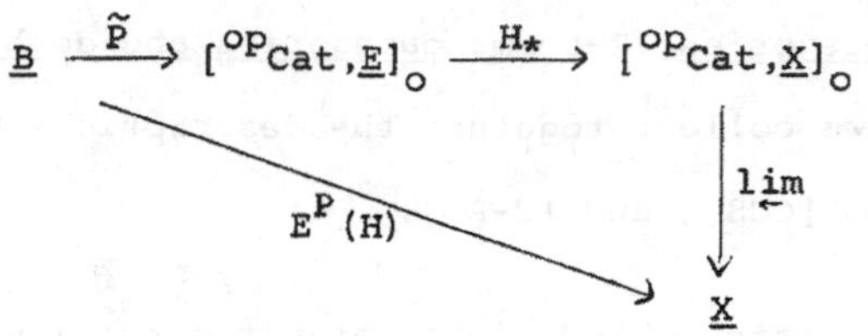

where H_* is given by composition with H .

iii) In general, given $F : \underline{A} \to \underline{B}$, replace F by the associated fibration P_F .

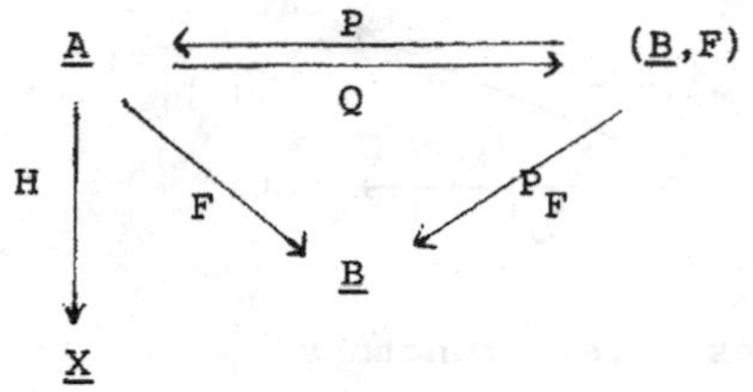

Then $E^F(H) = E^{P_F}(HP)$. (This was first observed in a different form by S. Mac Lane.)

In Part IV, these structural theorems for Cat will be studied in representable 2-categories.

I,2. 2-<u>categories</u>, 2-<u>comma</u> <u>categories</u> <u>and</u> <u>double</u> <u>categories</u>. In this section we collect together the descriptions we shall need, mostly from [CCS], and [2-A].

I,2.1. A 2-<u>category</u> $\mathcal{A}$ is a Cat-category; i.e., enriched in the cartesian closed category Cat . (See [E-K].). It therefore consists of an ordinary category $\mathcal{A}_o$ together with a factorization of the hom functor

$$\begin{array}{ccc} & & \mathrm{Cat} \\ & \overset{\mathcal{A}(-,-)}{\nearrow} & \big\downarrow |(-)| \\ \mathcal{A}_o^{op}\times\mathcal{A}_o & \xrightarrow[\mathcal{A}_o(-,-)]{} & \mathrm{Sets} \end{array}$$

and composition rules (i.e., functors)

$$\mathcal{A}(A,B)\times\mathcal{A}(B,C) \xrightarrow{\circ} \mathcal{A}(A,C)$$

for all A,B,C , which are natural in all variables, associative, unitary (this means there is an object $I_A \in \mathcal{A}(A,A)$ which is a two-sided unit for compositions) and agree with composition in $\mathcal{A}_o$ on objects. Objects of $\mathcal{A}_o$ are called 0-<u>cells</u>, morphisms of $\mathcal{A}_o$ are called 1-<u>cells</u> and morphisms of $\mathcal{A}(A,B)$ are called 2-<u>cells</u>. $\mathcal{A}_o$ is called the <u>underlying</u> (<u>locally discrete</u>) category of $\mathcal{A}$. If one forgets the 1-cells, one obtains the <u>total</u> <u>category</u> $\mathcal{A}_t$ of $\mathcal{A}$ whose hom functor is given by $|\mathcal{A}(-,-)\underline{2}|$. The composition "$\circ$" above will be denoted by juxtaposition and called <u>strong</u>, while composition within each $\mathcal{A}(A,B)$ will be denoted by "$\cdot$" and called <u>weak</u>. Both are associative and have units. 1-cells have domains and codomains, denoted by $\partial_o f$ and $\partial_1 f$ respectively.

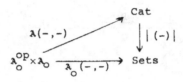

2-cells have <u>strong</u> domains and codomains with respect to juxtaposition composition, denoted by $\partial_o\mu$ and $\partial_1\mu$ respectively, and <u>weak</u> domains and codomains within the categories $\mathbb{A}(A,B)$, denoted by $\tilde{\partial}_o\mu$ and $\tilde{\partial}_1\mu$ respectively. Thus

(2.1)

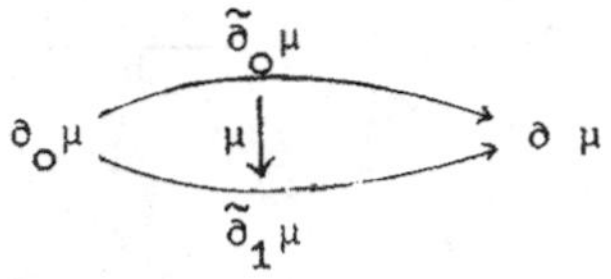

Note that $\partial_i\tilde{\partial}_j = \partial_i$.

Functorality of juxtaposition is equivalent to the preservation of units and, in the situation

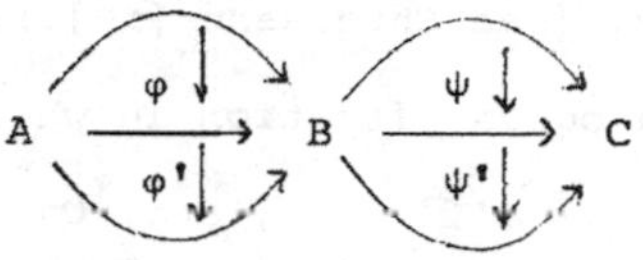

the validity of the <u>interchange</u> <u>law</u>

(2.2) $$(\psi'\varphi')\cdot(\psi\varphi) = (\psi'\cdot\psi)(\varphi'\cdot\varphi)$$

We shall often have occasion to compose squares with specified 2-cells between the two composed 1-cells. In the situation

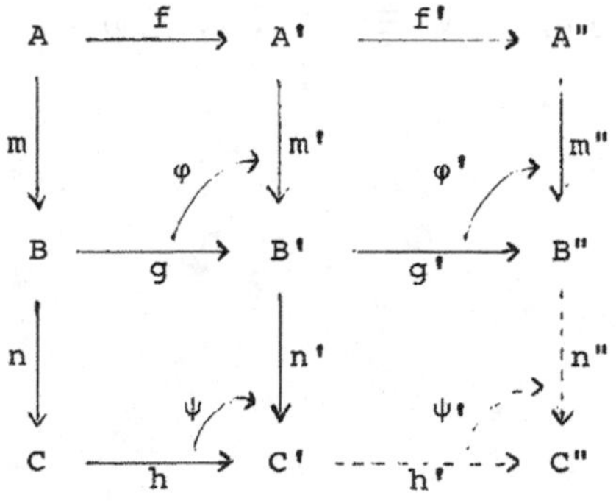

where $\varphi : gm \to m'f$, etc., there are composed squares

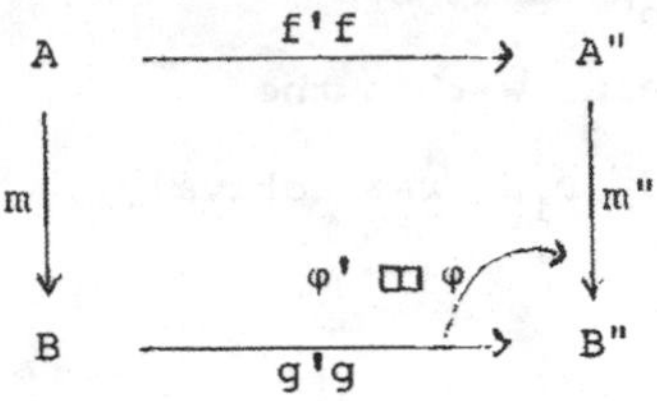

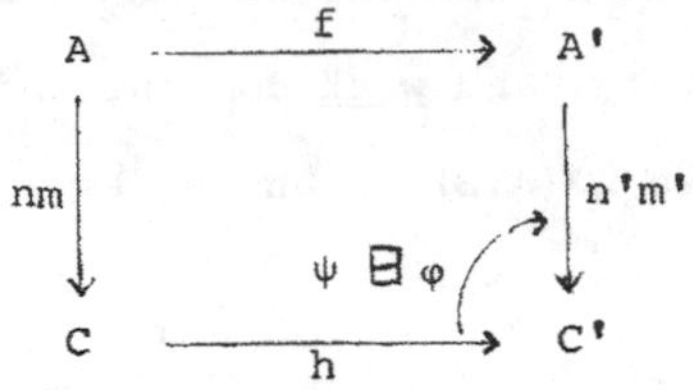

where

$$(2.3)\qquad \begin{aligned} \varphi' \boxbar \varphi &= (\varphi' f)\cdot(g'\varphi) \\ \psi \boxminus \varphi &= (n'\varphi)\cdot(\psi m) \end{aligned}$$

We refer to $\boxbar$ and $\boxminus$ as <u>horizontal</u> and <u>vertical</u> composition of squares, respectively. (See Ehresmann [15].). There is, of course, a completely analogous situation in which the 2-cells go the other way; i.e., $\varphi : m'f \to gm$, etc. One again has horizontal and vertical compositions given by

$$(2.3)'\qquad \begin{aligned} \varphi' \boxbar \varphi &= (g'\varphi)\cdot(\varphi' f) \\ \psi \boxminus \varphi &= (\psi m)\cdot(n'\varphi) \end{aligned}$$

It is easily verified that there is an interchange law for such squares:

$$(2.4)\qquad (\psi' \boxbar \psi) \boxminus (\varphi' \boxbar \varphi) = (\psi' \boxminus \varphi') \boxbar (\psi \boxminus \varphi)$$

(See [21], §1.)

A cube

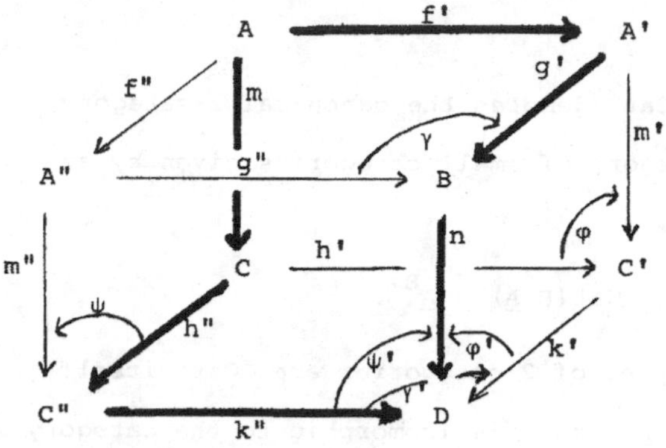

in which each face has a specified 2-cell between the two composed 1-cells, is called <u>commutative</u> if the two possible composed 2-cells between the compositions of the indicated heavily drawn 1-cells are equal; i.e.,

(2.5) $$(n\gamma)\cdot[\psi' \boxbar \psi] = [\varphi' \boxbar \varphi]\cdot(\gamma' m)$$

Note that by inserting dummy squares involving identities, this can be written just in terms of horizontal and vertical compositions. Also, the other possible expressions lead to nothing new since, for instance,

(2.6) $$(\psi' \boxminus \gamma)\cdot k''\psi = (n\gamma)\cdot[\psi' \boxbar \psi] .$$

In general, a 2-category is called <u>locally</u> <u>P</u> if all categories $\mathcal{A}(A,B)$ have property P ; e.g., <u>locally</u> <u>discrete</u> means $\mathcal{A}(-,-) = \mathcal{A}_o(-,-)$. If $\mathcal{A}$ is a 2-category, then it has three duals; the <u>strong</u> dual $\mathcal{A}^{op}$ in which $\mathcal{A}^{op}(A,B) = \mathcal{A}(B,A)$ (i.e , the 1-cells are reversed), the <u>weak</u> dual ${}^{op}\mathcal{A}$ in which ${}^{op}\mathcal{A}(A,B) = \mathcal{A}(A,B)^{op}$ (i.e., the 2-cells are reversed) and their combination ${}^{op}\mathcal{A}^{op}$ in which

${}^{op}\mathfrak{A}^{op}(A,B) = \mathfrak{A}(B,A)^{op}$ (i.e., 1-cells and 2-cells are reversed).

From now on, Cat denotes the canonical 2-category structure on the category of small categories given by exponentiation; i.e.,

$$\mathrm{Cat}(\underline{B},\underline{A}) = \underline{A}^{\underline{B}} .$$

Standard examples of 2-categories are Cat itself, $\mathrm{Cat}^{\underline{X}}$ for any small $\underline{X}$ (this is isomorphic to the category of split normal fibrations over $\underline{X}$ by the construction in I,2.9; see [FCC] also), and $\mathfrak{V}$-Cat where $\mathfrak{V}$ is a closed category or a monoidal category, as in [E-K]. There are various 2-categories which are analogues of the finite categories described in I,1.1. In particular, there is the 2-category $\underline{2}_2$ which looks like

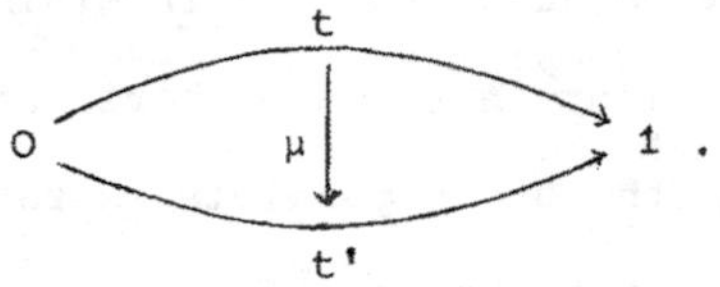

It is clearly a generator for 2-categories.

I,2.2. A 2-__functor__ $F : \mathfrak{B} \to \mathfrak{A}$ is a Cat-functor. Thus, it consists of an object function $F : |\mathfrak{B}_o| \to |\mathfrak{A}_o|$ together with functors

$$F_{B,C} : \mathfrak{B}(B,C) \to \mathfrak{A}(FB,FC)$$

for all B,C in $|\mathfrak{B}_o|$, which are compatible with units and composition. Alternatively, it is an ordinary functor $F_o : \mathfrak{B}_o \to \mathfrak{A}_o$ together with a natural transformation $F_{(-,-)}$ as illustrated

$$\begin{array}{ccc} \mathfrak{B}_o^{op}\times\mathfrak{B}_o & \xrightarrow{\quad F_o^{op}\times F_o \quad} & \mathfrak{A}_o^{op}\times\mathfrak{A}_o \\ & \mathfrak{B}(-,-) \searrow \quad \xRightarrow{F_{(-,-)}} \quad \swarrow \mathfrak{A}(-,-) & \\ & \mathrm{Cat} & \end{array}$$

which is compatible with units and composition; i.e.,

2F1 (naturality) if $f : B' \to B$, $g : C \to C'$, then

$$\begin{array}{ccc} \mathfrak{B}(B,C) & \xrightarrow{\mathfrak{B}(f,g)} & \mathfrak{B}(B',C') \\ {\scriptstyle F_{B,C}}\downarrow & & \downarrow{\scriptstyle F_{B',C'}} \\ \mathfrak{A}(FB,FC) & \xrightarrow[\mathfrak{A}(Ff,Fg)]{} & \mathfrak{A}(FB',FC') \end{array}$$

commutes.

2F2 (compatibility with units)

$$\begin{array}{ccc} & \nearrow^{\ulcorner I_B \urcorner} & \mathfrak{B}(B,B) \\ \underline{1} & & \downarrow{\scriptstyle F_{B,B}} \\ & \searrow_{\ulcorner I_{FB} \urcorner} & \mathfrak{A}(FB,FB) \end{array}$$

commutes.

2F3 (compatibility with composition)

$$\begin{array}{ccc} \mathfrak{B}(B,C)\times\mathfrak{B}(C,D) & \xrightarrow{\quad\circ\quad} & \mathfrak{B}(B,D) \\ {\scriptstyle F_{B,C}\times F_{C,D}}\downarrow & & \downarrow{\scriptstyle F_{B,D}} \\ \mathfrak{A}(FB,FC)\times\mathfrak{A}(FC,FD) & \xrightarrow{\quad\circ\quad} & \mathfrak{A}(FB,FD) \end{array}$$

commutes.

It is easily seen that the ordinary functor $\mathfrak{A}(-,-) : \mathfrak{A}_o^{op}\times\mathfrak{A}_o \to \mathrm{Cat}$ lifts to a 2-functor, denoted the same way,

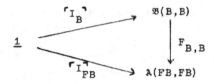

$$\mathfrak{A}(-,-) \; : \; \mathfrak{A}^{op}\times\mathfrak{A} \to \mathrm{Cat}$$

and that in the alternative description, $F_{(-,-)}$ becomes a Cat-natural transformation, as defined below. The partial functors

$$\mathfrak{A}(A,-) \; : \; \mathfrak{A} \to \mathrm{Cat}$$

$$\mathfrak{A}(-,A) \; : \; \mathfrak{A}^{op} \to \mathrm{Cat}$$

are called Cat-<u>representable</u> functors.

A 2-functor $F : \mathfrak{B} \to \mathfrak{A}$ is called <u>locally</u> P if all functors $F_{B,C}$ have properly P ; e.g., F is <u>locally full</u> (resp., <u>locally faithful</u>) if all $F_{B,C}$'s are full (resp., faithful).

I,2.3. A <u>Cat-natural transformation</u> is as defined in [E-K]. Thus $\sigma : F \to G$ is a family of morphisms $\sigma_B : F(B) \to G(B)$ such that the diagrams

$$\begin{array}{ccc} \mathfrak{B}(B,C) & \xrightarrow{F_{B,C}} & \mathfrak{A}(FB,FC) \\ {\scriptstyle G_{B,C}}\downarrow & & \downarrow{\scriptstyle \mathfrak{A}(1,\sigma_C)} \\ \mathfrak{A}(GB,GC) & \xrightarrow[\mathfrak{A}(\sigma_B,1)]{} & \mathfrak{A}(FB,GC) \end{array}$$

commute for all B and C . Alternatively, it is an ordinary natural transformation $\sigma_o : F_o \to G_o$ such that the diagram

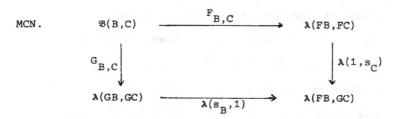

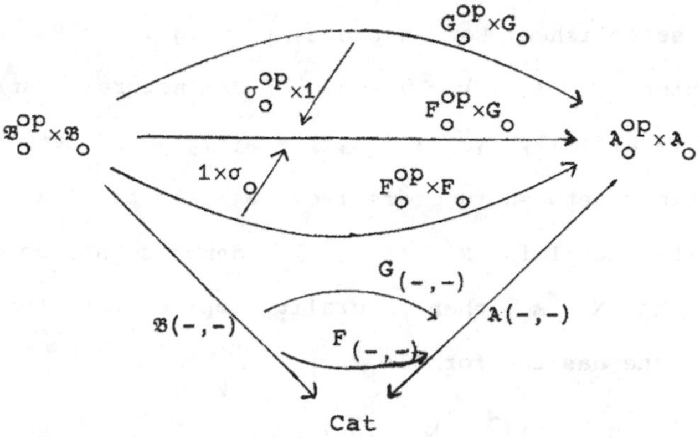

commutes; i.e.,

(2.7) $[\mathfrak{A}(-,-)(\sigma_o^{op} \times 1)] \cdot G_{(-,-)} = [\mathfrak{A}(-,-)(1 \times \sigma_o)] \cdot F_{(-,-)}$

$(\mathfrak{A}^{\mathfrak{B}})_o$ denotes the category of 2-functors and Cat-natural transformations from $\mathfrak{B}$ to $\mathfrak{A}$. It is the underlying category of a 2-category $\mathfrak{A}^{\mathfrak{B}}$ in which a 2-cell or modification (the term is from Benabou) $s : \sigma \to \sigma'$ is a family of 2-cells $s_B : \sigma_B \to \sigma'_B$ in $\mathfrak{A}$ such that the diagram

MCN.
$$\begin{array}{ccc} \mathfrak{B}(B,C) & \xrightarrow{F_{B,C}} & \mathfrak{A}(FB,FC) \\ {\scriptstyle G_{B,C}} \downarrow & & \downarrow {\scriptstyle \mathfrak{A}(1,s_C)} \\ \mathfrak{A}(GB,GC) & \xrightarrow[\mathfrak{A}(s_B,1)]{} & \mathfrak{A}(FB,GC) \end{array}$$

commutes. (Note that $\mathfrak{A}(1,s_C)$ is a natural transformation). Compositions are defined in the obvious fashion. This exponentiation yields a cartesian closed structure on the category, 2-Cat_o , of small 2-categories and 2-functors.

The usual full and faithful, locally full and faithful, Yoneda embeddings

$$A \to Cat^{A^{op}} \quad , \quad A^{op} \to Cat^{A}$$

are easily established. For instance, if $f,g : A \to B$ then a modification $\tilde{\sigma} : A(-,f) \to A(-,g)$ of Cat-natural transformations is a family $\{\tilde{\sigma}_X : A(X,f) \to A(X,g)\}$ of natural transformations between functors from $A(X,A)$ to $A(X,B)$ which is Cat-natural in X. If $(\tilde{\sigma}_X)_h$ denotes the component of $\tilde{\sigma}_X$ at $h : X \to A$, then naturality implies that for $k : Y \to X$, one has the formula

$$(2.8) \qquad (\tilde{\sigma}_X)_h k = (\tilde{\sigma}_Y)_{hk}$$

Thus, if $\sigma = (\sigma_A)_{id_A} : f \to g$, then $\tilde{\sigma} = A(-,\sigma)$, so Yoneda is locally full.

The collection of small 2-categories, 2-functors and Cat-natural transformations constitutes a 2-category which we hereafter denote by 2-Cat. If modifications are included, then one obtains a 3-category (see I,2.6), which we denote by $\widetilde{\text{2-Cat}}$.

The adjoint functors between Cat and Sets in I,1.5 c) give rise to analogous ones between 2-Cat and Cat. Here $(-)_o$ plays the role of $|(-)|$. Let $LD : Cat \to 2\text{-}Cat$ be the inclusion of categories as locally discrete 2-categories. Let $L\pi_o : 2\text{-}Cat \to Cat$ be "local π_o"; i.e., a bijection on objects and

$$[(L\pi_o)A](A,A') = \pi_o[A(A,A')]$$

Let $LG : Cat \to 2\text{-}Cat$ be "local G"; i.e., a bijection on objects and $[(LG)\underline{A}](A,A') = G[\underline{A}(A,A')]$. In each case one must check that these locally defined operations are compatible with compositions. Then

$$L\pi_o \dashv LD \dashv (-)_o \dashv LG$$

2-Cat is complete and cocomplete and limits and colimits are (2-Cat)-limits in the sense of closed categories, exactly as in I,1.3 . Limits are computed in the usual fashion, coproducts are disjoint unions and coequalizers are described by exactly the same construction as in I,1.3 except everything (especially the coequalizers between coproducts of finite products) is to be interpreted in Cat . (This is the point of Wolff's construction which holds in the general symmetric monoidal closed category situation). Note that since $(-)_o$ has a right adjoint, it preserves equal izers, so the underlying category of the coequalizer of two 2-functors is as expected. Note further, however, that while the coequalizer of two functors has a convenient description as equivalence classes of strings of morphisms in the codomain of the functors, a similar description of the 2-cells in the coequalizer of two 2-functors involves a very careful and lengthy discussion of equivalence classes of strings of strings of 2-cells. (See, for instance, the proof of I,4.9).

I,2.4. A <u>quasi-natural transformation</u> $\sigma : F \to G$. (Bunge, [7]) is what is called a 2-natural transformation in [CCS]. Thus it is a family of morphisms $\sigma_B : FB \to GB$ together with a family of 2-cells σ_f as illustrated

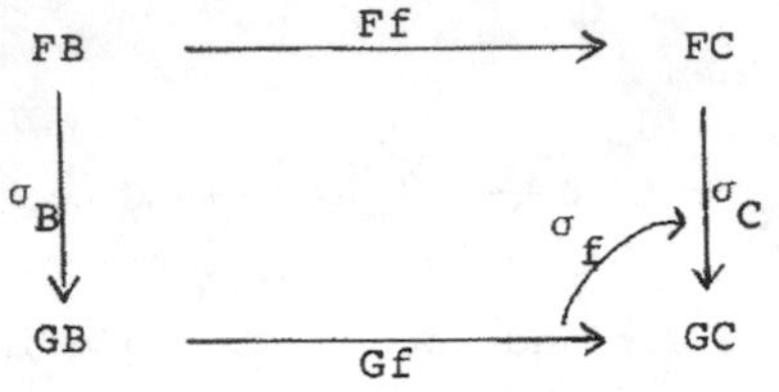

such that

QN1 if $\mu : f \to f'$, then $\sigma_{f'}\cdot(G\mu)\sigma_B = \sigma_C(F\mu)\cdot\sigma_f$

QN2 $\sigma_{I_B} = id_{\sigma_B}$

QN3 $\sigma_{gf} = \sigma_g \boxplus \sigma_f$

The composition of quasi-natural transformations is given by

(2.9) $(\sigma'\sigma)_B = \sigma'_B\sigma_B$, $(\sigma'\sigma)_f = \sigma'_f \boxminus \sigma_f$

Condition QN1 says that, for fixed B and C , the σ_f's constitute a natural transformation as illustrated

(2.10)

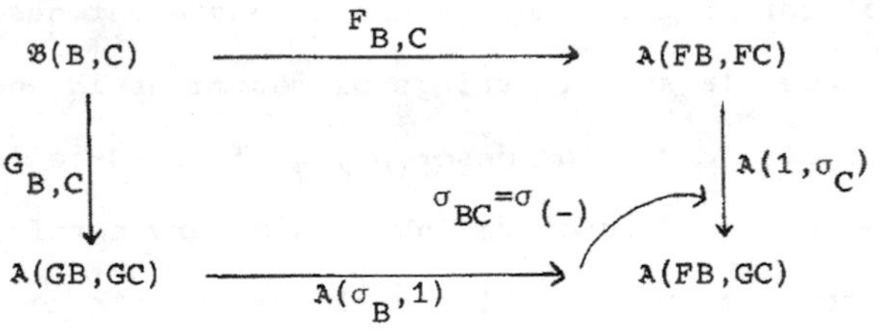

Note, however, that the $\sigma_{(-)}$'s do not behave naturally with respect to morphisms of B and C . Thus, taken together, they do not constitute a modification between the two compositions in the conical diagram (2.7) in I,2.3. Rather, the role of naturality is replaced by condition QN3 which says that the $\sigma_{(-)}$'s are "compatible with composition" in the sense that the diagram

$$\begin{array}{ccc}
\mathfrak{B}(B,C)\times\mathfrak{B}(C,D) & \xrightarrow{\{\ulcorner\sigma_{BC}\urcorner\times tG_{C,D},\, tF_{BC}\times\ulcorner\sigma_{CD}\urcorner\}} & \chi \\
 & & \Big\downarrow {\scriptstyle jux\ \times_{\mathfrak{A}(FB,GD)}\ jux} \\
\Big\downarrow {\scriptstyle jux} & & \mathfrak{A}(FB,GD)^{\underline{3}} \\
 & & \Big\downarrow {\scriptstyle c} \\
\mathfrak{B}(B,D) & \xrightarrow[\ulcorner\sigma_{B,D}\urcorner]{} & \mathfrak{A}(FB,GD)^{\underline{2}}
\end{array} \qquad (2.11)$$

commutes; where "jux" denotes juxtaposition composition, where natural transformations into (-) are regarded as functors into $(-)^{\underline{2}}$ (notations as in I,1.4) and where χ denotes the pullback of the two functors

$$d_1 jux : \mathfrak{A}(FB,GC)^{\underline{2}}\times\mathfrak{A}(GC,GD)^{\underline{2}} \to \mathfrak{A}(FB,GD)$$

$$d_o jux : \mathfrak{A}(FB,FC)^{\underline{2}}\times\mathfrak{A}(FC,GD)^{\underline{2}} \to \mathfrak{A}(FB,GD)$$

Similarly, condition QN2 says that the $\sigma_{(-)}$'s are "compatible with units" in the sense that the diagram

$$(2.12)\qquad \begin{array}{ccc}
 & \underline{1} & \\
{\scriptstyle \ulcorner I_B\urcorner}\swarrow & & \searrow{\scriptstyle \ulcorner\sigma_B\urcorner} \\
\mathfrak{B}(B,B) & \xrightarrow[\ulcorner\sigma_{B,B}\urcorner]{} & \mathfrak{A}(FB,GB)^{\underline{2}}
\end{array}$$

commutes.

In terms of components, these assert the commutativity of the diagrams

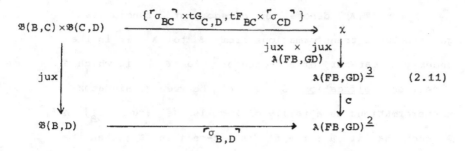

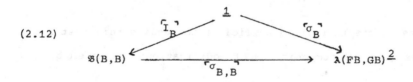

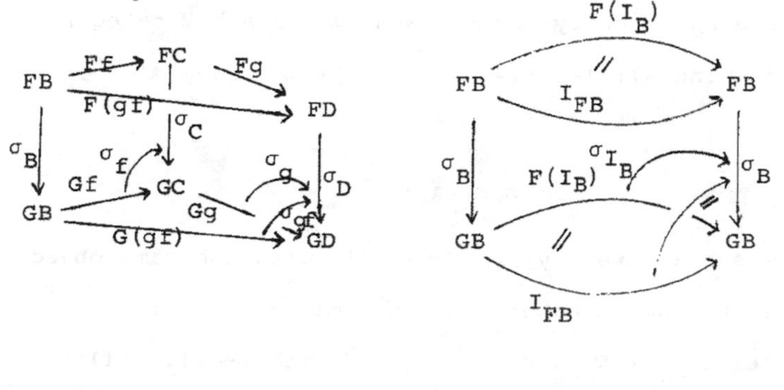

(2.13)

$Fun_o(\mathfrak{B},\mathfrak{A})$ denotes the category of 2-functors and quasi-natural transformations from $\mathfrak{B}$ to $\mathfrak{A}$. It is the underlying category of a 2-category $Fun(\mathfrak{B},\mathfrak{A})$ in which a 2-cell or <u>modification</u> $s : \sigma \to \sigma'$ between quasi-natural transformations is a family of 2-cells $\{s_B : \sigma_B \to \sigma'_B\}$ in $\mathfrak{A}$ such that if $\mu : f \to f'$ is a 2-cell in $\mathfrak{B}$, then the (degenerate) cube

MQN.

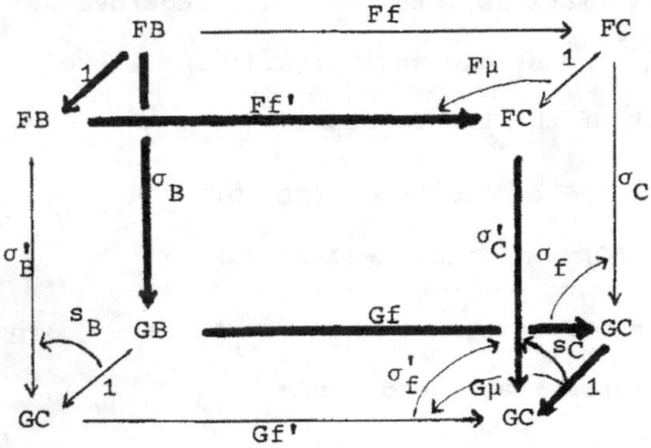

commutes. Note that it is sufficient to require this just for 1-cells. The two compositions of modifications are given by

(2.14) $$(s's)_A = s'_A s_A \ , \ (s' \cdot s)_A = s'_A \cdot s_A$$

Besides these 2-categories, there is an important type of non-full sub 2-category of $Fun(\mathfrak{B},\mathfrak{A})$ described as follows: Let $\mathfrak{B}'_o$ be a subcategory of $\mathfrak{B}_o$ and $\mathfrak{A}'$ a sub 2-category of $\mathfrak{A}$ containing all 1-cells; i.e., $\mathfrak{B}'_o \subset \mathfrak{B}_o$ and $\mathfrak{A}' \supset \mathfrak{A}_o$. Then

$$Fun(\mathfrak{B},\mathfrak{B}'_o;\mathfrak{A},\mathfrak{A}')$$

denotes the sub 2-category of $Fun(\mathfrak{B},\mathfrak{A})$ with the same objects, with 1-cells the quasi-natural transformations σ such that for all 1-cells $f \in \mathfrak{B}'_o$, $\sigma_f \in \mathfrak{A}'$, and with 2-cells all modifications of such quasi-natural transformations. We shall use only two special cases

i) $\mathrm{Fun}(\mathfrak{B},\mathfrak{B}'_o;\mathfrak{A},\mathfrak{A}_o)$; this is given by quasi-natural transformations whose restrictions to $\mathfrak{B}'_o$ are natural transformations. E.g., $\mathrm{Fun}(\mathfrak{B},\mathfrak{B}_o;\mathfrak{A},\mathfrak{A}_o) = \mathfrak{A}^{\mathfrak{B}}$.

ii) Let iso $\mathfrak{A}$ denote the sub 2-category of $\mathfrak{A}$ consisting of 2-cells which are isomorphisms. Then $\mathrm{Fun}(\mathfrak{B},\mathfrak{B}'_o;\mathfrak{A},\mathrm{iso}\ \mathfrak{A})$ consists of quasi-natural transformations such that if $f \in \mathfrak{B}'_o$, then the square involving σ_f commutes up to the isomorphism σ_f .

I,2.5. The 2-<u>comma category</u> $[F_1,F_2]$ of a pair of 2-functors $F_i : \mathfrak{A}_i \to \mathfrak{B}$, $i = 1,2$, is the 2-category with objects pairs of the form (A_1,f,A_2) where $f : F_1(A_1) \to F_2(A_2)$, morphisms triples (h_1,γ,h_2) in diagrams of the form

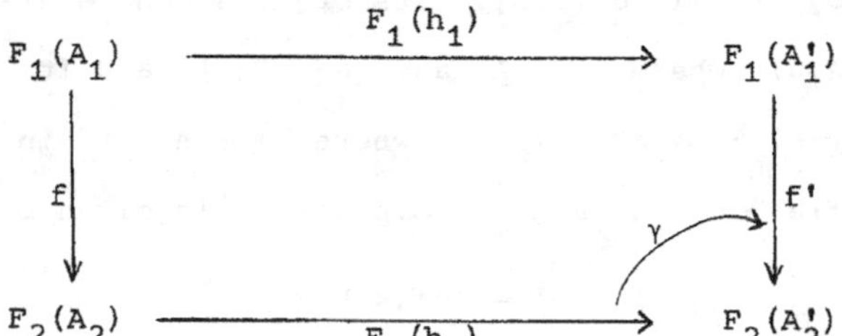

and 2-cells pairs $(\varphi_1 : h_1 \to \tilde{h}_1, \varphi_2 : h_2 \to \tilde{h}_2)$ of 2-cells such that $f'F(\varphi_1)\cdot\gamma = \tilde{\gamma}\cdot F(\varphi_2)f$. Composition of morphisms is given by

(2.16) $\quad (h'_1,\gamma',h'_2)(h_1,\gamma,h_2) = (h'_1h_1, \gamma' \boxdot \gamma, h'_2h_2)$,

while the two compositions of 2-cells are

$$(2.17)\quad \begin{aligned}(\psi_1,\psi_2)\cdot(\varphi_1,\varphi_2) &= (\psi_1\cdot\varphi_1,\psi_2\cdot\varphi_2)\\ (\psi_1,\psi_2)(\varphi_1,\varphi_2) &= (\psi_1\varphi_1,\psi_2\varphi_2)\ .\end{aligned}$$

Alternatively, $[F_1,F_2]$ is the inverse limit in 2-Cat of the diagram

$$\mathfrak{A}_1 \xrightarrow{F_1} \mathfrak{B} \xleftarrow{\delta_0} \mathrm{Fun}\ \mathfrak{B} \xrightarrow{\delta_1} \mathfrak{B} \xleftarrow{F_2} \mathfrak{A}_2$$

where $\mathrm{Fun}\ \mathfrak{B} = {}^{op}\mathrm{Fun}(\underline{2},{}^{op}\mathfrak{B})$ and where δ_i is induced by $\partial_i : \underline{1} \to \underline{2}$, $i = 0,1$. Hence there are projections $P_i : [F_1,F_2] \to \mathfrak{A}_i$, $i = 1,2$ and conditions QN 1,2,3 of I,2.4 say that a quasi-natural transformation $\sigma : F \to G$ can be identified with a 2-functor $\bar{\sigma} : \mathfrak{B} \to [F,G]$ such that $P_1\bar{\sigma} = F$ and $P_2\bar{\sigma} = G$.

An important example of a 2-comma category is one of the form $[1,F]$ where $F : \underline{A} \to \mathrm{Cat}$ and $1 : \underline{1} \to \mathrm{Cat}$ takes the value $\underline{1} \in \mathrm{Cat}$. $[1,F]$ is the opfibred category over $\underline{A}$ determined by F . (See [CSS].) Its objects can be described as pairs (A,a) where $A \in \underline{A}$ and $a \in F(A)$ and its morphisms as pairs $(f,\varphi) : (A,a) \to (B,b)$ where $f : A \to B$ in $\underline{A}$ and $\varphi : F(f)a \to b$ in $F(B)$. Composition is given by

$$(g,\psi)(f,\varphi) = (gf,\psi\cdot F(g)\varphi)$$

The class of morphisms of the form

$$(f,id_{F(f)a}) : (A,a) \to (B,F(f)a)$$

is the canonical choice of cocartesian morphisms for this opfibration.

Similarly, if $F : \underline{A}^{op} \to \mathrm{Cat}$ then the fibration over $\underline{A}$ corresponding to F is the category $[1,(-)^{op}F]^{op}$, where

$$(-)^{op} : \mathrm{Cat} \to \mathrm{Cat}$$

takes each category to its opposite category. The objects of this fibration are again pairs (A,a) as above, while morphisms are pairs

$$(f,\varphi) : (A,a) \to (B,b)$$

where $f : A \to B$ in $\underline{A}$ and $\varphi : a \to F(f)b$ in $F(A)$. Composition is given by

$$(g,\psi)(f,\varphi) = (gf, F(f)\psi \cdot \varphi)$$

The canonical cartesian morphisms are those of the form

$$(f, id_{F(f)b}) : (A, F(f)b) \to (B,b) \ .$$

I,2.6. A 3-<u>category</u> $\tilde{\mathfrak{A}}$ is a (2-Cat)-category; i.e., enriched in the cartesian closed category 2-Cat (see I,2.3). In this work all that is needed is a notation for the seven possible duals. They are combinations of

a) $\tilde{\mathfrak{A}}^{op}$, where $\tilde{\mathfrak{A}}^{op}(A,B) = \tilde{\mathfrak{A}}(B,A)$

b) ${}^{op}\tilde{\mathfrak{A}}$, where ${}^{op}\tilde{\mathfrak{A}}(A,B) = \tilde{\mathfrak{A}}(A,B)^{op}$

c) ${}_{op}\tilde{\mathfrak{A}}$, where ${}_{op}\tilde{\mathfrak{A}}(A,B) = {}^{op}[\tilde{\mathfrak{A}}(A,B)]$

A 3-functor is a (2-Cat)-functor. There are 2-Cat-natural transformations, modifications of such, and 3-cells between modifications, giving rise to a cartesian closed structure on the category, 3-Cat, of small 3-categories. Similarly, there are various kinds of quasi-natural transformations and constructions corresponding to $\mathrm{Fun}(\mathfrak{A},\mathfrak{B})$. Here we need only the analogous comma category.

I,2.7. The 3-<u>comma category</u> $[F_1,F_2]_3$ of a pair of 3-functors $F_i : \tilde{\mathfrak{A}}_i \to \tilde{\mathfrak{B}}$, $i = 1,2$ is the three category with objects and morphisms the same as in $[F_1,F_2]$. Following the notation of I,2.5, a 2-cell is a pair (φ_1,φ_2) of 2-cells together with a 3-cell ξ in $\tilde{\mathfrak{B}}$ as illustrated:

$$(2.18)\qquad \begin{array}{ccc} F_2(h_1)f & \xrightarrow{\;F_2(\varphi_2)f\;} & F_2(h_2')f \\ \Big\downarrow{\scriptstyle\gamma} & \overset{\xi}{\nearrow} & \Big\downarrow{\scriptstyle\gamma'} \\ f'F_1(h_1) & \xrightarrow[\;f'F_1(\varphi_1)\;]{} & f'F_1(h_1') \end{array}$$

The compositions of such 2-cells are defined in the evident fashion. Finally, a 3-cell is a pair $(\xi_1:\varphi_1\to\varphi_1',\xi_2:\varphi_2\to\varphi_2')$ of 3-cells such that

$$(2.19)\qquad \gamma'(F_2(\xi_2)f)\cdot\xi = \xi'\cdot(f'F_1(\xi_1))\gamma .$$

Alternatively, a pair of objects (A_1,f,A_i) and (A_1',f',A_2') determines 2-functors

$$\tilde{\mathfrak{A}}_2(A_2,A_2') \xrightarrow{\;F_2(-)f\;} \tilde{\mathfrak{B}}(F_1(A_1),F_2(A_2)) \xleftarrow{\;f'F_1(-)\;} \tilde{\mathfrak{A}}_1(A_1,A_1')$$

and the 2-Cat-valued hom object between these two objects is the 2-comma category

$$[F_2(-)f,f'F_1(-)]$$

It can also be described as an inverse limit

$$\tilde{\mathfrak{A}}_1 \to \tilde{\mathfrak{B}} \leftarrow 3\text{-Fun}_{\tilde{\mathfrak{B}}} \to \tilde{\mathfrak{B}} \leftarrow \tilde{\mathfrak{A}}_2$$

where 3-Fun is given by a "basic construction" for

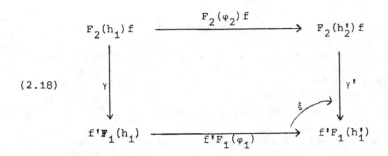

3-categories as in [CCS], §5, but replacing the comma category there by a 2-comma category as above.

I,2.8. A <u>double category</u> $\mathfrak{D}$ (Ehresmann [15]) is a class of "morphisms" carrying two different compatible category structures; i.e., a class M with two structures $(\partial_o, \partial_1, \cdot)$ and $(\tilde{\partial}_o, \tilde{\partial}_1, \tilde{\cdot})$ of domain, codomain, and composition such that

i) each is a category

ii) $\partial_i \tilde{\partial}_j = \tilde{\partial}_j \partial_i$ and the objects for one structure form a subcategory for the other structure.

iii) there is an interchange law

$$(\psi' \tilde{\cdot} \varphi') \cdot (\psi \tilde{\cdot} \varphi) = (\psi' \cdot \psi) \tilde{\cdot} (\varphi' \cdot \varphi)$$

We shall observe in II,1. that a small double category is the same thing as a category object in Cat . However, both interpretations are useful. A 2-category is a special case of a double category with the property that the objects for the first structure are also objects for the second structure. (Cf. also [CSS]); i.e., the category structure induced on the objects for the first structure by the second structure is discrete.

We generally think of the first category structure as the "strong" or "horizontal" structure and write $\mathfrak{D}^{op}$ for dualization with respect to this structure, while the second structure is considered as the "weak" or "vertical" one, with ${}^{op}\mathfrak{D}$ denoting dualization with respect to it.

Double functors and double natural transformations are described in the obvious way as having the required properties with respect to both structures.

It is clear from this what a triple category is; i.e., three category structures which form double category structures when taken two at a time. One defines an n-tuple category similarly. There are various special cases of triple categories which occur in this work - one, two, or all three of the double category structures may be 2-categories, some of which in turn may be locally discrete. A further refinement is to allow one or more of the double structures to be a bicategory structure, as in the next section, rather than a 2-category. This does not occur in our work, but "triple" functors which are only pseudo functors with respect to one of the 2-category structures do. The only kind of triple categories that occurs in this work is that in which two of the double category structures are 2-categories. This is the same as a category object (see II,1.) in 2-Cat, so we shall always call it this to avoid confusion.

The category $(\mathrm{Fun}\ \mathfrak{B})_o = {}^{op}\mathrm{Fun}_o(\underline{2}, {}^{op}\mathfrak{B})$ mentioned in I,2.5 and dealt with at greater length in §I,4 is a double category. As described, an object is a 1-cell $f : A \to B$ in $\mathfrak{B}$ and a morphism is a diagram

$$D:\quad \begin{array}{ccc} A & \xrightarrow{\;h\;} & A' \\ {\scriptstyle f}\big\downarrow & \overset{\varphi}{\nearrow} & \big\downarrow{\scriptstyle f'} \\ B & \xrightarrow[\;k\;]{} & B' \end{array} \quad ,$$

composition being horizontal composition of such diagrams. Another category structure on the same set of diagrams is given by defining $\tilde{\partial}_0 D = h$, $\tilde{\partial}_1 D = k$ and composition to be vertical composition. The interchange law (2.4) is exactly what is needed to show that this is a double category. If one includes 2-cells one obtains a double 2-category, which we prefer to discuss as a category object in 2-Cat.

I,2.9. 2- <u>and</u> 3-<u>Categorical</u> <u>Fibrations</u> can be described in exactly the same way as categorical fibrations, as in [CCS], §1. (See also [FCC] and the brief outline in I,1.11.) We use Cat-adjoint and 2-Cat-adjoint in the sense of closed categories; e.g., $F : \mathfrak{A} \to \mathfrak{B}$ and $U : \mathfrak{B} \to \mathfrak{A}$ are Cat-adjoint if there is a Cat-natural isomorphism

$$\mathfrak{B}(F(-),-) \xrightarrow{\sim} \mathfrak{A}(-,U(-))$$

between Cat-valued 2-functors (see also I,6.2, (4)). Thus, treating only the case of 2-categories for brevity, a 2-functor $P : \mathfrak{C} \to \mathfrak{B}$ is a 2-fibration if there exists a 2-functor $L : (\mathfrak{B},P) \to \mathfrak{C}^{\underline{2}}$ such that $SL = (\mathfrak{B},P)$ and S is the left Cat-adjoint to L . Here (F_1,F_2) is the comma 2-category; i.e., the limit in 2-Cat of the diagrams

$$\mathfrak{A}_1 \xrightarrow{F_1} \mathfrak{B} \xleftarrow{\mathfrak{B}^{\partial_0}} \mathfrak{B}^{\underline{2}} \xrightarrow{\mathfrak{B}^{\partial_1}} \mathfrak{B} \xleftarrow{F_2} \mathfrak{A}_2$$

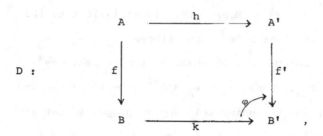

and $S = \{P^{\underline{2}}, \mathfrak{C}^{\partial_1}\} : \mathfrak{C}^{\underline{2}} \to (\mathfrak{B}, P)$. This is equivalent to the existence of 2-functors between the fibres, $f^* : \mathfrak{C}_B \to \mathfrak{C}_A$ for $f : A \to B$ in $\mathfrak{B}$, and Cat-natural transformations $\theta_f : J_B \circ f^* \to J_A$, where $J_A : \mathfrak{C}_A = P^{-1}(A) \to \mathfrak{C}$ is the inclusion, satisfying the universal mapping property that given $E \in \mathfrak{C}_B$ and 1-cells $m : D \to E$ in $\mathfrak{C}$ and $h : P(D) \to A$ with $fh = P(m)$ then there is a unique h' in $\mathfrak{C}$ with $(\theta_f)_E h' = m$ and $P(h') = h$

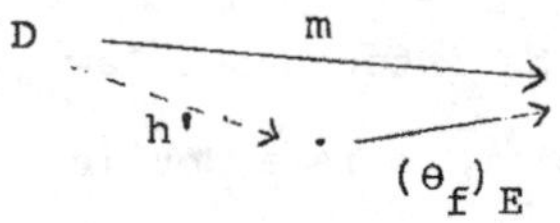

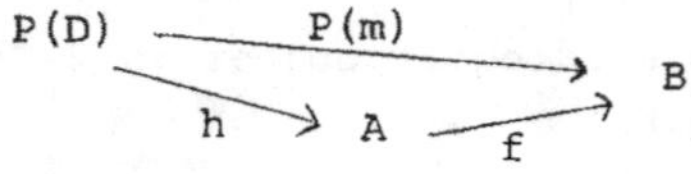

(resp., 2-cells, $\psi : m \to m'$ and $\varphi : h \to h'$ with $f\varphi = P(\psi)$, then there is a unique φ' with $(\theta_f)_E \varphi' = \psi$ and $P(\varphi') = \varphi$.) (Note that this property is incorrectly stated in [CCS]; cf., [FCC].) A choice of f^* and θ_f satisfying $(id_B)^* = \mathfrak{C}_B$, $\theta_{id_B} = id$, $(gf)^* = f^*g^*$ and $\theta_{gf} = \theta_g \theta_f$ (if possible) is called a split-normal 2-cleavage, and $P : \mathfrak{C} \to \mathfrak{B}$ together with such a choice is called a split-normal 2-fibration. The 2-category of cleavage preserving 2-functors and modifications of such between split-normal 2-fibrations over $\mathfrak{B}$ is isomorphic to the 2-category $(2\text{-Cat})^{\mathfrak{B}^{op}}$. If $F : \mathfrak{A} \to \mathfrak{B}$ is any 2-functor then it has an associated 2-fibration $P_F : (\mathfrak{B}, F) \to \mathfrak{B}$ such that $F = P_F Q_F$ where Q_F is right adjoint to the projection $(\mathfrak{B}, F) \to \mathfrak{A}$, and this is the best possible factorization through a split-normal 2-fibration.

Analogous results hold in the 3-category case and for opfibrations. The proofs are easily adapted from those in [FCC]. We note for later reference that if $\mathfrak{B}$ has enriched pullbacks, then the projection $(\mathfrak{B},F) \to \mathfrak{A}$ is also a 2-fibration, via pullbacks.

In I,1.13 ii) it was observed that a split normal fibration gave rise to a functor $\tilde{p} : B \to [{}^{op}Cat,\underline{E}]_o$, and this was used for Kan-extensions. In I,7.14, an analogous construction will be needed for quasi-Kan-extensions. We treat the dual case since that is what will be used there. Let $P : \mathfrak{E} \to \mathfrak{B}$ be a split normal 2-fibration. Then there is a 2-functor

$$\tilde{P} : \mathfrak{B} \to [2\text{-Cat},\mathfrak{E}]_3 \qquad \text{(Cf. I,2.7)}$$

where $\tilde{P}(B)$ and $\tilde{P}(f)$ are defined as in I,1.13 (except that φ_f goes the other way). If $\lambda : f \to f'$ is a 2-cell in $\mathfrak{B}$, then $\lambda_* : f_* \to f'_*$ is a Cat-natural transformation (i.e., 2-cell in 2-Cat) and $\tilde{P}(\lambda)$ is the 2-cell in $[2\text{-Cat},\xi]_3$ consisting of (λ_*,id) and the modification ξ_f as illustrated

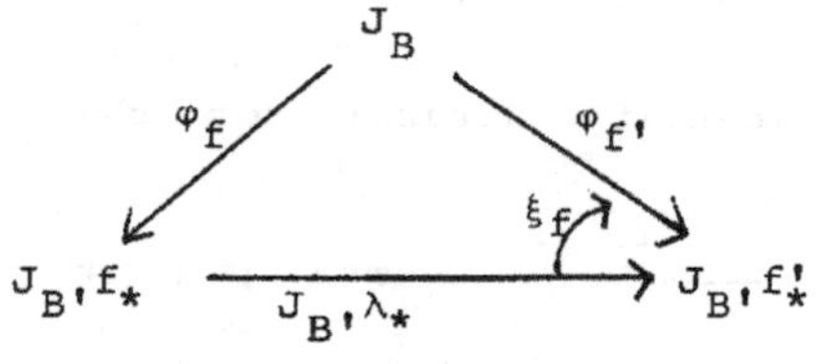

where $(\xi_f)_E$ is the unique 2-cell such that $P(\xi_f)_E = \lambda$. (See also I,7.13).

I,3. Bicategories. The notion of a bicategory is closely related to that of a 2-category and many of our main results involve notions from the theory of bicategories. The following description is essentially that of Benabou [BC].

I,3.1. A bicategory $\mathfrak{B}$ consists of a set (not necessarily small), Ob $\mathfrak{B}$, of objects together with

BC1 (small) categories $\mathfrak{B}(A,B)$ for each ordered pair of objects,

BC2 "composition" functors

$$\mathfrak{B}(A,B)\times\mathfrak{B}(B,C) \xrightarrow{\circ} \mathfrak{B}(A,C)$$

for each ordered triple of objects,

BC3 "identity" objects $I_A \in |\mathfrak{B}(A,A)|$ for each A,

BC4 "associativity" natural isomorphisms

$$\begin{array}{ccc} \mathfrak{B}(A,B)\times\mathfrak{B}(B,C)\times\mathfrak{B}(C,D) & \xrightarrow[\circ\times 1]{} & \mathfrak{B}(A,C)\times\mathfrak{B}(C,D) \\ {\scriptstyle 1\times\circ}\downarrow & \overset{\alpha_{ABCD}}{\curvearrowright} & \downarrow{\scriptstyle\circ} \\ \mathfrak{B}(A,B)\times\mathfrak{B}(B,D) & \xrightarrow[\circ]{} & \mathfrak{B}(A,D) \end{array},$$

BC5 "left and right identity" natural isomorphisms

$$\begin{array}{ccc} \underline{1}\times\mathfrak{B}(A,B) & \xrightarrow{\ulcorner I_A \urcorner \times 1} & \mathfrak{B}(A,A)\times\mathfrak{B}(A,B) \\ & {\scriptstyle\cong}\searrow \ \overset{r_{AB}}{\curvearrowright} \ \swarrow{\scriptstyle\circ} & \\ & \mathfrak{B}(A,B) & \end{array},$$

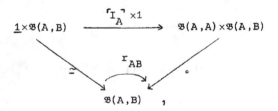

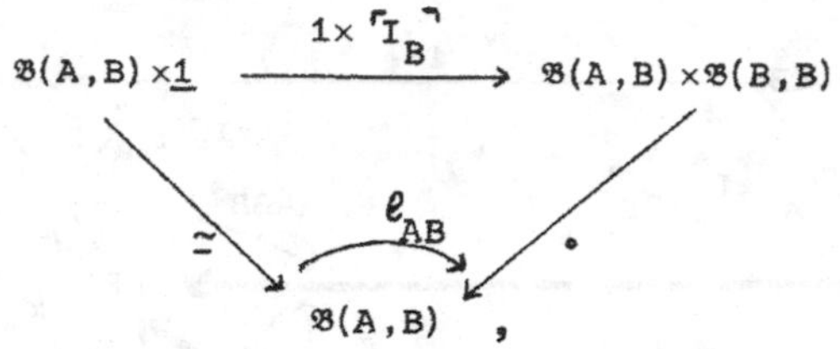

,

subject to two conditions:

BC6 The cube (in which we write AB for 𝔅(A,B) and 1 for all identity maps)

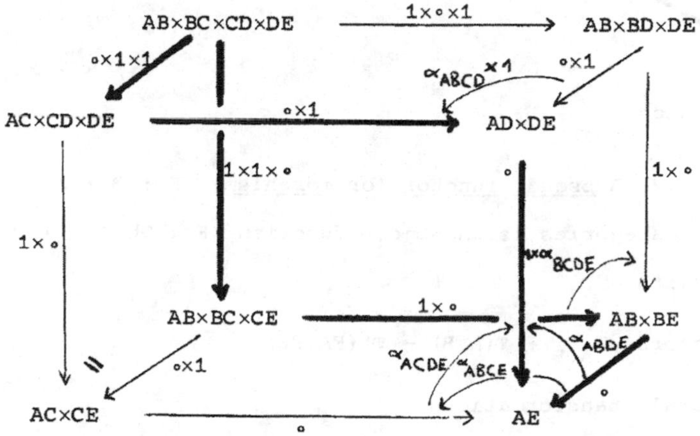

commutes. (This is the "pentagon" condition for coherence of the associativity isomorphisms.)

BC7 The (degenerate) cube

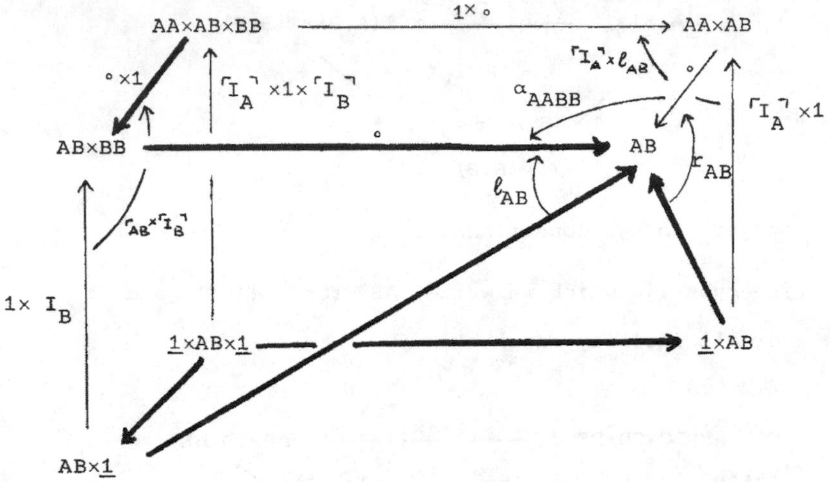

commutes.

I,3.2. A <u>pseudo</u>-<u>functor</u> (or <u>morphism</u>) $F : \mathfrak{B} \to \mathfrak{B}'$ between bicategories is an object function $F : \mathrm{Ob}\ \mathfrak{B} \to \mathrm{Ob}\ \mathfrak{B}'$ together with

PF1 functors $F_{A,B} : \mathfrak{B}(A,B) \to \mathfrak{B}'(FA,FB)$,

PF2 natural transformations

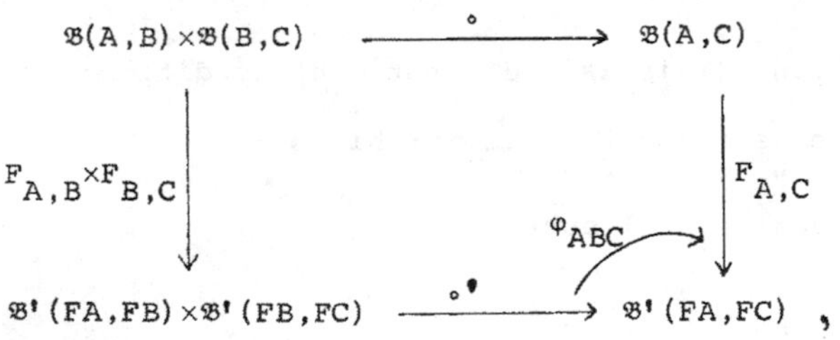

PF3 1-cells $\varphi_A : I'_{FA} \to F(I_A)$ in $\mathfrak{B}'(FA,FA)$, subject to two conditions:

PF4 The cube

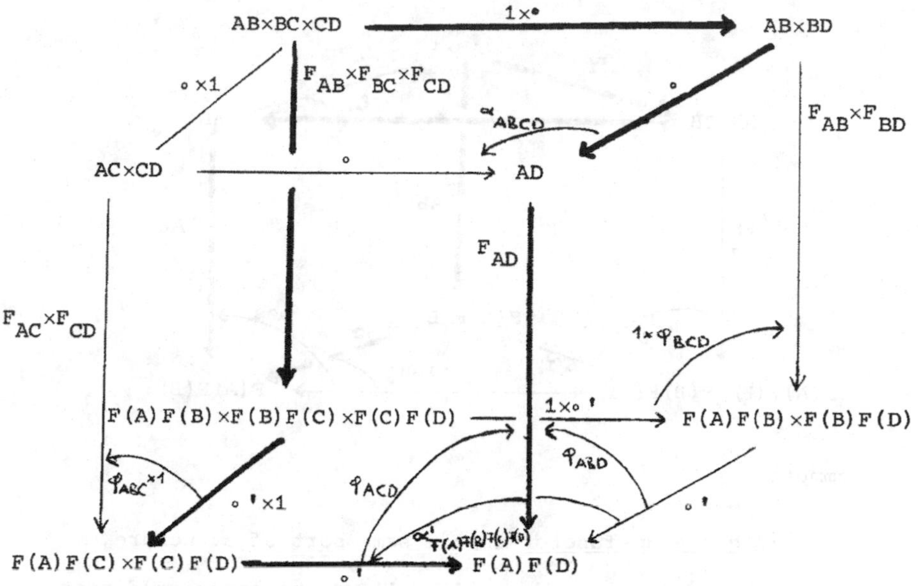

commutes

PF5 The (degenerate) cubes

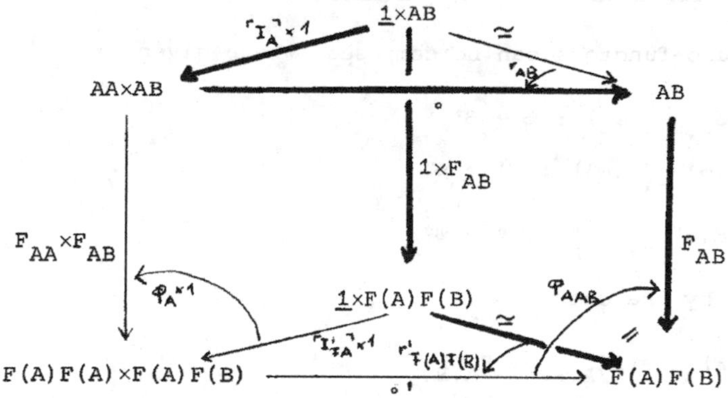

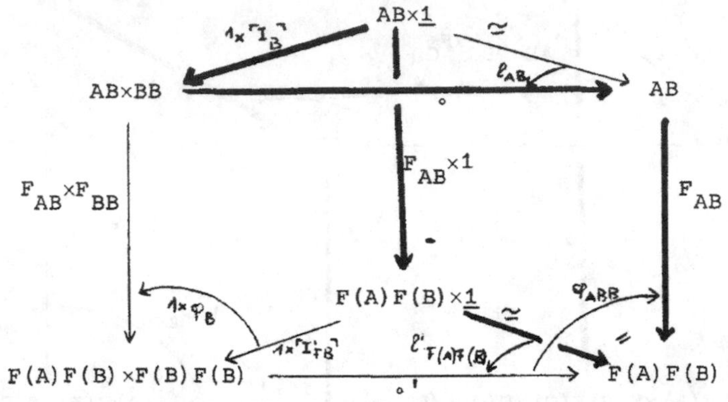

commute.

A <u>copseudo-functor</u> is the same sort of structure, except the φ_A's and $\varphi_{A,B,C}$'s all go the opposite direction. This can also be described by suitable duals. If all the φ's are isomorphisms (resp., identities), then a pseudo-functor is called a <u>homomorphic</u> (resp., <u>strict</u>) pseudo-functor.

Pseudo-functors can be composed, where given

$(F,\varphi_{-,-,-},\varphi_{-}) : \mathfrak{B} \to \mathfrak{B}'$

$(F',\varphi'_{-,-,-},\varphi'_{-}) : \mathfrak{B}' \to \mathfrak{B}''$

then $(F'F,\varphi''_{-,-,-},\varphi''_{-}) : \mathfrak{B} \to \mathfrak{B}''$

is defined by the data

$$(F'F)_{AB} = F'_{FA,FB} \circ F_{A,B}$$

$$\varphi''_{ABC} = \varphi'_{FA,FB,FC} \boxminus \varphi_{A,B,C}$$

$$\varphi''_{A} = F'(\varphi_A) \circ \varphi'_{FA}$$

Vertically stacking the diagrams in PF4 and PF5 shows that this is a pseudo-functor and that composition is associative. We denote the category of small bicategories and pseudo-functors by Bicat .

I,3.3. A <u>quasi-natural transformation</u> between pseudo-functors F and F' from $\mathfrak{B}$ to $\mathfrak{B}'$ consists of

QNP1 a family of 1-cells $\sigma_B : FB \to F'B$,

QNP2 a family of natural transformations

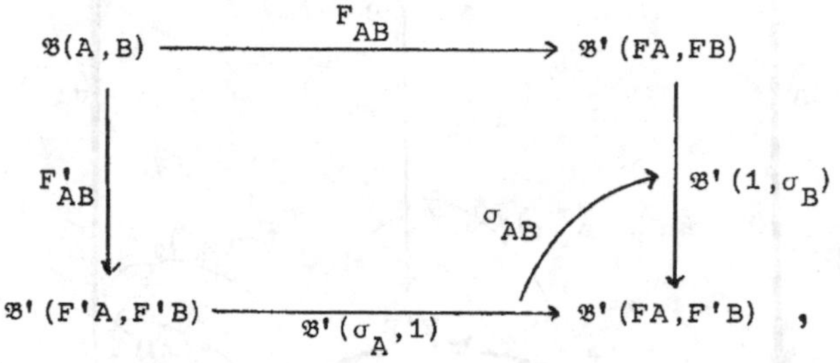

,

subject to two conditions:

QNP3 the (degenerate) cube

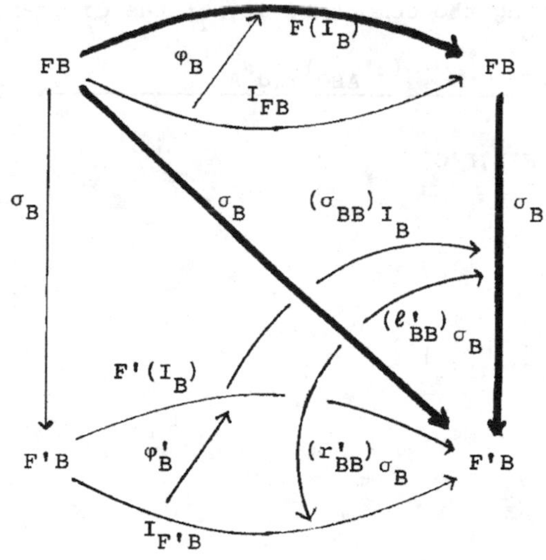

commutes for all B .

QNP4 The degenerate cube

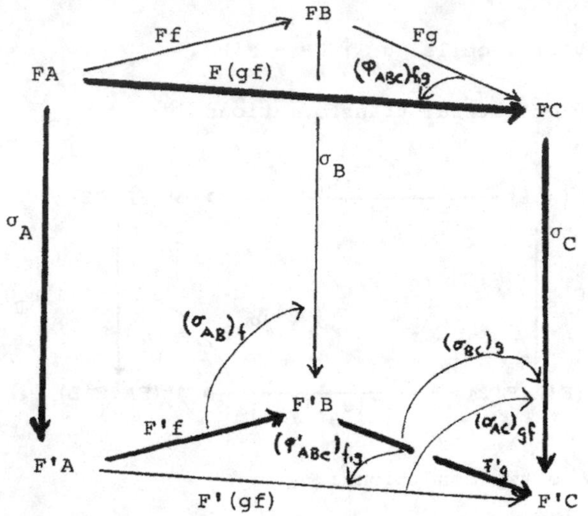

commutes for all f and g . Because of the non-strict associativity of composition, this must be understood as asserting the commutativity of the diagram

$$
\begin{array}{ccc}
[(F'g)(F'f)]\sigma_A & \xrightarrow{(\varphi'_{ABC})_{f,g}\sigma_A} & F'(gf)\sigma_A \\
\downarrow \alpha'_{FA,F'A,F'B,F'C} & & \\
(F'g)[(F'f)\sigma_A] & & \\
\downarrow (F'g)\sigma_f & & \\
(F'g)[\sigma_B(Ff)] & & \downarrow \sigma_{gf} \\
\downarrow (\alpha'_{FA,FB,F'B,F'C})^{-1} & & \\
[(F'g)\sigma_B](Ff) & & \\
\downarrow \sigma_g(Ff) & & \\
[\sigma_C(Fg)](Ff) & & \\
\downarrow \alpha'_{FA,FB,FC,F'C} & & \\
\sigma_C[(Fg)(Ff)] & \xrightarrow[\sigma_C(\varphi_{ABC})_{f,g}]{} & \sigma_C F(gf)
\end{array}
$$

The composition of quasi-natural transformations is defined by composing such squares vertically, again inserting all the required instances of α' .

Modifications are defined as in I,2.4; i.e., $s : \sigma \to \sigma'$ is a family of 2-cells $\{s_A : \sigma_A \to \sigma'_A\}$ such that

MQNP. $$(\sigma'_{AB})_f \cdot [(F'\mu) s_A] = [s_B (F\mu)] \cdot (\sigma_{AB})_f$$

The two compositions are defined component-wise, and yield a bicategory Pseud($\mathfrak{B}$,$\mathfrak{B}'$) . Note that, since the compositions only use the composition structure of $\mathfrak{B}'$, if $\mathfrak{B}'$ is a 2-category then so is Pseud ($\mathfrak{B}$,$\mathfrak{B}'$) . There are faithful Yoneda embeddings

$$\mathfrak{B} \to \text{Pseud}\ (\mathfrak{B}^{op}, \text{Cat})$$
$$\mathfrak{B}^{op} \to \text{Pseud}\ (\mathfrak{B}, \text{Cat})$$

but they are __not__ full in general.

The corresponding constructions for copseudo-functors yield a bicategory coPseud ($\mathfrak{B}$,$\mathfrak{B}'$) .

I,3.4. __Examples__.

1) Examples of bicategories in [BC] that concern us are

a) multiplicative categories = bicategories with one object. The category of abelian groups with $\otimes$ as multiplication provides a simple example of a non-full Yoneda embedding.

b) Bim = the bicategory of bimodules where objects are rings with unit, where 1-cells $R \to S$ are R-S-bimodules $_R M_S$ and where 2-cells are homomorphisms of bimodules. Composition is tensor product over the middle ring.

c) Spans χ , where χ is a category with pullbacks. An object of Spans χ is an object of χ , a 1-cell from V to V' is a diagram $V \leftarrow X \rightarrow V'$ and a 2-cell is a commutative diagram

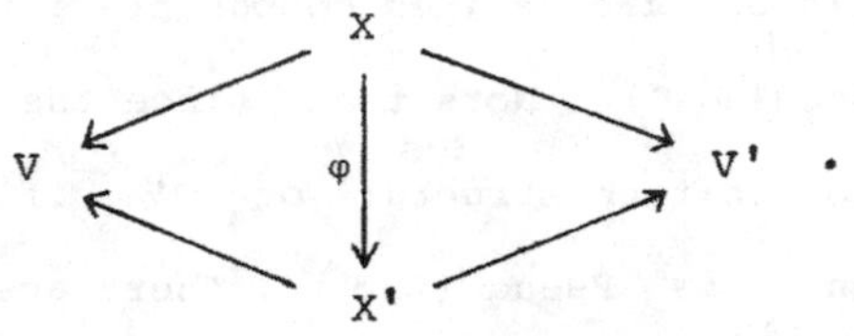

Composition is given by pullbacks.

2) The construction of Bim above can be generalized to an arbitrary bicategory $\mathcal{B}$ such that all categories $\mathcal{B}(A,B)$ have coequalizers which are preserved by composition with fixed 1-cells, giving a bicategory $\mathrm{Bim}(\mathcal{B})$ as follows: The composition in $\mathcal{B}$ makes each category $\mathcal{B}(A,A)$ a multiplicative category. An object $R \in \mathcal{B}(A,A)$ is a monoid if one is given morphisms

$$R \circ R \xrightarrow{m_R} R \ , \ I_A \xrightarrow{e_R} R$$

which are associative and unitary, up to the isomorphisms involving "$\circ$" and I_A .

a) An object of $\mathrm{Bim}(\mathcal{B})$ is a pair (A,R) where $A \in \mathrm{Ob}\ B$ and R is a monoid in $\mathcal{B}(A,A)$.

Similarly, composition determines actions of $\mathcal{B}(A,A)$ and $\mathcal{B}(B,B)$ on $\mathcal{B}(A,B)$,

$$\mathcal{B}(A,A) \times \mathcal{B}(A,B) \rightarrow \mathcal{B}(A,B)$$

$$\mathcal{B}(A,B) \times \mathcal{B}(B,B) \rightarrow \mathcal{B}(A,B)$$

If $S \in \mathcal{B}(A,A)$ and $R \in \mathcal{B}(B,B)$ are monoids then

$M \in \mathfrak{B}(A,B)$ is a left R - right S-bimodule if there are given morphisms

$$R \circ M \xrightarrow{\varphi} M \ , \quad M \circ S \xrightarrow{\psi} M$$

such that the diagrams

$$\begin{array}{ccc} I_B \circ M & & \\ {\scriptstyle e_R \circ M}\downarrow & \searrow {\scriptstyle (\ell_{AB})^{-1}_M} & \\ R \circ M & \xrightarrow{\varphi} & M \end{array} \qquad \begin{array}{ccc} M \circ I_A & & \\ {\scriptstyle M \circ e_S}\downarrow & \searrow {\scriptstyle (r_{AB})^{-1}_M} & \\ M \circ S & \xrightarrow{\psi} & M \end{array}$$

$$\begin{array}{ccc} R \circ R \circ M & \xrightarrow{m_R \circ M} & R \circ M \\ {\scriptstyle R \circ \varphi}\downarrow & & \downarrow{\scriptstyle \varphi} \\ R \circ M & \xrightarrow{\varphi} & M \end{array} \qquad \begin{array}{ccc} M \circ S \circ S & \xrightarrow{M \circ m_S} & M \circ S \\ {\scriptstyle \psi \circ S}\downarrow & & \downarrow{\scriptstyle \psi} \\ M \circ S & \xrightarrow{\psi} & M \end{array}$$

$$\begin{array}{ccc} R \circ M \circ S & \xrightarrow{R \circ \psi} & R \circ M \\ {\scriptstyle \varphi \circ S}\downarrow & & \downarrow{\scriptstyle \varphi} \\ M \circ S & \xrightarrow{\psi} & M \end{array}$$

commute, where we have left out α, ℓ, and r. A morphism of bimodules is a morphism $f : M \to M'$ such that the diagrams

$$\begin{array}{ccc} R \circ M & \longrightarrow & M \\ {\scriptstyle R \circ f}\downarrow & & \downarrow{\scriptstyle f} \\ R \circ M' & \longrightarrow & M' \end{array} \qquad \begin{array}{ccc} M \circ S & \longrightarrow & M \\ {\scriptstyle f \circ S}\downarrow & & \downarrow{\scriptstyle f} \\ M' \circ S & \longrightarrow & M' \end{array}$$

commute.

b) A 1-cell from (A,S) to (B,R) in $\mathrm{Bim}(\mathfrak{B})$ is a bimodule ${}_R M_S$. A 2-cell is a morphism of such bimodules.

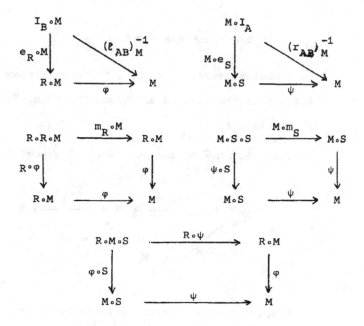

c) If ${}_R M_S : (A,S) \to (B,R)$ and ${}_S N_T : (C,T) \to (A,S)$ then the composition of M and N is the tensor product

$$ {}_R M_S \underset{S}{\otimes} {}_S N_T $$

where

$$ M \circ S \circ N \underset{M \circ \varphi}{\overset{\psi \circ N}{\rightrightarrows}} M \circ N \longrightarrow M \underset{S}{\otimes} N $$

is a coequalizer in $\mathcal{B}(A,C)$. The composition of 2-cells is the induced morphism. The identity for (B,R) is ${}_R R_R$.

That $R \underset{R}{\otimes} M \simeq M$ follows as usual since

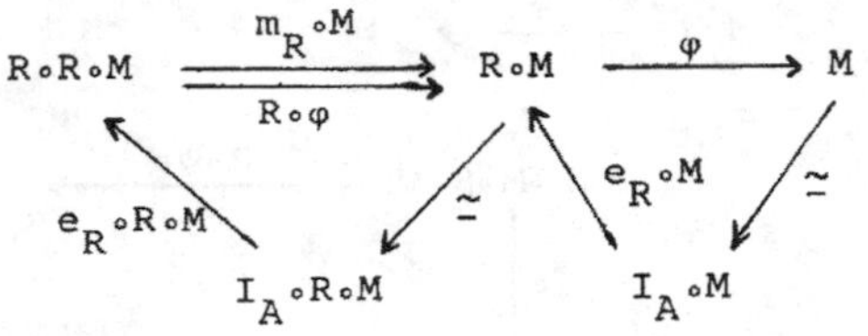

is a split coequalizer. Associativity follows since composition with a fixed 1-cell preserves coequalizers.

Note that if the category of abelian groups, Ab, is treated as a multiplicative category via $\otimes$; i.e., as a bicategory with one object, then

$$ \text{Bim(Ab)} = \text{Bim} $$

3) In this work we shall be concerned with Bim(Spans χ) . If $V \in \chi$, then Spans $\chi(V,V)$ is the multiplicative category whose objects look like $E \underset{d_1}{\overset{d_0}{\rightrightarrows}} V$, and whose morphisms are diagrams

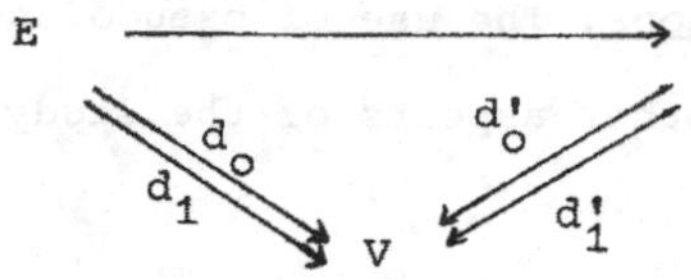

where both triangles commute. The product of two objects as above is given by

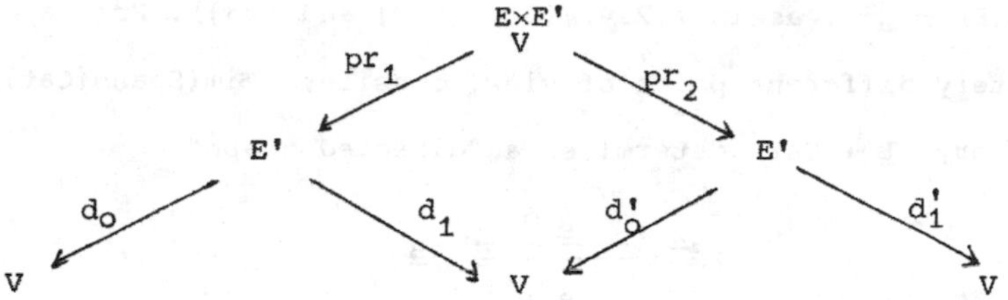

and the unit is $V \rightrightarrows V$. One can think of Spans $\chi(V,V)$ as directed graphs with vertices V ; monoids will be discussed in II,1 and bimodules in II,2 and the interpretation in terms of Bim(Spans χ) in II,2.2.

The condition about the preservation of coequalizers in 2) above only affects the associativity of the composition in Bim($\mathfrak{B}$) . We shall frequently ignore this when associativity plays no role. However, we do note the following easily verified result:
If χ has universal coequalizers (as well as pullbacks) then it satisfies the conditions for associativity of composition in Bim(Spans χ) . Here, universal coequalizers means that if $f : X \to Y$, then the functor given by pullbacks

$$f^* : (\chi,Y) \to (\chi,X) ,$$

preserves coequalizers. This is always satisfied in a topos since there f^* has a right adjoint.

I,3.5. <u>Fibrations</u>. The use of pseudofunctors and spans illuminates various other aspects of the study of fibrations. (Cf., I,2.9.)

i) Homomorphic pseudo-functors from $\underline{B}^{op}$ to Cat are in bijective correspondence with fibrations $P : \underline{E} \to \underline{B}$ with chosen cleavages; equivalently, with chosen lifting functors $L : (\underline{B},P) \to \underline{E}^2$, as in I,2.9. (Cf., [BC] and [23]). From a completely different point of view, consider $\mathfrak{Bim}$(Spans(Cat)). A category $\underline{B} \in$ Cat determines a "directed graph"

$$\underline{B}^2 \underset{\underline{B}^{\partial_1}}{\overset{\underline{B}^{\partial_o}}{\rightrightarrows}} \underline{B}$$

and the structure described in I,1.6 makes this a monoid in Spans Cat$(\underline{B},\underline{B})$. Consider a functor $P : \underline{E} \to \underline{B}$ as a span $(P,Id_{\underline{E}})$ from $\underline{B}$ to $\underline{E}$. An "action" of this monoid on this span is a commutative diagram

$$\begin{array}{ccc} \underline{B}^2 \times_{\underline{B}} \underline{E} & \xrightarrow{\quad M \quad} & \underline{E} \\ & {\scriptstyle \underline{B}^{\partial_o} pr_1} \searrow \quad \swarrow {\scriptstyle P} & \\ & \underline{B} & \end{array}$$

where the pullback is formed from $\underline{B}^{\partial_o}$ and P; i.e., $\underline{B}^2 \times_{\underline{B}} \underline{E} = (\underline{B},P)$. There is a bijective correspondence between strict actions in the sense of 3) above and split normal cleavages L for P. The correspondence from cleavages to modules is given by setting $M = \underline{B}^{\partial_o} L$. Conversely, given M, $L(f,E) : M(f,E) \to E$ is defined as the value of M on the morphism

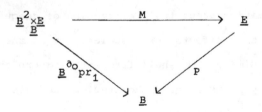

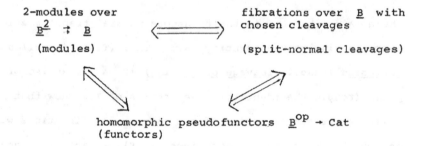

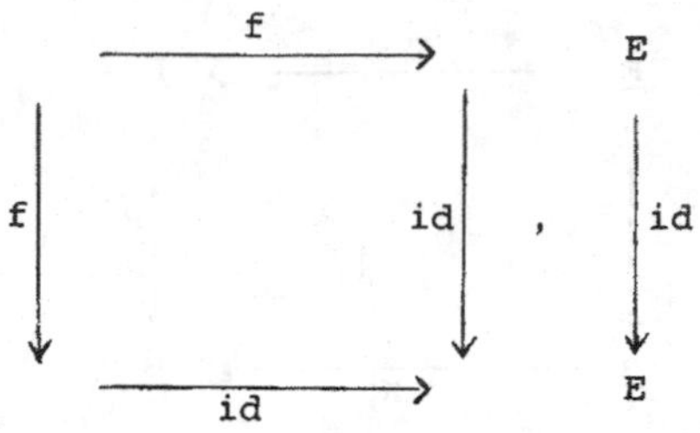

in $\underline{B}^2 \times_{\underline{B}} \underline{E}$. Suitable commutative squares give the extension of L to a functor $(\underline{B},P) \to \underline{E}^2$ with the desired properties. The same correspondence gives a bijection between "2-actions", (i.e., actions which are associative and unitary up to given coherent isomorphisms) and arbitrary cleavages. Such actions are models of the 2-theory of monoids; see I,8. We thus obtain a bijection between the following classes and indicated subclasses:

2-modules over $\underline{B}^2 \rightrightarrows \underline{B}$ (modules) $\Longleftrightarrow$ fibrations over $\underline{B}$ with chosen cleavages (split-normal cleavages)

$\Updownarrow$ homomorphic pseudofunctors $\underline{B}^{op} \to \mathrm{Cat}$ (functors)

The top line is described in greater generality in the sections on fibrations in the later chapters of this work.

ii) The correspondences extend to isomorphisms of categories. In each case, there are three natural choices of morphisms. Treating fibrations first, a commutative diagram

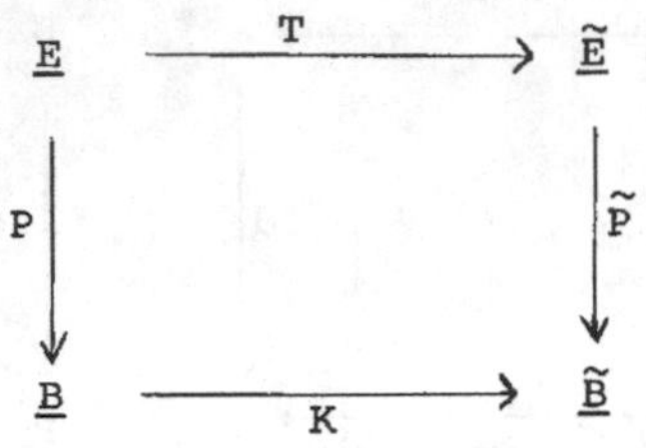

where P and $\tilde{P}$ are fibrations with chosen cleavages L and $\tilde{L}$ gives rise to an adjoint square (cf., I,6)

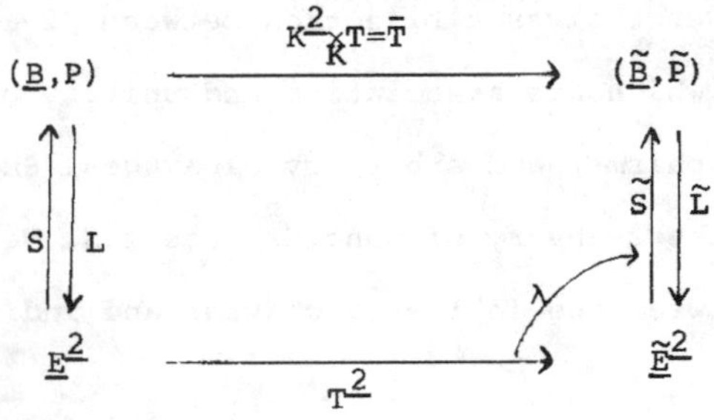

where $\lambda : T^2L \to \tilde{L}\bar{T}$ is the transpose natural transformation corresponding to the identity $\tilde{S}T^2 = \bar{T}S$. (T,K) is called <u>cartesian</u> (resp., <u>cleavage preserving</u>) if λ is an isomorphism (resp., the identity). The results of I,6 show that these define subcategories of the category of fibrations with chosen cleavages and all morphisms (T,K) as above. We denote the possibilities by

$$\begin{array}{ccccc} (\text{Fib};\text{all}) & \supset & (\text{Fib};\text{cart}) & \supset & (\text{Fib};\text{cleap}) \\ \cup & & \cup & & \cup \\ (\text{Split};\text{all}) & \supset & (\text{Split};\text{cart}) & \supset & (\text{Split};\text{cleap}) \end{array}$$

(Cf., [FCC], where different notation is used).

Turning now to modules, defining $\mu = \underline{B}^{\partial_0}\lambda$ gives rise to a diagram

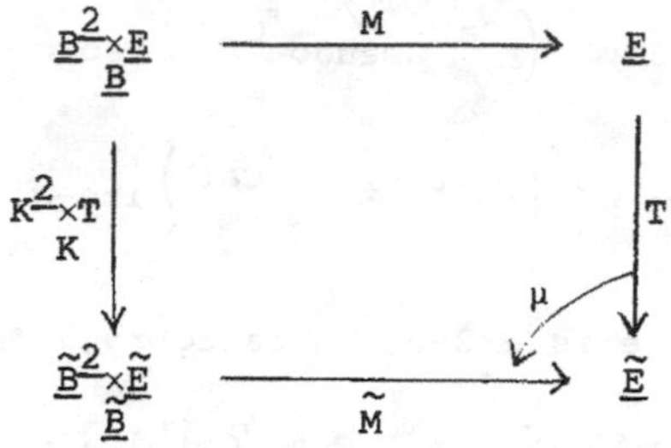

This μ is compatible with associativity and units and such a pair (T,μ) is called a quasi-homomorphism of 2-modules or modules, as the case may be. (Cf., I,8). From μ one can reconstruct λ and μ is an isomorphism or the identity if and only if the corresponding λ is. In these cases we speak of iso-quasi-homomorphisms and homomorphisms. All three cases form categories where the μ's compose via vertical composition, $\boxminus$, of squares. We denote the possibilities by

$$\begin{array}{ccccc} (\text{2-mod,quasi}) & \supset & (\text{2-mod,iso-}) & \supset & (\text{2-mod,homo}) \\ \cup & & \cup & & \cup \\ (\text{mod,quasi}) & \supset & (\text{mod,iso-}) & \supset & (\text{mod,homo}) \end{array}$$

Finally, turning to the interpretation in terms of functors from $\underline{B}^{op}$ to Cat , one can consider diagrams

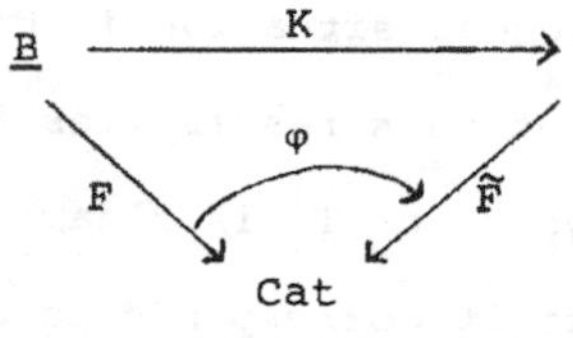

where K is a functor, F and $\tilde{F}$ can be homomorphic pseudo-functors or functors and φ can be quasi-natural, iso-quasi-natural (i.e., all φ_f's are isomorphisms), or natural (i.e., all φ_f's are identities). We denote the possibilities by

$$\begin{array}{ccccc}
\left({}_{\ell}\mathrm{Cat}_{\mathrm{pseudo}} /\!/ \ulcorner\mathrm{Cat}\urcorner\right)_{\mathrm{quasi}} & \supset & \left({}_{\ell}\mathrm{Cat}_{\mathrm{pseudo}} /\!/ \ulcorner\mathrm{Cat}\urcorner\right)_{\mathrm{iso\text{-}q}} & \supset & \left({}_{\ell}\mathrm{Cat}_{\mathrm{pseudo}} /\!/ \ulcorner\mathrm{Cat}\urcorner\right) \\
\cup & & \cup & & \cup \\
\left({}_{\ell}\mathrm{Cat} /\!/ \ulcorner\mathrm{Cat}\urcorner\right)_{\mathrm{quasi}} & \supset & \left({}_{\ell}\mathrm{Cat} /\!/ \ulcorner\mathrm{Cat}\urcorner\right)_{\mathrm{iso\text{-}q}} & \supset & \left({}_{\ell}\mathrm{Cat} /\!/ \ulcorner\mathrm{Cat}\urcorner\right)
\end{array}$$

Note that the last case is a 2-comma category,

$$({}_{\ell}\mathrm{Cat} /\!/ \ulcorner\mathrm{Cat}\urcorner) = [{}_{\ell}\mathrm{Cat}, \ulcorner\mathrm{Cat}\urcorner]$$

If we restrict to F being a functor and $K = \underline{B}$, then the second line becomes

$$\mathrm{Fun}(\underline{B},\mathrm{Cat}) \supset \mathrm{Fun}(\underline{B},\underline{B};\mathrm{Cat},\mathrm{iso\text{-}Cat}) \supset \mathrm{Cat}^{\underline{B}} .$$

__Theorem__: There are canonical isomorphisms

$$\begin{array}{ccc}
(2\text{-mod},\mathrm{quasi}) & \longleftrightarrow & (\mathrm{Fib};\mathrm{all}) \\
 & \nwarrow\!\!\searrow \qquad \swarrow\!\!\nearrow & \\
 & ({}_{\ell}\mathrm{Cat}_{\mathrm{pseudo}} /\!/ \ulcorner\mathrm{Cat}\urcorner)_{\mathrm{quasi}} &
\end{array}$$

which induce isomorphism between the five corresponding subcategories.

The proof is left to the reader since the details are straightforward. Aspects of the top line are discussed in later chapters. An analogous situation is treated in I,5. The main intuitive content of this result lies in the most restrictive case, where it says that T is cleavage preserving iff it is a homomorphism iff it corresponds to a natural transformation between Cat-valued functors.

I,4. Properties of $\mathrm{Fun}(\mathfrak{A},\mathfrak{B})$ and $\mathrm{Pseud}(\mathfrak{A},\mathfrak{B})$. In operating with quasi-natural transformations as though they were natural, completely new phenomena appear. The reason for this is best explained by the fact that if 2-Cat is regarded as a closed category with $\mathrm{Fun}(\mathfrak{A},\mathfrak{B})$ as the internal hom object, then 2-Cat is neither a 2-category nor a bicategory, the trouble being that composition is not a functor of two variables in this situation (see I,4.5.) In [CCS], the 2-categories $\mathrm{Fun}_{\mathfrak{A}}$ were discussed first, being given by the basic construction of that paper; then 2-comma categories were constructed as in I,2.5 of this paper, and these were used to construct the 2-categories $\mathrm{Fun}(\mathfrak{A},\mathfrak{B})$. The reason for this procedure is the analogy with what is done in representable 2-categories in Chapters III and IV of this paper. However, the analogy is misleading since $\mathrm{Fun}_{\mathfrak{A}}$ does not have the properties of a representing functor Φ as in this paper. We hope in a later paper to describe in an abstract fashion the properties that it does have. For our purposes here it is more useful to investigate the closed structure given by $\dot{\mathrm{Fun}}(\mathfrak{A},\mathfrak{B})$ directly.

We show first that there is a notion of a "quasi-functor of two variables" which is the analogue of a "bilinear mapping" and that evaluation going from $\mathfrak{A}\times\mathrm{Fun}(\mathfrak{A},\mathfrak{B}) \to \mathfrak{B}$ is the universal such thing (I,4.2). This characterizes $\mathrm{Fun}(\mathfrak{A},\mathfrak{B})$ and allows one to derive some properties. The existence of a composition (I,4.5)

$$\mathrm{Fun}(\mathfrak{A},\mathfrak{B})\times\mathrm{Fun}(\mathfrak{B},\mathfrak{C}) \to \mathrm{Fun}(\mathfrak{A},\mathfrak{C})$$

requires either a development of the analogue of "multilinear mappings", which is described briefly in I,4.6 and I,4.7, or, equivalently, the existence of a tensor product. We give a direct construction in I,4.9 and an indirect one in an appendix, I,4.23. This tensor product is <u>not</u> symmetric, but it does have right adjoints in each variable giving a biclosed structure on 2-Cat_o (I,4.14). This involves introducing much broader classes of quasi-functors and quasi-natural transformations and a careful study of their interrelations (I,4.10-I,4.13). One can then define Fun $\mathfrak{A}$ and deduce its properties (I,4.16).

In I,4.20 and I,4.21 we indicate briefly what happens with Pseud($\mathfrak{A}$,$\mathfrak{B}$) for bicategories $\mathfrak{A}$ and $\mathfrak{B}$. The situation is not analogous, but since we need little more than the terminology, we do not attempt a systematic analysis of this more complex topic.

A few examples are given in I,4.22. A more extended application is given in I,5 and this is utilized in I,7.

I,4.1. <u>Definition</u>. Let $\mathfrak{A}$, $\mathfrak{B}$ and $\mathfrak{C}$ be 2-categories

i) A <u>quasi-functor of two variables</u> $H : \mathfrak{A}\times\mathfrak{B} \to \mathfrak{C}$ consists of families of 2-functors

$$\{H(A,-) : \mathfrak{B} \to \mathfrak{C} \mid \text{for all } A \in \text{Ob } \mathfrak{A}\}$$
$$\{H(-,B) : \mathfrak{A} \to \mathfrak{C} \mid \text{for all } B \in \text{Ob } \mathfrak{B}\}$$

such that

$$H(A,-)(B) = H(-,B)(A) =_{def} H(A,B)$$

together with 2-cells $\gamma_{f,g}$ for each $f : A \to A'$, $g : B \to B'$ as illustrated

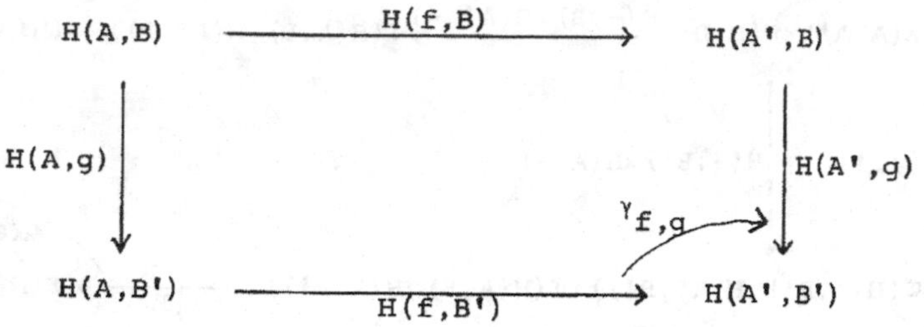

satisfying

$QF_2 1$ $\gamma_{id,g} = id$, $\gamma_{f,id} = id$

$QF_2 2$ $\gamma_{f'f,g} = \gamma_{f',g} \boxbar \gamma_{f,g}$

$\gamma_{f,g'g} = \gamma_{f,g'} \boxminus \gamma_{f,g}$

$QF_2 3$ if $\mu : f \to f'$ and $\nu : g \to g'$ are 2-cells, then the (degenerate) cube

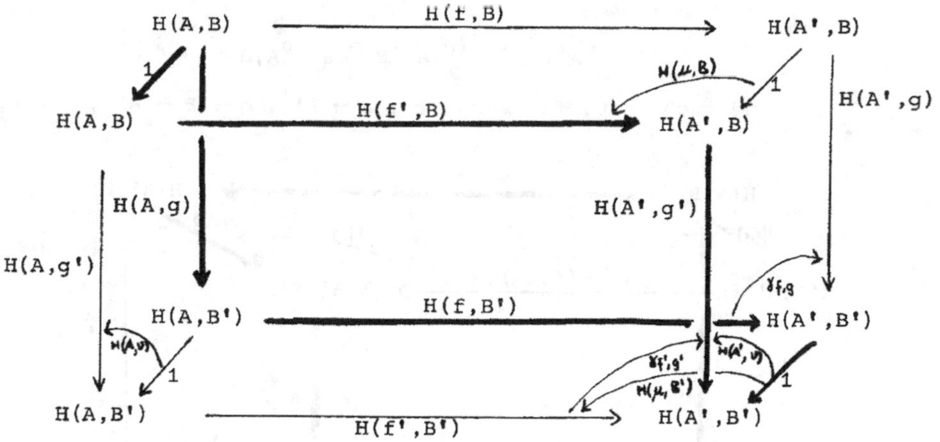

commutes.

Conditions $QF_2 1$ and $QF_2 3$ say that the $\gamma_{f,g}$'s are the components of a natural transformation $\gamma_{AA'BB'}$, as illustrated.

$$\begin{array}{ccc} \mathfrak{A}(A,A')\times\mathfrak{B}(B,B') & \xrightarrow{H(-,B)\times H(A',-)} & \mathfrak{C}(H(A,B),H(A',B))\times\mathfrak{C}(H(A',B),H(A',B')) \\ \Big\downarrow{\scriptstyle H(-,B')\times H(A,-)} & \overset{\gamma_{AA'BB'}}{\nearrow} & \Big\downarrow{\scriptstyle \circ} \\ \mathfrak{C}(H(A,B'),H(A',B'))\times\mathfrak{C}(H(A,B),H(A,B')) & \xrightarrow[\circ(Tw)]{} & \mathfrak{C}(H(A,B),H(A',B')) . \end{array} \tag{4.1}$$

Condition $QF_2{}^2$ corresponds to diagrams like (2.11) in I,2.4.

ii) A <u>quasi-natural transformation</u>, σ, between quasi-functors of two variables H and $H' : \mathfrak{A}\times\mathfrak{B} \to \mathfrak{C}$, consists of families of quasi-natural transformations

$$\{(\sigma_A)_{(-)} : H(A,-) \to H'(A,-) \mid \text{for all } A \in \text{Ob } \mathfrak{A}\}$$
$$\{(\sigma_{(-)})_B : H(-,B) \to H'(-,B) \mid \text{for all } B \in \text{Ob } \mathfrak{B}\}$$

such that

$$(\sigma_A)_{(B)} = (\sigma_{(A)})_B =_{\text{def}} \sigma_{A,B}$$

and such that, for each $f : A \to A'$, $g : B \to B'$, the cube

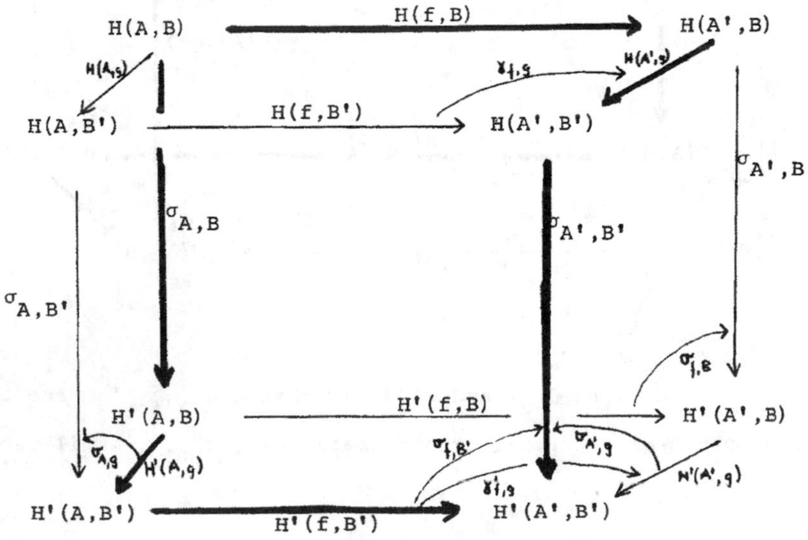

QN_2.

commutes.

iii) A <u>modification</u> $s : \sigma \to \sigma'$ between such quasi-natural transformations is a family of 2-cells $\{s_{A,B} : \sigma_{A,B} \to \sigma'_{A,B}\}$ such that for each fixed A and each fixed B ,

$$\mathrm{MQN}_2 \quad \begin{array}{l} (s_A)_{(-)} : (\sigma_A)_{(-)} \to (\sigma'_A)_{(-)} \\ (s_{(-)})_B : (\sigma_{(-)})_B \to (\sigma'_{(-)})_B \end{array}$$

are modifications (see I,2.3.)

The composition of quasi-natural transformations is given by vertical composition of squares, as in I,2.4, and the two compositions of modifications are defined as before. The resulting 2-category is denoted by $\text{q-Fun}(\mathfrak{A}\times\mathfrak{B},\mathfrak{C})$.

I,4.2. <u>Theorem</u>. $\mathrm{Fun}(\mathfrak{A},\mathfrak{B})$ is the unique 2-category (up to isomorphism) such that

i) there exists a quasi-functor of two variables $ev : \mathfrak{A}\times\mathrm{Fun}(\mathfrak{A},\mathfrak{B}) \to \mathfrak{B}$,

ii) given any quasi-functor of two variables $H : \mathfrak{A}\times\chi \to \mathfrak{B}$, there is a unique 2-functor $H^t : \chi \to \mathrm{Fun}(\mathfrak{A},\mathfrak{B})$ such that the diagram

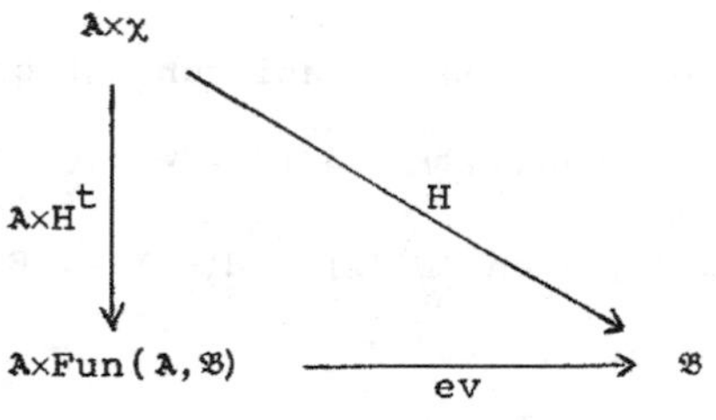

commutes.

Furthermore ,

iii) the correspondence in part ii) extends to an isomorphism

$$\text{q-Fun}(\mathcal{A}\times\mathcal{X},\mathcal{B}) \cong \text{Fun}(\mathcal{X},\text{Fun}(\mathcal{A},\mathcal{B}))$$

for every $\mathcal{A},\mathcal{X}$ and $\mathcal{B}$. These are the components of a natural isomorphism between functors

$$(2\text{-Cat}_o)^{op}\times(2\text{-Cat}_o)^{op}\times(2\text{-Cat}_o) \to 2\text{-Cat}_o$$

<u>Proof</u>: i) $ev : \mathcal{A}\times\text{Fun}(\mathcal{A},\mathcal{B}) \to \mathcal{B}$ is the quasi-functor of two variables such that $ev(A,F) = FA$, $ev(f,F) = Ff$, $ev(\mu,F) = F\mu$, $ev(A,\sigma) = \sigma_A$ and $ev(A,s) = s_A$. Clearly, if one variable is fixed, then it is a 2-functor in the other. A pair of morphisms (f,σ) gives rise to a square

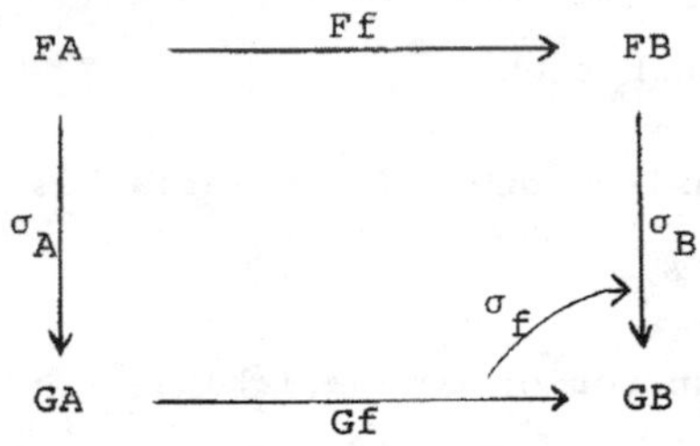

and it follows immediately from the definition of $\text{Fun}(\mathcal{A},\mathcal{B})$ that this satisfies the required relations.

ii) Let $H : \mathcal{A}\times\mathcal{X} \to \mathcal{B}$ be a quasi-functor of two variables. Define $H^t : \mathcal{X} \to \text{Fun}(\mathcal{A},\mathcal{B})$ as follows:

a) If $X \in \text{Ob}\,\mathcal{X}$, then $H^t(X) = H(-,X)$. By definition, this is a 2-functor from $\mathcal{A}$ to $\mathcal{B}$.

b) If $g : X \to X'$, then $H^t(g) : H^t(X) \to H^t(X')$ is the quasi-natural transformation whose components are

$$[H^t(g)]_A = H(A,g)$$

and, for a morphism $f : A \to A'$,

$$
\begin{array}{ccc}
(H^t X)A & \xrightarrow{(H^t X)f} & (H^t X)A' \\
\Big\downarrow{\scriptstyle (H^t g)_A} & \nearrow{\scriptstyle (H^t g)_f} & \Big\downarrow{\scriptstyle (H^t g)_{A'}} \\
(H^t X')A & \xrightarrow[(H^t X')f]{} & (H^t X')A'
\end{array}
\quad = \quad
\begin{array}{ccc}
H(A,X) & \xrightarrow{H(f,X)} & H(A',X) \\
\Big\downarrow{\scriptstyle H(A,g)} & \nearrow{\scriptstyle \gamma_{f,g}} & \Big\downarrow{\scriptstyle H(A',g)} \\
H(A,X') & \xrightarrow[H(f,X')]{} & H(A',X')
\end{array}
$$

i.e., $[H^t(g)]_f = \gamma_{f,g}$. It is easily seen that this gives a functor from χ_o to $Fun_o(\mathfrak{A},\mathfrak{B})$.

c) If $\nu : g \to g'$ is a 2-cell in χ , then

$$[H^t(\nu)]_A = H(A,\nu)$$

Clearly H^t is a 2-functor, and is the unique one such that $ev(\mathfrak{A}\times H^t) = H$.

The uniqueness of $Fun(\mathfrak{A},\mathfrak{B})$, up to an isomorphism, then follows from general principles. It is interesting, instructive, and amusing that the uniqueness can also be shown by deriving the internal structure of $Fun(\mathfrak{A},\mathfrak{B})$ from this property, as in Lawvere [CCFM], p.9. That is, set $\chi = \underline{1}$, $\underline{2}$, $\underline{3}$, and $\underline{2}_2$, and consider the structure of quasi-functors of two variables from $\mathfrak{A}\times\chi$ to $\mathfrak{B}$ in these cases.

iii) Part ii) gives the desired isomorphism at the level of objects. If $\sigma : H \to H'$ is a quasi-natural transformation between quasi-functors of two variables, define $\sigma^t : H^t \to H'^t$ as follows:

$$\begin{array}{ccccc} (\sigma^t)_X & : & H^t(X) & \to & H'^t(X) \\ \| & & \| & & \| \\ (\sigma_{(-)})_X & : & H(-,X) & \to & H'(-,X) \end{array} \;;$$

i.e., σ^t_X is a morphism in $\mathrm{Fun}(\mathfrak{A},\mathfrak{B})$.

If $g : X \to X'$, then σ^t_g is the modification between composed quasi-natural transformations given by

$$\begin{array}{ccc} H^tX & \xrightarrow{H^tg} & H^tX' \\ {\scriptstyle \sigma^t_X}\downarrow & \sigma^t_g & \downarrow{\scriptstyle \sigma^t_{X'}} \\ H'^tX & \xrightarrow[H'^tg]{} & H'^tX' \end{array} \quad = \quad \begin{array}{ccc} H(-,X) & \xrightarrow{H(-,g)} & H(-,X') \\ {\scriptstyle (\sigma_{(-)})_X}\downarrow & \sigma_{(-),g} & \downarrow{\scriptstyle (\sigma_{(-)})_{X'}} \\ H'(-,X) & \xrightarrow[H'(-,g)]{} & H'(-,X') \end{array} \;;$$

i.e., the components are given by $(\sigma^t_g)_A = \sigma_{A,g}$. The cube in I,4.1, QN_2, can be interpreted precisely as asserting that σ^t_g is a modification (I,2.4, MQN). Thus $\sigma^t : H^t \to H'^t$ is a quasi-natural transformation.

Finally, let $s : \sigma \to \sigma'$ be a modification as in I,4.1, iii). Then $s^t : \sigma^t \to \sigma'^t$ is the modification (as in I,2.4) whose components $s^t_X : H^tX \to H'^tX$ are the modifications in $\mathrm{Fun}(\mathfrak{A},\mathfrak{B})$ with components $(s^t_X)_A = s_{A,X}$.

The naturality of this correspondence is easily established, once one has observed that Fun(-,-) and q-Fun(-×-,-) are functors (but <u>not</u> 2-functors) on suitable products of 2-Cat_o and its dual. A proof of this for Fun(-,-) is suggested in [CCS], §6. A direct proof is immediate, since if, in the situation

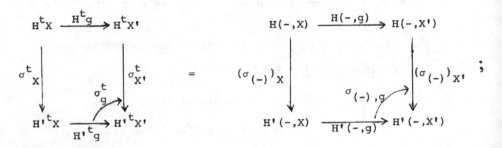

$$(4.2)\qquad \mathfrak{A}' \xrightarrow{K} \mathfrak{A} \underset{G}{\overset{F}{\underset{\sigma\Downarrow}{\rightrightarrows}}} \mathfrak{B} \xrightarrow{H} \mathfrak{B}' ,$$

σ is a quasi-natural transformation, then so is $H\sigma K$, where $(H\sigma K)_{A'} = H(\sigma_{K(A')})$ and $(H\sigma K)_f = H(\sigma_{Kf})$. q-Fun(-×-,-) is a functor in the analogous fashion; i.e., in the situation

$$(4.3)\qquad \mathfrak{A}'\times\mathfrak{B}' \xrightarrow{F\times G} \mathfrak{A}\times\mathfrak{B} \underset{H'}{\overset{H}{\underset{\sigma\Downarrow}{\rightrightarrows}}} \mathfrak{C} \xrightarrow{K} \mathfrak{C}'$$

where F , G , and K are 2-functors, while σ is a quasi-natural transformation between quasi-functors of two variables, it follows that $KH(F\times G)$ is a quasi-functor of two variables and $K\sigma(F\times G)$ is a quasi-natural transformation.

Alternatively, using this last fact, the functorial behavior of $\mathrm{Fun}(\mathfrak{A},\mathfrak{B})$ is determined since, given 2-functors $F : \mathfrak{A}' \to \mathfrak{A}$ and $G : \mathfrak{B} \to \mathfrak{B}'$, then $\mathrm{Fun}(F,G)$ is the unique 2-functor such that the diagram

$$(4.4)\qquad \begin{array}{ccccc} \mathfrak{A}'\times\mathrm{Fun}(\mathfrak{A},\mathfrak{B}) & \xrightarrow{F\times 1} & \mathfrak{A}\times\mathrm{Fun}(\mathfrak{A},\mathfrak{B}) & \xrightarrow{ev} & \mathfrak{B} \\ \downarrow{\scriptstyle \mathfrak{A}'\times\mathrm{Fun}(F,G)} & & & & \downarrow{\scriptstyle G} \\ \mathfrak{A}'\times\mathrm{Fun}(\mathfrak{A}',\mathfrak{B}') & & \xrightarrow{\quad ev \quad} & & \mathfrak{B}' \end{array}$$

commutes. It follows directly from this description that Fun(-,-) is a functor.

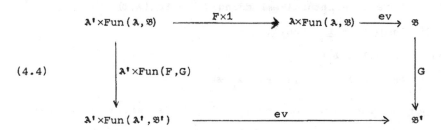

I,4.3. <u>Remark</u>. The naturality of the isomorphism above usually is used in the following three situations, in which the diagram on the left commutes if and only if the diagram on the right does.

a)

$$\begin{array}{ccc} \bar{\mathfrak{A}}\times\chi & \xrightarrow{\bar{H}} & \\ {\scriptstyle G\times\chi}\big\downarrow & & \mathfrak{B} \\ \mathfrak{A}\times\chi & \xrightarrow{H} & \end{array} \quad \text{iff} \quad \begin{array}{ccc} & \xrightarrow{H^t} & \mathrm{Fun}(\mathfrak{A},\mathfrak{B}) \\ \chi & & \big\downarrow{\scriptstyle \mathrm{Fun}(G,\mathfrak{B})} \\ & \xrightarrow{\bar{H}^t} & \mathrm{Fun}(\bar{\mathfrak{A}},\mathfrak{B}) \end{array} \qquad (4.5)$$

b)

$$\begin{array}{ccc} & \xrightarrow{H} & \mathfrak{B} \\ \mathfrak{A}\times\chi & & \big\downarrow{\scriptstyle G} \\ & \xrightarrow{\bar{H}} & \bar{\mathfrak{B}} \end{array} \quad \text{iff} \quad \begin{array}{ccc} & \xrightarrow{H^t} & \mathrm{Fun}(\mathfrak{A},\mathfrak{B}) \\ \chi & & \big\downarrow{\scriptstyle \mathrm{Fun}(\mathfrak{A},G)} \\ & \xrightarrow{\bar{H}^t} & \mathrm{Fun}(\mathfrak{A},\bar{\mathfrak{B}}) \end{array} \qquad (4.6)$$

c)

$$\begin{array}{ccc} \mathfrak{A}\times\chi & \xrightarrow{H} & \\ {\scriptstyle \mathfrak{A}\times G}\big\downarrow & & \mathfrak{B} \\ \mathfrak{A}\times\bar{\chi} & \xrightarrow{\bar{H}} & \end{array} \quad \text{iff} \quad \begin{array}{ccc} \chi & \xrightarrow{H^t} & \\ {\scriptstyle G}\big\downarrow & & \mathrm{Fun}(\mathfrak{A},\mathfrak{B}) \\ \bar{\chi} & \xrightarrow{\bar{H}^t} & \end{array} \qquad (4.7)$$

I,4.4. <u>Corollary</u>.

i) There is a natural embedding $\mathfrak{B}^{\mathfrak{A}} \to \mathrm{Fun}(\mathfrak{A},\mathfrak{B})$

ii) $\mathrm{Fun}(\underline{0},\mathfrak{B}) \simeq \underline{1}$, $\mathrm{Fun}(\underline{1},\mathfrak{B}) \simeq \mathfrak{B}$

iii) $\mathrm{Fun}(\mathfrak{A},\varprojlim \mathfrak{B}_i) \simeq \varprojlim \mathrm{Fun}(\mathfrak{A},\mathfrak{B}_i)$

$\mathrm{Fun}(\varinjlim \mathfrak{A}_i,\mathfrak{B}) \simeq \varprojlim \mathrm{Fun}(\mathfrak{A}_i,\mathfrak{B})$

<u>Proof</u>: i) The 2-functor $ev : \mathfrak{A}\times\mathfrak{B}^{\mathfrak{A}} \to \mathfrak{B}$ is certainly a quasi-functor of two variables and hence determines a 2-functor $\mathfrak{B}^{\mathfrak{A}} \to \mathrm{Fun}(\mathfrak{A},\mathfrak{B})$ which is easily seen to be an embedding by considering 2-functors from $\underline{2}_2$ into $\mathfrak{B}^{\mathfrak{A}}$.

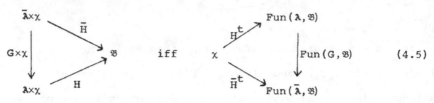

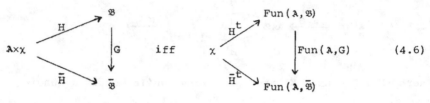

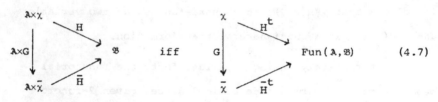

ii) Since $\underline{O}$ is not only initial, but empty, $\underline{O}\times\chi$ is empty for all χ , so there is a unique quasi-functor of two variables $\underline{O}\times\chi \to \mathfrak{B}$ for any $\mathfrak{B}$. Thus $\underline{1}$ satisfies the universal mapping property characterizing $\mathrm{Fun}(\underline{O},\mathfrak{B})$. A similar argument shows that $\mathrm{Fun}(\underline{1},\mathfrak{B}) \simeq \mathfrak{B}$.

iii) Since limits in 2-Cat can be represented as suitable subobjects of products, it is evident that a quasi-functor of two variables

$$H : \mathfrak{A}\times\chi \to \varprojlim \mathfrak{B}_i$$

corresponds to a uniquely determined cone of quasi-functors of two variables

$$H_i : \mathfrak{A}\times\chi \to \mathfrak{B}_i$$

This shows that the unique quasi-functor of two variables, $\overline{ev}$, making the diagram

$$\begin{array}{ccc}
\mathfrak{A}\times\varprojlim \mathrm{Fun}(\mathfrak{A},\mathfrak{B}_i) & \xrightarrow{\overline{ev}} & \varprojlim \mathfrak{B}_i \\
\Big\downarrow {\scriptstyle \mathfrak{A}\times pr_i} & & \Big\downarrow {\scriptstyle pr_i} \\
\mathfrak{A}\times\mathrm{Fun}(\mathfrak{A},\mathfrak{B}_i) & \xrightarrow[ev_i]{} & \mathfrak{B}_i
\end{array}$$

commute satisfies the same universal property as $ev : \mathfrak{A}\times\mathrm{Fun}(\mathfrak{A},\varprojlim \mathfrak{B}_i) \to \varprojlim \mathfrak{B}_i$, and hence $\mathrm{Fun}(\mathfrak{A},\varprojlim \mathfrak{B}_i) \simeq \varprojlim \mathrm{Fun}(\mathfrak{A},\mathfrak{B}_i)$.

Colimits are a bit more difficult. Since $-\times\chi$ has a right adjoint, it preserves colimits. The proof consists in showing that there are natural correspondences between the following types of morphisms and cones:

a) $\mathfrak{X} \to \mathrm{Fun}(\varinjlim \mathfrak{A}_i, \mathfrak{B})$

b) $(\varinjlim \mathfrak{A}_i) \times \mathfrak{X} \to \mathfrak{B}$

c) $\varinjlim (\mathfrak{A}_i \times \mathfrak{X}) \to \mathfrak{B}$

d) cones $\{\mathfrak{A}_i \times \mathfrak{X} \to \mathfrak{B}\}$

e) cones $\{\mathfrak{X} \to \mathrm{Fun}(\mathfrak{A}_i, \mathfrak{B})\}$

f) $\mathfrak{X} \to \varprojlim \mathrm{Fun}(\mathfrak{A}_i, \mathfrak{B})$

In b), c), and d), quasi-functors of two variables are intended. From I,4.2 and I,4.3 and the properties of ordinary limits, one has

$$\text{a)} \leftrightarrow \text{b)} \ , \ \text{d)} \leftrightarrow \text{e)} \leftrightarrow \text{f)} \ ,$$

so we must show that b), c), and d) are equivalent. This holds trivially for coproducts since they are disjoint unions, so it is sufficient to consider coequalizers. In this case, it is easiest to bypass c) and show b) $\leftrightarrow$ d). Suppose

$$\mathfrak{A} \underset{G}{\overset{F}{\rightrightarrows}} \mathfrak{B} \xrightarrow{P} \mathfrak{Q}$$

is a coequalizer in 2-Cat (see I,2.11, 1).) Let $H : \mathfrak{B}\times\mathfrak{X} \to \mathfrak{C}$ be a quasi-functor of two variables such that $H(F\times\mathfrak{X}) = H(G\times\mathfrak{X})$. We must show there is a unique $K : \mathfrak{Q}\times\mathfrak{X} \to \mathfrak{C}$ with $KP = H$. Define

$K(Q,-) = H(B,-) : \mathfrak{X} \to \mathfrak{C}$ where $PB = Q$

$K(-,X) : \mathfrak{Q} \to \mathfrak{C}$ is the functor induced by $H(-,X)$.

Now if $q : Q \to Q'$ and $g : X \to X'$, then q can be represented as a string $(f_1,\ldots,f_n)$ of morphisms in $\mathfrak{B}$ such that $H(f_1,X),\ldots,H(f_n,X)$ is composable in $\mathfrak{C}$. We set

$$\gamma_{q,g} = \gamma_{f_n,g} \boxbar \gamma_{f_{n-1},g} \boxbar \cdots \boxbar \gamma_{f_1,g}$$

Using the explicit description of coequalizers in 2-Cat and Cat, one shows that this is well-defined and gives the desired unique quasi-functor of two variables.

I,4.5. Theorem: There is a strictly associative, strictly unitary composition quasi-functor of two variables which is natural in all three variables

$$\mathrm{Fun}(\mathfrak{A},\mathfrak{B}) \times \mathrm{Fun}(\mathfrak{B},\mathfrak{C}) \xrightarrow{\circ} \mathrm{Fun}(\mathfrak{A},\mathfrak{C})$$

Proof: This is the main fact we want about $\mathrm{Fun}(\mathfrak{A},\mathfrak{B})$ since it is responsable for all of the complexities surrounding quasi-adjunctions (see I,7). "Natural" means as functors on 2-Cat_0 . There are three ways to describe this composition.

i) An explicit formula for the composition can be given as follows: Consider the situation

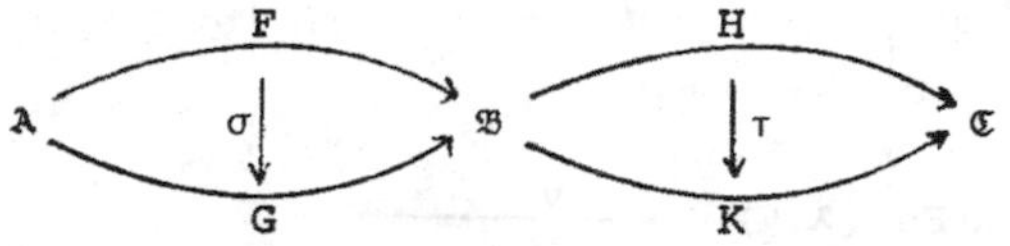

If H is fixed, then it is easily seen that $H\sigma : HF \to HG$ is a quasi-natural transformation and that composition with H on the right gives a 2-functor

$$H\circ- : \mathrm{Fun}(\mathfrak{A},\mathfrak{B}) \to \mathrm{Fun}(\mathfrak{A},\mathfrak{C})$$

Similarly, composition with F on the left gives a 2-functor

$$-\circ F : \mathrm{Fun}(\mathfrak{B},\mathfrak{C}) \to \mathrm{Fun}(\mathfrak{A},\mathfrak{C})$$

But given σ and τ as indicated, instead of a commutative square one has, for any $A \in \mathfrak{A}$, a diagram

$$
(3.8) \qquad
\begin{array}{ccc}
HF(A) & \xrightarrow{H(\sigma_A)} & HG(A) \\
\Big\downarrow \tau_{FA} & \nearrow \tau_{\sigma_A} & \Big\downarrow \tau_{GA} \\
KF(A) & \xrightarrow[K(\sigma_A)]{} & KG(A)
\end{array}
$$

One checks easily that

$$\gamma_{\sigma,\tau} = \tau_{\sigma_{(-)}} \quad : K\sigma \cdot \tau F \to \tau G \cdot H\sigma$$

is a modification and satisfies the required properties.

ii) Composition is the unique quasi-functor of two variables making the diagram

$$
(3.9) \qquad
\begin{array}{ccc}
\mathfrak{A}\times \mathrm{Fun}(\mathfrak{A},\mathfrak{B})\times \mathrm{Fun}(\mathfrak{B},\mathfrak{C}) & \xrightarrow{ev\times 1} & \mathfrak{B}\times \mathrm{Fun}(\mathfrak{B},\mathfrak{C}) \\
\Big\downarrow \mathfrak{A}\times\circ & & \Big\downarrow ev \\
\mathfrak{A}\times \mathrm{Fun}(\mathfrak{A},\mathfrak{C}) & \xrightarrow{ev} & \mathfrak{C}
\end{array}
$$

commute. From this, the strict associativity and strict unitaryness as well as naturality are easily derived from uniqueness. The unit is the unique functor $\underline{1} \to \mathrm{Fun}(\mathfrak{A},\mathfrak{A})$ corresponding to the isomorphism $\mathfrak{A}\times\underline{1} \cong \mathfrak{A}$. However, to do this, one must discuss quasi-functors of three variables. (Actually this must be done to make any sense out of associativity in any case).

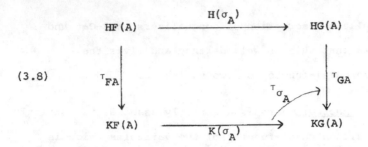

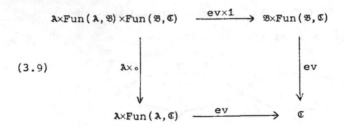

iii) The third approach is to show that there is a tensor product with

$$\mathrm{Fun}(\mathfrak{A}\otimes\mathfrak{B},\mathfrak{C}) \simeq \mathrm{Fun}(\mathfrak{A},\mathrm{Fun}(\mathfrak{B},\mathfrak{C}))$$

and then proceed as in [E-K], 5.10. We shall briefly describe both of these procedures, omitting detailed proofs.

I,4.6. Definition. i) A quasi-functor of n-variables, $n \geq 2$, $H : \prod_{i=1}^{n} \mathfrak{A}_i \to \mathfrak{C}$ consists of quasi-functors of two variables

$$H(A_1,\ldots,A_{i-1},-,A_{i+1},\ldots,A_{j-1},-,A_{j+1},\ldots,A_n) : \mathfrak{A}_i \times \mathfrak{A}_j \to \mathfrak{C}$$

for all $i < j$ and all choices of indicated objects $A_k \in \mathfrak{A}_k$, which agree on objects and as 2-functors of 1-variable, such that for all triples of indices $i < j < k$ and morphisms

$$f_i : A_i \to A_i', \quad f_j : A_j \to A_j', \quad f_k : A_k \to A_k'$$

the diagram (in which extraneous variables are omitted)

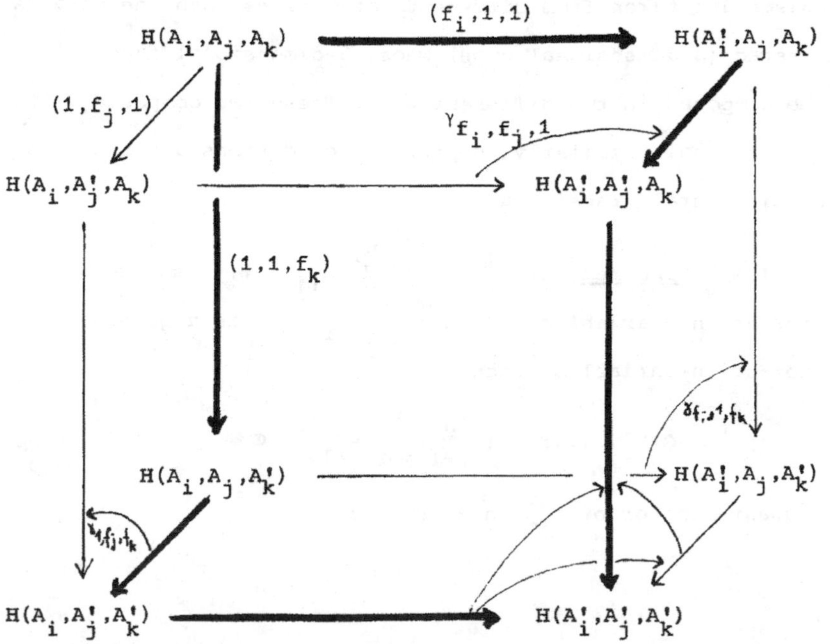

commutes.

ii) A <u>quasi-natural transformation</u> $\sigma : H \to H'$ is a family of quasi-natural transformations

$$\{\sigma_{A_1,\ldots,A_{i-1},-,A_{i+1},\ldots,A_n}\}$$

for all i and all choices of objects $A_k \in \mathbb{A}_k$, which are quasi-natural transformations of two variables for all choices of indices $i < j$.

iii) A <u>modification</u> $s : \sigma \to \sigma'$ is a family of modifications for each variable. The resulting 2-category is denoted by

$$q_n\text{-Fun}(\prod_{i=1}^{n} \mathbb{A}_i, \mathbb{C})$$

<u>Remark</u>. For the preceeding to be justified, it must and can be shown that there are no "higher commutativity relations". For instance, in the description of quasi-functors, if four morphisms are taken from different categories, then one obtains a tesserac (4-dimensional cube) whose 3-dimensional faces can be composed in two different ways. These two composed cubes are equal. Similarly, no further conditions are required on quasi-natural transformations.

I,4.7. <u>Theorem</u>. i) If $F_i : \prod_{j=1}^{n_i} A_{ij} \to \mathfrak{B}_i$ is a quasi-functor of n_i-variables and $G : \prod_{i=1}^{n} \mathfrak{B}_i \to \mathbb{C}$ is a quasi-functor of n-variables, then

$$G(F_1,\ldots,F_n) : \prod_{i=1}^{n} (\prod_{j=1}^{n_i} A_{ij}) \to \mathbb{C}$$

is a quasi-functor of $\sum_{i=1}^{n} n_i$-variables.

ii) If $H : \mathbb{A} \times \prod_{i=1}^{n} \mathbb{A}_i \to \mathbb{C}$ is a quasi-functor of n+1-variables, then there is a unique quasi-functor $H^t : \prod_{i=1}^{n} \mathbb{A}_i \to \mathrm{Fun}(\mathbb{A},\mathbb{C})$ of n-variables such that the diagram

$$\begin{array}{ccc} \mathbb{A} \times \prod_{i=1}^{n} \mathbb{A}_i & & \\ \Big\downarrow{\scriptstyle \mathbb{A} \times H^t} & \searrow^{H} & \\ \mathbb{A} \times \mathrm{Fun}(\mathbb{A},\mathbb{C}) & \xrightarrow[\mathrm{ev}]{} & \mathbb{C} \end{array}$$

commutes.

iii) This correspondence extends to a natural isomorphism

$$q_{n+1}\text{-}\mathrm{Fun}(\mathbb{A} \times \prod_{i=1}^{n} \mathbb{A}_i, \mathbb{C}) \cong q_n\text{-}\mathrm{Fun}(\prod_{i=1}^{n} \mathbb{A}_i, \mathrm{Fun}(\mathbb{A},\mathbb{C})) .$$

Proof. The proof is a straightforward bookkeeping exercise based on the proof of I,4.2. This now justifies the second approach to the proof of Theorem I,4.5.

We now turn to the third approach using tensor products. This is based on the observation of S. Mac Lane that quasi-functors should be special pseudo-functors, together with the information from J. Duskin that Benabou has given a universal construction to "straighten out" pseudo-functors. The idea of the construction was then worked out in conversation with Duskin and Mac Lane. Although it is not needed later, we begin with Mac Lane's suggestion since it can be used together with I,4.21 to give an alternative construction for the tensor product which follows.

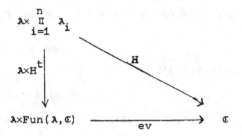

I,4.8. <u>Proposition</u>. There is an inclusion (see I,3.3)

$$\text{q-Fun}(\mathfrak{A}\times\mathfrak{B},\mathfrak{C}) \hookrightarrow \text{coPseud}(\mathfrak{A}\times\mathfrak{B},\mathfrak{C})$$

<u>Proof</u>. Let $H : \mathfrak{A}\times\mathfrak{B} \to \mathfrak{C}$ be a quasi-functor of two variables. Define a copseudo-functor $\bar{H} : \mathfrak{A}\times\mathfrak{B} \to \mathfrak{C}$ by the following data:

a)

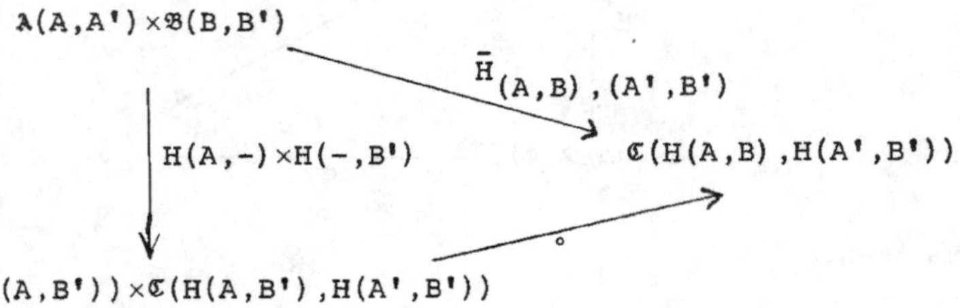

i.e., $\bar{H}(f,g) = H(f,B')H(A,g)$

b) $\varphi_{(f',B)(f,B)} = \text{id}$

$\varphi_{(A,g')(A,g)} = \text{id}$

$\varphi_{(f,B')(A,g)} = \text{id}$

$\varphi_{(A',g)(f,B)} = \gamma_{f,g}$

i.e., the only non-preservation of composition is in the situation

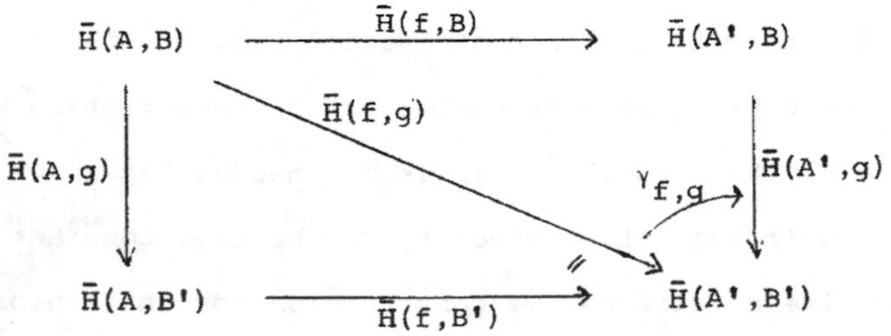

We have chosen to represent this as a copseudo-functor rather than a pseudo-functor with $\gamma_{f,g}$ in the lower triangle because

it then looks like a commutation relation

$$\gamma_{f,g} : \bar{H}(f,g) \to \bar{H}(A',g)\bar{H}(f,B)$$

which helps to keep things straight. One shows by explicit calculations that the conditions on γ are exactly equivalent to $(\bar{H},\varphi)$ being a pseudo-functor. The correspondence for quasi-natural transformations and modifications is then clear.

I,4.9. Theorem. There exists a tensor product for 2-categories, $\mathfrak{A} \otimes \mathfrak{B}$ together with a natural isomorphism

$$\mathrm{Fun}(\mathfrak{A}\otimes\mathfrak{B},\mathfrak{C}) \cong \mathrm{Fun}(\mathfrak{B},\mathrm{Fun}(\mathfrak{A},\mathfrak{C}))$$

satisfying the conditions of [E-K], II,5.10.

Proof. (See I,4.23 also). We construct $\mathfrak{A} \otimes \mathfrak{B}$ so that

$$\text{q-Fun}(\mathfrak{A}\times\mathfrak{B},\mathfrak{C}) \cong \mathrm{Fun}(\mathfrak{A}\otimes\mathfrak{B},\mathfrak{C})$$

and apply I,4.2, iii).

The objects of $\mathfrak{A} \otimes \mathfrak{B}$ are pairs (A,B) of objects where $A \in |\mathfrak{A}|$ and $B \in |\mathfrak{B}|$. The morphisms are equivalence classes of "approximations of the diagonal"; i.e., strings

$$(f_1,g_1)(f_2,g_2) \ldots (f_n,g_n)$$

where

i) $f_i \in \mathfrak{A}$, $g_i \in \mathfrak{B}$ and the compositions

$$f_1 f_2 \ldots f_n$$

$$g_1 g_2 \ldots g_n$$

exist; i.e., $\partial_0 f_i = \partial_1 f_{i+1}$, $\partial_0 g_i = \partial_1 g_{i+1}$.

ii) for all i, either f_i or g_i is an identity morphism. Two strings are equivalent if they are made so by the smallest equivalence relation compatible with composition such that

$$(f_1,1)(f_2,1) \sim (f_1 f_2,1)$$

$$(1,g_1)(1,g_2) \sim (1,g_1 g_2)$$

Composition of morphisms is induced by juxtaposition of strings. Note that 1 always denotes whatever identity map is appropriate.

The 2-cells of $\mathfrak{A} \otimes \mathfrak{B}$ are constructed as follows. First of all, for all non-identity 1-cells $f \in \mathfrak{A}$ and $g \in \mathfrak{B}$, there is a 2-cell

$$\gamma_{f,g} : (f,1)(1,g) \to (1,g)(f,1)$$

Now consider equivalence classes of well-formed strings

$$\Lambda = [\lambda_1,\ldots,\lambda_n]$$

where λ_i is either $\gamma_{f,g}$ for some f and g, or $\lambda_i = (\tau_i,\sigma_i)$ where τ_i is a 2-cell in $\mathfrak{A}$, σ_i is a 2-cell in $\mathfrak{B}$ and either τ_i or σ_i is a strong identity.

iii) A string is well-formed if whenever

$$\lambda_i\lambda_{i+1} = \begin{cases} (\tau_i,\sigma_i)(\tau_{i+1},\sigma_{i+1}) & \text{then } \tau_i\tau_{i+1} \text{ and } \sigma_i\sigma_{i+1} \\ (\tau_i,\sigma_i)\gamma_{f,g} & \text{then } \tau_i f \text{ and } \sigma_i g \\ \gamma_{f,g}(\tau_{i+1},\sigma_{i+1}) & \text{then } f\tau_{i+1} \text{ and } g\sigma_{i+1} \\ \gamma_{f,g}\gamma_{f',g'} & \text{then } ff' \text{ and } gg' \end{cases}$$

are defined in $\mathfrak{A}$ and $\mathfrak{B}$ respectively.

iv) Two well-formed strings are equivalent if they are made so by the smallest equivalence relation compatible with juxtaposition of strings such that

$$(\tau,1)(\tau',1) \sim (\tau\tau',1)$$

$$(1,\sigma)(1,\sigma') \sim (1,\sigma\sigma')$$

Strong composition is induced by juxtaposition of strings. The strong domain and codomain are given by $\partial_o\Lambda = \partial_o\lambda_n$ and $\partial_1\Lambda = \partial_1\lambda_1$ where

$$\partial_i(\tau_i,\sigma_i) = (\partial_i\tau_i,\partial_i\sigma_i)$$

$$\partial_i\gamma_{f,g} = (\partial_i f,\partial_i g)$$

The weak domain and codomain are given by

$$\tilde{\partial}_i\Lambda = [\tilde{\partial}_i\lambda_1,\ldots,\tilde{\partial}_i\lambda_n]$$

where $\tilde{\partial}_i(\tau_j,\sigma_j) = (\tilde{\partial}_i\tau_j,\tilde{\partial}_i\sigma_j)$ and

$$\tilde{\partial}_o\gamma_{f,g} = (f,1)(1,g) \;,\; \tilde{\partial}_1\gamma_{f,g} = (1,g)(f,1)$$

Square brackets denote the equivalence class of the indicated 1-cell.

Finally, the 2-cells of $\mathfrak{A} \otimes \mathfrak{B}$ are equivalence classes of strings of well-formed strings

$$\Gamma = [\Lambda_1\cdot\Lambda_2\cdot\ldots\cdot\Lambda_n]$$

such that $\tilde{\partial}_o\Lambda_i = \tilde{\partial}_1\Lambda_{i+1}$. Two such strings are equivalent if they are made so by the smallest equivalence relation "compatible with composition" such that

v) $\gamma_{f',g}(f,1)\cdot(f',1)\gamma_{f,g} \sim \gamma_{f'f,g}$

$(1,g')\gamma_{f,g}\cdot\gamma_{f,g'}(1,g) \sim \gamma_{f,g'g}$

vi) $(1,g')(f',1)\gamma_{f,g}\cdot\gamma_{f',g'}(f,1)(1,g)$

$\sim \gamma_{f',g'}(1,g)(f,1)\cdot(f',1)(1,g')\gamma_{f,g}$

when $f'f$ and $g'g$ are defined.

vii) If $\tau : f \to f'$, $\sigma : g \to g'$ then

$$\gamma_{f',g'} \cdot (\tau,1)(1,\sigma) \sim (1,\sigma)(\tau,1) \cdot \gamma_{f,g}$$

viii) $(\tau',1) \cdot (\tau,1) \sim (\tau' \cdot \tau,1)$

$(1,\sigma') \cdot (1,\sigma) \sim (1,\sigma' \cdot \sigma)$

when $\tau' \cdot \tau$ and $\sigma' \cdot \sigma$ are defined in $\mathfrak{A}$ and $\mathfrak{B}$, respectively. The weak domain and codomain are given by $\tilde{\partial}_o \Gamma = \tilde{\partial}_o \Lambda_n$, $\tilde{\partial}_1 \Gamma = \tilde{\partial}_1 \Lambda_1$, and the weak composition is induced by "dot juxtaposition" of strings; i.e., $\Gamma \cdot \Gamma'$ is defined if $\tilde{\partial}_1 \Gamma' = \tilde{\partial}_o \Gamma$ and then it is represented by

$$\Lambda_1 \cdot \Lambda_2 \cdot \ldots \cdot \Lambda_n \cdot \Lambda'_1 \cdot \ldots \cdot \Lambda'_m$$

To define the strong domain and codomain, one observes that $\partial_i \Lambda_j$ is independent of j so we can set $\partial_i \Gamma = \partial_i \Lambda_1$. If Γ and Γ' are represented by

$$\Gamma = [\Lambda_1 \cdot \ldots \cdot \Lambda_n] \ , \ \Gamma' = [\Lambda'_1 \cdot \ldots \cdot \Lambda_m]$$

and $\partial_1 \Gamma' = \partial_o \Gamma$, then the strong composition $\Gamma \Gamma'$ is represented as follows: if $n \leq m$, let

$$\tilde{\Lambda}_1 \cdot \tilde{\Lambda}_2 \cdot \ldots \cdot \tilde{\Lambda}_m$$

be any string of length m constructed by inserting suitable weak identities into Γ . This still represents Γ by viii). Then

$$\Gamma \Gamma' = [(\tilde{\Lambda}_1 \Lambda'_1) \cdot \ldots \cdot (\tilde{\Lambda}_m \Lambda'_m)]$$

If $n > m$ make the analogous construction for Γ' .
Finally, we can explain that the equivalence relation being "compatible with composition" means that it is to be compatible with both compositions described here.

We leave most of the details of checking that this is a 2-category to the reader. There is a non-trivial case of the interchange law (I,2.1, (2.2)) corresponding to the interchange law for squares described in I,2.1, (2.4). This is covered by requirement vi) above. Using this , one shows that it doesn't matter how identities are inserted in giving the definition of $\Gamma\Gamma'$, which is needed to show associativity of strong composition.

There is an obvious quasi-functor of two variables

$$J : \mathfrak{A} \times \mathfrak{B} \longrightarrow \mathfrak{A} \otimes \mathfrak{B}$$

given by $J(-,B)(\sigma) = (\sigma, 1_B)$, $J(A,-)(\tau) = (1_A, \tau)$, with $\gamma_{f,g}$ the unique so labeled 2-cell. Clearly for any H , there is a unique 2-functor $\bar{H} : \mathfrak{A} \otimes \mathfrak{B} \to \mathfrak{C}$ with $H = \bar{H}J$. The commutativity of the cube QN_2 in I,4.1, ii) shows that quasi-natural transformations $\sigma : H \to H'$ correspond bijectively to quasi-natural transformations $\bar{\sigma} : \bar{H} \to \bar{H}'$. Similarly, modifications correspond to each other.

The conditions in [E-K], II,5.10 require the commutativity of the diagrams (in their terminology but adapted to our order of variables)

$$\begin{array}{ccccc}
((A\otimes B)\otimes C, D) & & \xrightarrow{\quad p \quad} & & (C, (A\otimes B, D)) \\
\Big\downarrow{\scriptstyle (a,1)} & & & & \Big\downarrow{\scriptstyle (1,p)} \\
(A\otimes(B\otimes C), D) & \xrightarrow[p]{} & (B\otimes C, (A,D)) & \xrightarrow[p]{} & (C, (B, (A,D)))
\end{array}$$

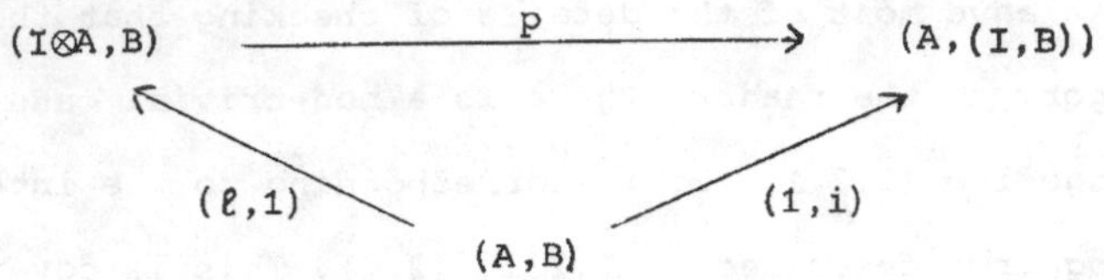

where a and ℓ are induced by the corresponding diagrams at the level of underlying categories. Here (X,Y) means Fun(X,Y) . The second diagram is immediate from the easily derived explicit expression for ℓ . The first diagram is equivalent ([E-K], II,4.1) to the coherence of the associativity isomorphism a . This, in turn, is equivalent in this case to the fact that the various ways of writing a tensor product of four variables all represent quasi-functors of four variables; and this, finally, is equivalent to the non-existence of "higher commutativity relations" (Remark after I,4.6), whose proof we have omitted, mostly because the relevant diagram is too big. Alternatively, the first diagram above can be shown to commute as a diagram of 2-categories by reducing everything to quasi-functors of several variables; e.g., one easily establishes isomorphisms compatible with "p" :

$$
\begin{aligned}
\mathrm{Fun}(\mathfrak{A}\otimes(\mathfrak{B}\otimes\mathfrak{C}),\mathfrak{D}) &\simeq q\text{-}\mathrm{Fun}(\mathfrak{A}\times(\mathfrak{B}\otimes\mathfrak{C}),\mathfrak{D}) \\
&\simeq q_3\text{-}\mathrm{Fun}(\mathfrak{A}\times\mathfrak{B}\times\mathfrak{C},\mathfrak{D}) \\
&\simeq q\text{-}\mathrm{Fun}((\mathfrak{A}\otimes\mathfrak{B})\times\mathfrak{C},\mathfrak{D}) \\
&\simeq \mathrm{Fun}((\mathfrak{A}\otimes\mathfrak{B})\otimes\mathfrak{C},\mathfrak{D}) \ .
\end{aligned}
$$

This finishes the construction of the tensor product.

The final question to be treated is the lack of symmetry of $\mathfrak{A} \otimes \mathfrak{B}$. One wants a hom-functor strongly adjoint to $- \otimes \mathfrak{B}$. Note that what we have is

$$\mathfrak{A} \otimes (-) \dashv \mathrm{Fun}(\mathfrak{A},-)$$

and this is an enriched adjunction with respect to "hom"$(\mathfrak{A},\mathfrak{C}) = \mathrm{Fun}(\mathfrak{A},\mathfrak{C})$. The discussion of this other adjoint and its relation to $\mathrm{Fun}(\mathfrak{A},\mathfrak{C})$ is rather intricate, and has, in fact, been incorrectly described by me on several occasions. This accounts for the seemingly excessive generality in what follows.

The basic consideration is connected with our consistent choice of squares with the 2-cell directed upwards (u) as on the left.

A 2-cell directed downwards (d) looks like the diagram on the right. This is closely related to forming duals. For completeness, we can also include the case of commuting squares (e for equal). We first treat the case of $\mathrm{Fun}(-,-)$. In what follows we always set

$$\bar{u} = d \ , \ \bar{d} = u \ , \text{ and } \bar{e} = e \ .$$

I,4.10. <u>Definition</u>. Let $\mathfrak{A}$ and $\mathfrak{B}$ be 2-categories. We set $\mathrm{Fun}_e(\mathfrak{A},\mathfrak{B}) = \mathfrak{B}^{\mathfrak{A}}$ and $\mathrm{Fun}_u(\mathfrak{A},\mathfrak{B}) = \mathrm{Fun}(\mathfrak{A},\mathfrak{B})$. We define $\mathrm{Fun}_d(\mathfrak{A},\mathfrak{B})$ to be the 2-category whose objects are 2-functors from $\mathfrak{A}$ to $\mathfrak{B}$, whose morphisms are <u>quasi$_d$-natural</u>

<u>transformations</u>; i.e., $\sigma : F \to G$ consists of a family of 1-cells $\{\sigma_A : FA \to GA\}$ together with a family of 2-cells

Q_dN

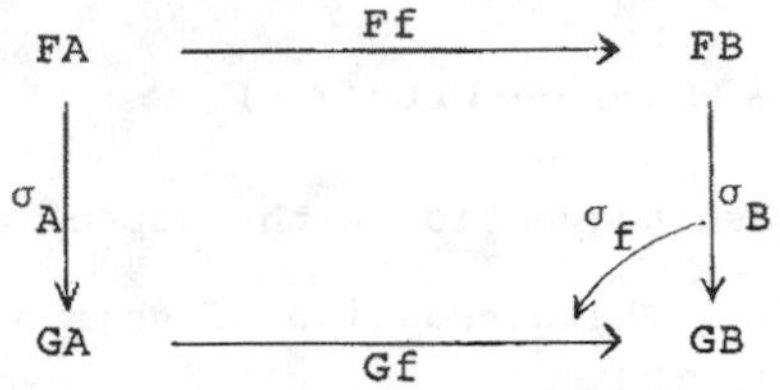

satisfying conditions analogous to I,2.1, QN 1,2,3, and whose 2-cells are modifications $s : \sigma \to \sigma'$ analogous to those in I,2.4, MQN.

I,4.11. <u>Proposition</u>. There are natural isomorphisms

$$ {}^{op}\mathrm{Fun}_x({}^{op}\mathcal{A}, {}^{op}\mathcal{B}) \simeq \mathrm{Fun}_x(\mathcal{A}^{op}, \mathcal{B}^{op})^{op} \simeq \mathrm{Fun}_{\bar{x}}(\mathcal{A}, \mathcal{B}) $$

where $x = u$, d , e . In particular,

$$ \mathrm{Fun}_x({}^{op}\mathcal{A}^{op}, {}^{op}\mathcal{B}^{op}) \simeq {}^{op}\mathrm{Fun}_x(\mathcal{A}, \mathcal{B})^{op} $$

<u>Proof</u>. The case $x = e$ is trivial. In all cases, there is an obvious bijection between the objects in the various 2-categories. We treat the case $x = u$ in detail. If $F : \mathcal{A} \to \mathcal{B}$, we write $F^{op} : \mathcal{A}^{op} \to \mathcal{B}^{op}$ and ${}^{op}F : {}^{op}\mathcal{A} \to {}^{op}\mathcal{B}$ for the corresponding 2-functors. Similarly, if $f : A \to A'$ is a 1-cell in $\mathcal{A}$ and $\tau : f \to f'$ a 2-cell, we write $f^{op} : A' \to A$ and $\tau^{op} : f' \to f$ for the corresponding 1-cell in $\mathcal{A}^{op}$ and 2-cell in ${}^{op}\mathcal{A}$. It is then clear that a quasi-natural transformation $\sigma : F \to G$ has the same description as a quasi$_d$-natural transformation ${}^{op}\sigma : {}^{op}F \to {}^{op}G$, but that modifications - being families of 2-cells in the codomain category - go the wrong way. Hence ${}^{op}\mathrm{Fun}_u({}^{op}\mathcal{A}, {}^{op}\mathcal{B}) \simeq \mathrm{Fun}_d(\mathcal{A}, \mathcal{B})$.

For the case of strong dualization, a morphism $\sigma^{op} : F^{op} \to G^{op}$ in $\mathrm{Fun}(\mathfrak{A}^{op},\mathfrak{B}^{op})^{op}$ is the same as a morphism $\sigma : G^{op} \to F^{op}$ in $\mathrm{Fun}(\mathfrak{A}^{op},\mathfrak{B}^{op})$. This consists of a family of 1-cells $\{(\sigma_A)^{op} : G^{op}(A) \to F^{op}(A)\}$ in $\mathfrak{B}^{op}$ - i.e., 1-cells $\{\sigma_A : FA \to GA\}$ in $\mathfrak{B}$ - together with 2-cells $\sigma_{f^{op}}$ for $f^{op} : B \to A$ in $\mathfrak{A}^{op}$ as illustrated

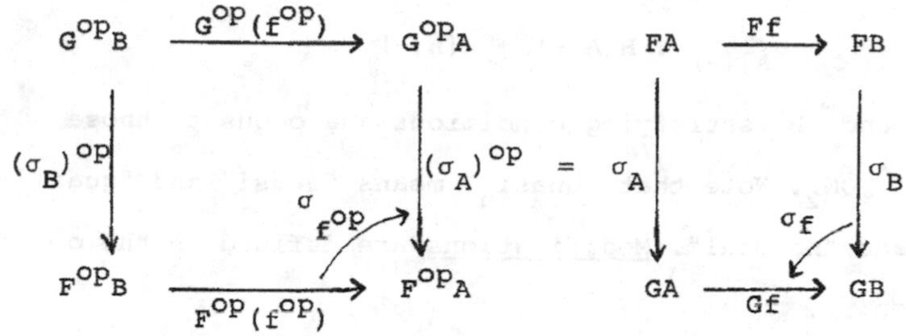
$G^{op}B$
$G^{op}(f^{op})$
$G^{op}A$
$(\sigma_B)^{op}$
$(\sigma_A)^{op}$
$\sigma_{f^{op}}$
$F^{op}B$
$F^{op}(f^{op})$
$F^{op}A$
=
FA
Ff
FB
σ_A
σ_B
σ_f
GA
Gf
GB

Modifications are unaffected by this dualization, and hence

$$\mathrm{Fun}_u(\mathfrak{A}^{op},\mathfrak{B}^{op})^{op} \simeq \mathrm{Fun}_d(\mathfrak{A},\mathfrak{B})$$

We next study the various possibilities for quasi-functors of two variables and quasi-natural transformations between them. There are eight logical possibilities depending on whether the $\gamma_{f,g}$'s, the σ_A's, and the σ_B's are "up" or "down". In fact only six occur. We again include the commuting case for completeness.

I,4.12. Definition.

i) A quasi$_u$-functor of two variables $H : \mathfrak{A}\times\mathfrak{B} \to \mathfrak{C}$ is a quasi-functor of two variables. A quasi$_d$-functor of two variables $H : \mathfrak{A}\times\mathfrak{B} \to \mathfrak{C}$ is defined as in I,4.1 i) except that $\gamma_{f,g}$ is "down"; i.e., $\gamma_{f,g} : H(A',g)H(f,B) \to H(f,B')H(A,g)$, and these γ's satisfy equations analogous to QF_2 of I,4.1 i).

A $\underline{\text{quasi}_e}$-$\underline{\text{functor}}$ $\underline{\text{of}}$ $\underline{\text{two}}$ $\underline{\text{variables}}$ $H : \mathfrak{A}\times\mathfrak{B} \to \mathfrak{C}$ is a 2-functor.

ii) A $\underline{\text{quasi}_{y,z}}$-$\underline{\text{natural}}$ $\underline{\text{transformation}}$ between quasi$_x$-functors of two variables H and $\bar{H}$ consists of

a) quasi$_y$-natural transformations

$$\sigma_{(-),B} : H(-,B) \to \bar{H}(-,B)$$

b) quasi$_z$-natural transformations

$$\sigma_{A,(-)} : H(A,-) \to \bar{H}(A,-)$$

for all A and B satisfying conditions analogous to those in I,4.1 ii), QN_2. Note that "quasi$_u$" means "quasi" and "quasi$_e$-natural" means "natural". $\underline{\text{Modifications}}$ are defined in the obvious fashion.

iii) $q\text{-Fun}_{x;y,z}(\mathfrak{A}\times\mathfrak{B},\mathfrak{C})$ denotes the resulting 2-category. The cases $u;d,u$ and $d;u,d$ are excluded. (I.e., there are 25 rather than 27 cases).

I,4.13. $\underline{\text{Proposition}}$. There are natural isomorphisms

i) $q\text{-Fun}_{x;x,z}(\mathfrak{A}\times\mathfrak{B},\mathfrak{C}) \cong \text{Fun}_z(\mathfrak{B},\text{Fun}_x(\mathfrak{A},\mathfrak{C}))$

ii) $q\text{-Fun}_{x;y,\bar{x}}(\mathfrak{A}\times\mathfrak{B},\mathfrak{C}) \cong \text{Fun}_y(\mathfrak{A},\text{Fun}_{\bar{x}}(\mathfrak{B},\mathfrak{C}))$

iii) $q\text{-Fun}_{x;y,z}(\mathfrak{A}\times\mathfrak{B},\mathfrak{C}) \cong q\text{-Fun}_{\bar{x};z,y}(\mathfrak{B}\times\mathfrak{A},\mathfrak{C})$

$\cong {}^{op}[q\text{-Fun}_{\bar{x};\bar{y},\bar{z}}({}^{op}\mathfrak{A}\times{}^{op}\mathfrak{B},{}^{op}\mathfrak{C})$

$\cong [q\text{-Fun}_{\bar{x};\bar{y},\bar{z}}(\mathfrak{A}^{op}\times\mathfrak{B}^{op},\mathfrak{C}^{op})]^{op}$

iv) $\text{Fun}_x(\mathfrak{A}\otimes\mathfrak{B},\mathfrak{C}) \cong q\text{-Fun}_{u;x,x}(\mathfrak{A}\times\mathfrak{B},\mathfrak{C})$

$\underline{\text{Proof}}$. The proofs of i), ii) and iii) are straightforward adaptations of the proofs of I,4.2 iii) and I,4.11, using numerous diagrams like the one in I,4.1 ii). The proof of iv) follows from the observation that objects on the left are

2-functors from $\mathfrak{A} \otimes \mathfrak{B}$ to $\mathfrak{C}$ which correspond to quasi$_u$-functors of two variables from $\mathfrak{A}\times\mathfrak{B}$ to $\mathfrak{C}$, while morphisms on the left are quasi$_x$-natural transformations between such 2-functors which are easily seen to correspond to quasi$_{x,x}$-natural transformations between the corresponding quasi$_u$-functors, using the explicit structure of $\mathfrak{A} \otimes \mathfrak{B}$.

I,4.14. Theorem. The tensor product of 2-categories is part of a monoidal closed category structure on 2-Cat$_o$ in each variable (a biclosed category in the sense of Lambek [28]), i.e.,

$$\mathfrak{A} \otimes (-) \dashv \mathrm{Fun}_u(\mathfrak{A},-) \; ; \; (-) \otimes \mathfrak{B} \dashv \mathrm{Fun}_d(\mathfrak{B},-)$$

and there are natural isomorphisms

i) $\mathrm{Fun}_u(\mathfrak{A}\otimes\mathfrak{B},\mathfrak{C}) \cong \mathrm{Fun}_u(\mathfrak{B},\mathrm{Fun}_u(\mathfrak{A},\mathfrak{C}))$

ii) $\mathrm{Fun}_d(\mathfrak{A}\otimes\mathfrak{B},\mathfrak{C}) \cong \mathrm{Fun}_d(\mathfrak{A},\mathrm{Fun}_d(\mathfrak{B},\mathfrak{C}))$.

Furthermore, the enriched representable functors for these two structures commute; i.e.,

iii) $\mathrm{Fun}_u(\mathfrak{A},\mathrm{Fun}_d(\mathfrak{B},\mathfrak{C})) \cong \mathrm{Fun}_d(\mathfrak{B},\mathrm{Fun}_u(\mathfrak{A},\mathfrak{C}))$

Proof. i) is the same as I,4.9. ii) follows from I,4.13 iv) with $x = d$ together with the special case $u;d,d$ of I,4.13 ii). The adjunctions follow by considering these isomorphisms at the level of objects. Finally, iii) is immediate from I,4.13 i) and ii), taking $(x;x,z) = (u;u,d) = (x;y,\bar{x})$. Note that i), ii) and iii) also follow from adjointness plus the associativity of the tensor product (end of proof of I,4.9).

I,4.15. Definition. Let $\mathfrak{A}$ be a 2-category. Then

$$\mathrm{Fun}\,\mathfrak{A} = \mathrm{Fun}_d(\underline{2},\mathfrak{A})$$

I,4.16. Corollary.

i) $\mathrm{Fun}_u(\mathfrak{A},\mathrm{Fun}\,\mathfrak{B}) \cong \mathrm{Fun}_d(\underline{2},\mathrm{Fun}_u(\mathfrak{A},\mathfrak{B}))$

$\cong \mathrm{Fun}(\mathrm{Fun}_u(\mathfrak{A},\mathfrak{B}))$

ii) $\mathrm{Fun}_u(\mathfrak{A}\otimes 2,\mathfrak{B}) \cong \mathrm{Fun}_u(\underline{2},\mathrm{Fun}_u(\mathfrak{A},\mathfrak{B}))$

$\cong {}^{op}\mathrm{Fun}({}^{op}\mathrm{Fun}_u(\mathfrak{A},\mathfrak{B}))$

iii) $\mathrm{Fun}_d(\underline{2}\otimes\mathfrak{A},\mathfrak{B}) \cong \mathrm{Fun}(\mathrm{Fun}_d(\mathfrak{A},\mathfrak{B}))$

iv) $\mathrm{Fun}(\mathrm{Fun}\,\mathfrak{B}) \cong \mathrm{Fun}_d(\underline{2}\otimes\underline{2},\mathfrak{B})$

Note. In particular, morphisms in $\mathrm{Fun}(\mathfrak{A},\mathfrak{B})$ (i.e., quasi-natural transformations) can be identified with 2-functors from $\mathfrak{A}$ to $\mathrm{Fun}\,\mathfrak{B}$ or from $\mathfrak{A}\otimes\underline{2}$ to $\mathfrak{B}$. A similar analysis can be given for the 2-cells in $\mathrm{Fun}(\mathfrak{A},\mathfrak{B})$ by replacing $\underline{2}$ with $\underline{2}_2$ in the formulas above. Thus 2-cells can be identified with 2-functors from $\mathfrak{A}$ to $\mathrm{Fun}_d(\underline{2}_2,\mathfrak{B})$ or from $\mathfrak{A}\otimes\underline{2}_2$ to $\mathfrak{B}$.

If one wishes, by using I,4.11, all subscripts d can be eliminated from I,4.14 and I,4.16.

Proof. The first equation in i) is a special case of I,4.14 iii) and the second is the definition of $\mathrm{Fun}(-)$. The first equation in ii) is a special case of I,4.14 i) and the second follows from I,4.14 ii), the definition of $\mathrm{Fun}(-)$ and the property that ${}^{op}\underline{2} = \underline{2}$ since $\underline{2}$ is locally discrete. iii) and iv) are special cases of I,4.14 ii).

I,4.17. Corollary. There are isomorphisms of 2-categories:

$${}^{op}\mathfrak{A}\otimes{}^{op}\mathfrak{B} \cong {}^{op}(\mathfrak{B}\otimes\mathfrak{A})$$

$$\mathfrak{A}^{op}\otimes\mathfrak{B}^{op} \cong (\mathfrak{B}\otimes\mathfrak{A})^{op}$$

$${}^{op}\mathfrak{A}^{op}\otimes{}^{op}\mathfrak{B}^{op} \cong {}^{op}(\mathfrak{A}\otimes\mathfrak{B})^{op}$$

Proof. It follows from I,4.11, I,4.14, and I,4.13 that there are natural isomorphisms

$$
\begin{aligned}
{}^{op}\mathrm{Fun}_u({}^{op}(\mathfrak{A}\otimes\mathfrak{B}), {}^{op}\mathfrak{C}) &\cong \mathrm{Fun}_d(\mathfrak{A}\otimes\mathfrak{B}, \mathfrak{C}) \\
&\cong \text{q-}\mathrm{Fun}_{u;d,d}(\mathfrak{A}\times\mathfrak{B}, \mathfrak{C}) \\
&\cong {}^{op}[\text{q-}\mathrm{Fun}_{u;u,u}({}^{op}\mathfrak{B}\times{}^{op}\mathfrak{A}, {}^{op}\mathfrak{C})] \\
&\cong {}^{op}\mathrm{Fun}_u({}^{op}\mathfrak{B}\otimes{}^{op}\mathfrak{A}, {}^{op}\mathfrak{C})
\end{aligned}
$$

Setting ${}^{op}\mathfrak{C} = {}^{op}(\mathfrak{A}\otimes\mathfrak{B})$ or ${}^{op}\mathfrak{B}\otimes{}^{op}\mathfrak{A}$ produces 2-functors

$$ {}^{op}(\mathfrak{A}\otimes\mathfrak{B}) \rightleftarrows {}^{op}\mathfrak{B}\otimes{}^{op}\mathfrak{A} $$

such that the effect of the above isomorphisms on objects is given by composition with these functors. Hence they are inverse to each other. (I.e., the isomorphism part of the Yoneda argument works for natural transformations between $\mathfrak{D}$-representable functors). The second isomorphism is proved similarly, and the third follows from the first two.

I,4.18. Remark. The various versions of the $\mathfrak{D}$-Yoneda lemma are available here; e.g., [E-K], I,8.6; I,10.8; II,7.4, where it is called the representation theorem. Parts of Dubuc [9], Kelly [24], and Day-Kelly [8] are also available; namely those parts which only use the biclosed structure and not symmetry.

I,4.19. Proposition. For all 2-categories $\mathfrak{A}$ (resp., $\mathfrak{A}$ and $\mathfrak{B}$), $\mathrm{Fun}_x(\mathfrak{A},-)$ (resp., $\text{q-}\mathrm{Fun}_{x;y,z}(\mathfrak{A}\times\mathfrak{B},-)$ is a 3-functor from 2-Cat to 2-Cat , and the isomorphisms in I,4.9, 11, 13, 14, and 15 are 3-natural. (Cf. I,2.6.)

Proof. It is easily checked, using the explicit formula in part 1) of the proof of I,4.5, that the transpose of composition yields a factorization

$$\begin{array}{ccc}
\mathfrak{C}^{\mathfrak{B}} & \dashrightarrow & \mathrm{Fun}(\mathfrak{A},\mathfrak{C})^{\mathrm{Fun}(\mathfrak{A},\mathfrak{B})} \\
\cap & & \cap \\
\mathrm{Fun}(\mathfrak{B},\mathfrak{C}) & \longrightarrow & \mathrm{Fun}(\mathrm{Fun}(\mathfrak{A},\mathfrak{B}),\mathrm{Fun}(\mathfrak{A},\mathfrak{C}))
\end{array}$$

which gives the 3-functor structure on $\mathrm{Fun}(\mathfrak{A},-)$. The other cases are handled analogously. Note, however, that nothing beyond naturality can be said about the behavior of these functors in their first variables with respect to the standard 2-category and 3-category structure in 2-Cat_0 . Thus the Proposition on page 281 of [CCS] must be corrected accordingly.

It would be nice to be able to carry out a similar analysis for bicategories, pseudo-functors and quasi-natural transformations between them (see I,3). In fact, it is essential for our purposes to do so. However, an examination of the analogue of the explicit formula for I,4.5 discloses a fatal flaw. Suppose that in the notation there, $\mathfrak{A}$, $\mathfrak{B}$, and $\mathfrak{C}$ are bicategories, F, G, H, and K are pseudo-functors and σ and τ are quasi-natural transformations. Then already in trying to define

$$H\circ(-) : \mathrm{Pseud}(\mathfrak{A},\mathfrak{B}) \to \mathrm{Pseud}(\mathfrak{A},\mathfrak{C})$$

there is a problem since, when H is applied to a diagram

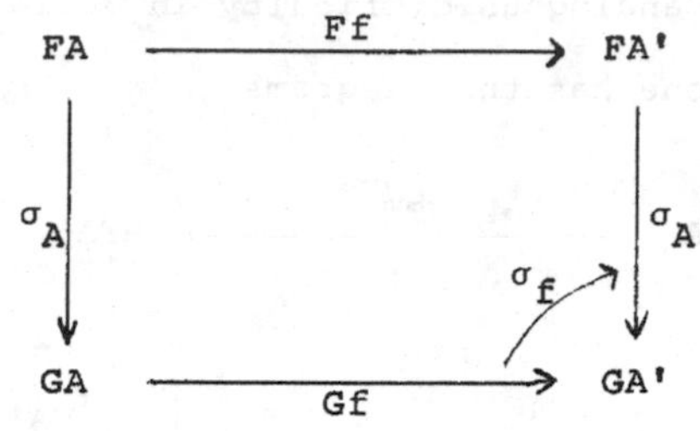

one gets a diagram

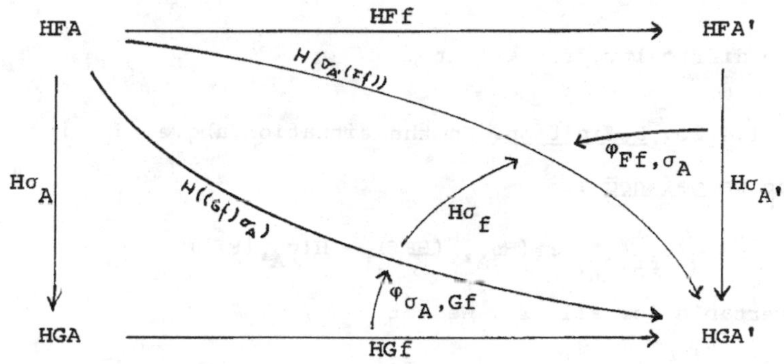

In general, there is no way to put together these 2-cells to get a quasi-natural transformation from HF to HG . It does no good to try to generalize the notion of quasi-natural transformations to allow diagrams like

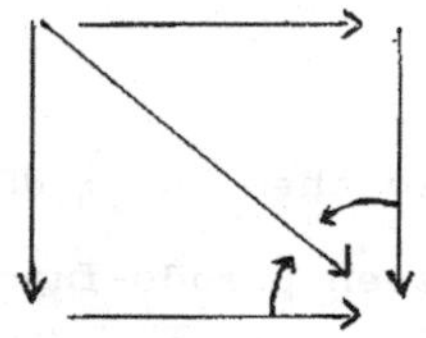

or even a whole path of 2-cells from one composition to the other, since there is no way to compose such diagrams.

There is no analogous difficulty in defining τF where $\tau : H \to K$, since one has the diagrams

$$
\begin{array}{ccc}
HFA & \xrightarrow{HFf} & HFA' \\
\Big\downarrow{\scriptstyle \tau_{FA}} & \overset{\tau_{Ff}}{\nearrow} & \Big\downarrow{\scriptstyle \tau_{FA'}} \\
KFA & \xrightarrow[KFf]{} & KFA'
\end{array}
$$

with no difficulty, for all f .

I,4.20. <u>Definition</u>. In the situation above, $H\sigma$ is said to be <u>defined</u> if

$$\varphi_{Ff,\sigma_{A'}} : (H\sigma_{A'})(HFf) \to H(\sigma_{A'}(Ff))$$

is invertable for all f . We set

$$(H\sigma)_f = (\varphi_{Ff,\sigma_{A'}})^{-1}(H\sigma_f)(\varphi_{\sigma_A,Gf})$$

Note that if H is homomorphic then $H\sigma$ is defined for all σ .

I,4.21. <u>Proposition</u>. In the situation above,

i) $\tau F : HF \to KF$ is a quasi-natural transformation between pseudo-functors.

ii) If $H\sigma$ is defined then $H\sigma : HF \to HG$ is a quasi-natural transformation between pseudo-functors.

iii) $(-)F : \mathrm{Pseud}(\mathfrak{B},\mathfrak{C}) \to \mathrm{Pseud}(\mathfrak{A},\mathfrak{C})$ is a strictly homomorphic pseudo-functor.

iv) $H(-) : \mathrm{Pseud}(\mathfrak{A},\mathfrak{B}) \to \mathrm{Pseud}(\mathfrak{A},\mathfrak{C})$ is a pseudo-functor on any sub-bicategory of $\mathrm{Pseud}(\mathfrak{A},\mathfrak{B})$ on which it is defined.

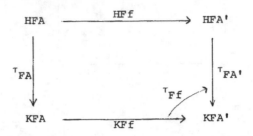

Proof. These are all exercises using the naturality of $\varphi : M(q)M(q') \rightarrow M(qq')$ and the compatibility of φ with the associativity isomorphisms (see I,3.2, PF2 and PF4). i) and iii) are easy, iv) is moderate and ii) is a long calculation.

I,4.22. Examples

1) In general, the tensor product of two 2-categories is very complicated. The simplest, non-trivial case is $\underline{2} \otimes \underline{2}$ which looks like

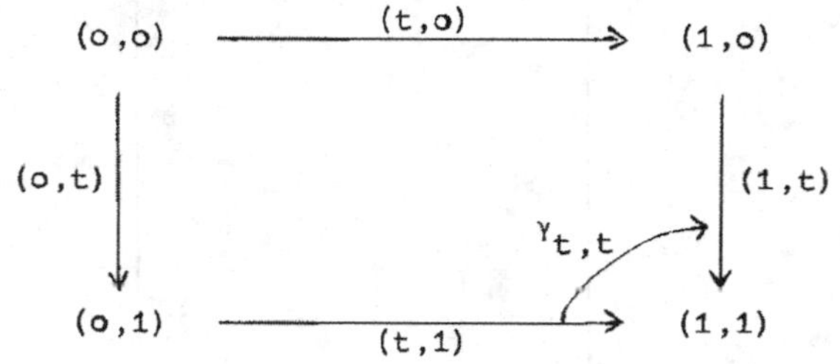

The next case is $\underline{2}_2 \otimes \underline{2}_2$ (cf. I,2.11, 1)). Here the only hom-category which is neither empty nor $\underline{2}$ is $\underline{2}_2 \otimes \underline{2}_2((o,o),(1,1))$, which turns out to be the commutative cube represented by

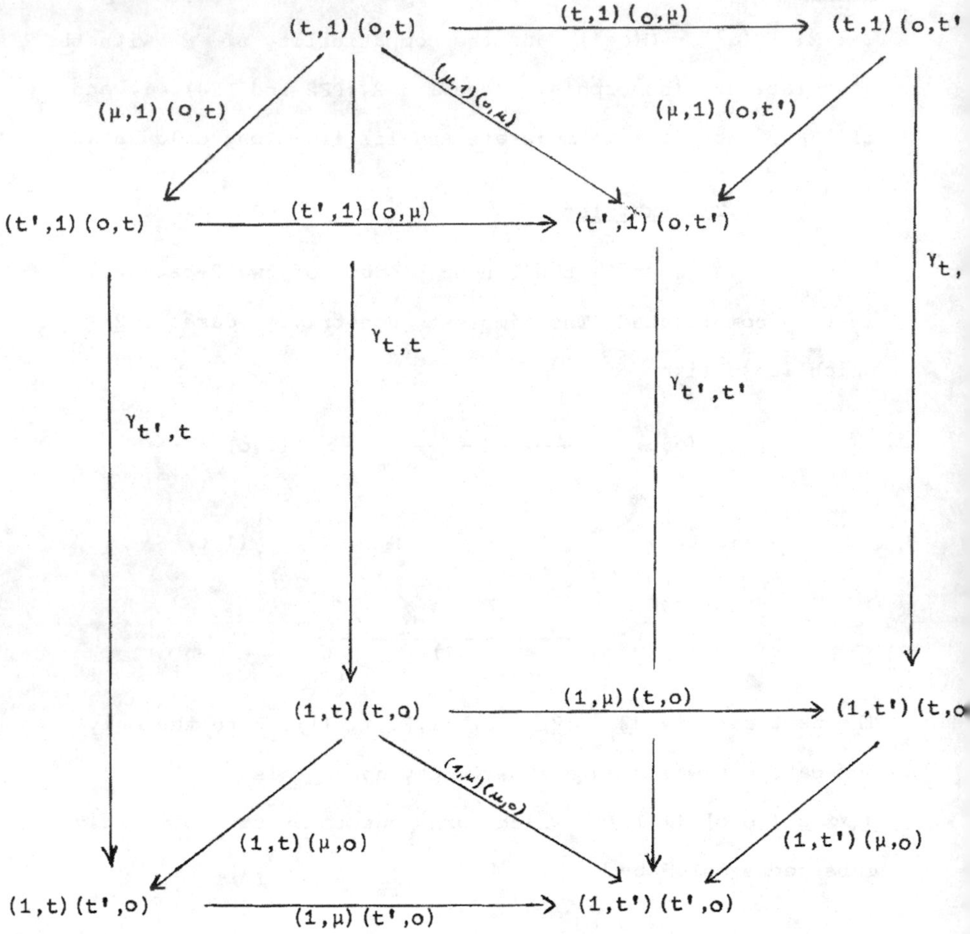

Only relation vii) of I,4.9. is needed here. We have omitted most of the diagonal compositions in this picture of $\underline{2}\times\underline{2}\times\underline{2}$.

From these two cases one can more or less visualize $A \otimes \underline{2}$ and $A \otimes \underline{2}_2$ for arbitrary A . More complicated cases seem out of reach at the moment. For instance, what is Cat ⊗ Cat ; i.e., describe

$$\text{Cat} \otimes \text{Cat}((\underline{A},\underline{B}),(\underline{A}',\underline{B}'))$$

at least in special cases.

Another interesting case is that of a pair of strictly associative and unitary monoidal categories, $\underline{A}$ and $\underline{B}$, regarded as 2-categories $\bar{\underline{A}}$ and $\bar{\underline{B}}$ with a single object. Then $\bar{\underline{A}} \otimes \bar{\underline{B}}$ has a single object and its morphisms are represented by strings of objects

$$X_1 B_1 A_2 B_2 \ldots A_n Y_n$$

where X_1 is empty or A_1 and Y_n is empty or B_n, and $A_i \in \underline{A}$, $B_i \in \underline{B}$. The 2-cells are generated by morphisms $A_i \to A_i'$ or $B_i \to B_i'$ together with interchange 2-cells $AB \to BA$ subject to the relations in I,4.9. There has not been time to calculate specific examples of the monoidal category that arises from this construction.

2) The construction Fun(-) is, of course, analogous to $(-)^{\underline{2}}$. In I,1.5, we remarked that $(-)^{\underline{2}}$ is a triple, or equivalently $\underline{2}\times(-)$ is a cotriple, and Cat_t is isomorphic to the Kleisli category for this triple. Fun(-) has a much more complicated structure; namely, two triples and a Cat-natural transformation between them. It is easier to discuss this in terms of $\underline{2} \otimes (-)$, the structure transfering properly, by I,4.15, iii). $\underline{2} \otimes (-)$ has two cotriple structures given by the two functors $\underline{2} \to \underline{2} \otimes \underline{2}$ represented, in the terminology above, by

$$\delta_o : \underline{2} \to \underline{2} \otimes \underline{2} : t \longmapsto (t,1)(o,t)$$

$$\delta_1 : \underline{2} \to \underline{2} \otimes \underline{2} : t \longmapsto (1,t)(t,o) .$$

γ_{tt} represents a quasi-natural transformation between these. This induces a Cat-natural transformation between $\delta_o \otimes (-)$ and $\delta_1 \otimes (-)$ or equivalently between the two corresponding functors from Fun(Fun(-)) to Fun(-) . This means precisely: 2-Cat is a 2-category; Fun(-) and Fun(Fun(-)) are 2-functors (by I,4.19); a morphism Fun(Fun(-)) → Fun(-) is a Cat-natural transformation - i.e., for each $\mathfrak{B}$, a 2-functor Fun(Fun $\mathfrak{B}$) → Fun $\mathfrak{B}$, natural in $\mathfrak{B}$; a 2-cell between two such is a Cat-natural transformation for each $\mathfrak{B}$ which is natural in $\mathfrak{B}$. γ_{tt} yields such a thing (cf. I,5.5).

I,4.23 Appendix A. Let $\mathfrak{A}$ be a 2-category. The problem of finding a 2-category $\tilde{\mathfrak{A}}$ and a copseudo-functor (I,3.2) $P : \mathfrak{A} \to \tilde{\mathfrak{A}}$ such that if $F : \mathfrak{A} \to \mathfrak{B}$ is any copseudo-functor, then there is a unique 2-functor $\tilde{F} : \tilde{\mathfrak{A}} \to \mathfrak{B}$ with $\tilde{F}P = F$, has a relatively simple solution which is presumably a special case of the corresponding (unavailable) construction of Benabou for bicategories. It is as follows. Let $I : \underline{\Delta} \to$ Sets take the ordinal n to the set $[n] = \{1,2,\ldots,n\}$ and increasing functions to the corresponding maps. (Note: $[0] = \emptyset$). A set X determines a functor

$$\underline{\Delta}^{op} \to \text{Sets} : n \longmapsto \text{Sets}(I(n),X)$$

and hence a fibred category Seq X over $\underline{\Delta}$ whose objects are finite sequences of elements of X and such that $\alpha : m \to n$ determines a unique map

$$(x_{\alpha(1)},\ldots,x_{\alpha(m)}) \to (x_1,\ldots,x_n)$$

for each sequence $(x_1,\ldots,x_n)$ (cf. I,1.11, and I,2.9.)

If A and B are objects in the 2-category $\mathbb{A}$, let $\mathrm{Seq}(\mathrm{Ob}\ \mathbb{A})_{A,B}$ be the full subcategory of $\mathrm{Seq}(\mathrm{Ob}\ \mathbb{A})$ determined by sequences $(X_1,\ldots,X_n)$ such that $X_1 = A$ and $X_n = B$. Note that if $A \neq B$ then there are no such sequences with $n = 0$ or 1 and a unique sequence (A,B) for $n = 2$, while if $A = B$ then there is a unique sequence (A) for $n = 1$.

If $\mathbb{A}$ is a small 2-category, then $\tilde{\mathbb{A}}$ is the small 2-category in which the objects are the same as those in $\mathbb{A}$ and such that $\tilde{\mathbb{A}}(A,B)$ is the fibred category over $\mathrm{Seq}(\mathrm{Ob}\ \mathbb{A})_{A,B}$ corresponding to the functor

$$[\mathrm{Seq}(\mathrm{Ob}\ \mathbb{A})_{A,B}]^{op} \to \mathrm{Cat}$$

taking a sequence $(X_1,X_2,\ldots,X_n)$ to $\prod_{i=1}^{n-1} \mathbb{A}(X_i,X_{i+1})$ for $n \geq 1$ and, if $A = B$, taking (A) to $\underline{1}$. It is sufficient to describe the values of this functor on morphisms of the form

i) $(X_1,\ldots,\hat{X}_i,\ldots,X_n) \to (X_1,\ldots,X_i,\ldots,X_n)$

in which case it is the product of identity functors with composition

$$\mathbb{A}(X_{i-1},X_i)\times\mathbb{A}(X_i,X_{i+1}) \xrightarrow{\circ} \mathbb{A}(X_{i-1},X_{i+1}) \ ;$$

and

ii) $(X_1,\ldots,X_i,X_i,\ldots,X_n) \to (X_1,\ldots,X_i,\ldots,X_n)$

in which case it is given by inserting $\underline{1}$ between the i'th and (i+1)'st factors in the product for $(X_1,\ldots,X_i,\ldots,X_n)$ and then taking the product of identity functors with the functor

$$I_{X_i} : \underline{1} \to \mathbb{A}(X_i,X_i)$$

In particular, $(A,A) \to (A)$ gives $I_A : \underline{1} \to \mathbb{A}(A,A)$. The relations satisfied by composition and units I_X, show that this extends to a functor.

In the corresponding fibred category, $\tilde{A}(A,B)$ there are "cartesian" morphisms

$$(f_1,\ldots,f_{i+1}f_i,\ldots,f_n) \rightarrow (f_1,\ldots,f_i,f_{i+1},\ldots,f_n)$$

$$(f_1,\ldots,f_i,I_{X_i},f_{i+1},\ldots,f_n) \rightarrow (f_1,\ldots,f_i,f_{i+1},\ldots,f_n)$$

for each composable sequence of 1-cells from A to B.

Given a triple of objects A,B,C in A, there is a functor

$$\mathrm{Seq}(\mathrm{Ob}\ A)_{A,B} \times \mathrm{Seq}(\mathrm{Ob}\ A)_{B,C} \rightarrow \mathrm{Seq}(\mathrm{Ob}\ A)_{A,C}$$

$$((X_1,\ldots,X_n),(Y_1,\ldots,Y_m)) \mapsto (X_1,\ldots,X_n=Y_1,\ldots,Y_m)$$

which induces a unique "cartesian" functor between fibred categories

$$\tilde{A}(A,B) \times \tilde{A}(B,C) \xrightarrow{\circ} \tilde{A}(A,C)$$

which is the inclusion on each fibre and preserves the given cartesian morphisms. This is the composition in $\tilde{A}$. There is an obvious copseudo-functor $P : A \rightarrow \tilde{A}$ given by the inclusion of $A(A,B)$ in $\tilde{A}(A,B)$ as the fibre over the unique sequence of length 2 in which, for $f : A \rightarrow B$, $g : B \rightarrow C$, $(\varphi_{A,B,C})_{f,g} : gf \rightarrow (f,g)$ is the given cartesian morphism. The unit I_A in $\tilde{A}(A,A)$ is the inclusion of $\underline{1}$ as the fibre over the sequence (A), so $\varphi_A : P(I_A) \rightarrow \tilde{I}_A$ is the given such cartesian morphism.

Since composition above is a cartesian functor, all of the cartesian morphisms can be expressed in terms of the $\varphi_{A,B,C}$'s and φ_A's by composition. It follows from this that P has the desired universal property.

Now let Σ be any family of composable pairs of 1-cells in $\mathfrak{A}$. Let $\tilde{\Sigma}_{A,B}$ be the set of all cartesian morphisms in $\tilde{\mathfrak{A}}(A,B)$ of the form

$$(f_1,\ldots,f_{i+1}f_i,\ldots,f_n) \to (f_1,\ldots,f_i,f_{i+1},\ldots,f_n)$$

where $(f_i,f_{i+1}) \in \Sigma$ and let $\tilde{\mathfrak{A}}(A,B)[[\tilde{\Sigma}_{A,B}^{-1}]]$ be the category in which all of these cartesian morphisms are made identity maps (i.e., coequalized with their domains). This is stable with respect to the composition defined above and yields a 2-category $\tilde{\mathfrak{A}}[[\tilde{\Sigma}^{-1}]]$ which is universal with respect to copseudo-functors $F : \mathfrak{A} \to \mathfrak{B}$ such that $F(f,f') = F(f)F(f')$ for all $(f,f') \in \Sigma$.

In particular, $\mathfrak{A} \otimes \mathfrak{B}$ is given by taking in $\mathfrak{A} \times \mathfrak{B}$ the class Σ of all pairs of the forms $((f,1),(f',1))$, $((1,g),(1,g'))$ and $((f,1),(1,g))$. Then

$$\mathfrak{A} \otimes \mathfrak{B} = \widetilde{\mathfrak{A}\times\mathfrak{B}}[[\tilde{\Sigma}^{-1}]] .$$

A simple example of this construction is $\tilde{\underline{1}} = \underline{\Delta}^{op}$. This is checked by the observation of Benabou that copseudo-functors from $\underline{1}$ to Cat are the same as cotriples.

I,4.24 <u>Appendix B</u>. Then is another biclosed structure between $\mathfrak{B}^{\mathfrak{A}}$ and $\mathrm{Fun}(\mathfrak{A},\mathfrak{B})$ given by considering quasi-natural transformations in which all 2-cells are isomorphisms. We shall use the prefix <u>iso</u> (or, later, simply i) for this case. Thus $\varphi : F \to G$ is an <u>iso-quasi$_x$-natural transformation</u> if φ_f is an isomorphism for all f.
$\mathrm{Iso\text{-}Fun}_x(\mathfrak{A},\mathfrak{B})$ is the locally full sub 2-category of $\mathrm{Fun}_x(\mathfrak{A},\mathfrak{B})$ determined by these. Note that $\mathrm{Iso\text{-}Fun}_u(\mathfrak{A},\mathfrak{B}) = $
$= \mathrm{Fun}(\mathfrak{A},\mathfrak{A}_o;\mathfrak{B},\mathrm{iso}\ \mathfrak{B})$ (I,2.4). Iso-quasi-functors of n-variables and iso-quasi-natural transformations between them are defined

analogously. $\mathfrak{A} \underset{\text{iso}}{\otimes} \mathfrak{B}$ is constructed by inverting all the 2-cells $\gamma_{f,g}$.

If $\varphi : F \to G$ is an iso-quasi$_u$-natural transformation, then $\tilde{\varphi}$ is an iso-quasi$_d$-natural transformation, where $\tilde{\varphi}_A = \varphi_A$ and $\tilde{\varphi}_f = \varphi_f^{-1}$. This yields an isomorphism

$$\text{iso-Fun}_u(\mathfrak{A},\mathfrak{B}) \cong \text{iso-Fun}_d(\mathfrak{A},\mathfrak{B})$$

which is compatible with the commutativity isomorphism

$$\mathfrak{A} \underset{\text{iso}}{\otimes} \mathfrak{B} \cong \mathfrak{B} \underset{\text{iso}}{\otimes} \mathfrak{A}$$

in which $\gamma_{f,g}$ is sent to $(\gamma_{g,f})^{-1}$, and shows that this construction gives a symmetric monoidal closed category structure on 2-Cat_o.

I,4.25 <u>Appendix C</u>. <u>Categories enriched in $2\text{-Cat}_{\otimes}$</u>. Let $2\text{-Cat}_{\otimes}$ denote the monoidal (non-symmetric) closed category whose objects and morphisms are 2-categories and 2-functors, respectively; and whose internal hom-functor is given by $\text{Fun}(-,-)$ with its associated tensor product. Categories enriched in $2\text{-Cat}_{\otimes}$ are defined in the usual way as in [E-K]. Note that since $\mathfrak{B}^{\mathfrak{A}} \subset \text{Fun}(\mathfrak{A},\mathfrak{B})$, or, equivalently, since there is a 2-functor $\mathfrak{A} \otimes \mathfrak{B} \to \mathfrak{A}\times\mathfrak{B}$ taking the 2-cells $\gamma_{f,g}$ of I,4.9 to identities, an ordinary 3-category may always be considered as a $2\text{-Cat}_{\otimes}$-category. In I,7 we shall need the notion of a <u>quasi-enriched functor</u> between $2\text{-Cat}_{\otimes}$-categories. Let $\mathfrak{A}$ and $\mathfrak{B}$ be such, with enriched hom-functors denoted by $\mathfrak{A}(-,-)$ and $\mathfrak{B}(-,-)$. Then a quasi-enriched functor $F : \mathfrak{A} \to \mathfrak{B}$ is an object function, denoted by F, together with 2-functors

$$F_{A,B} : \mathfrak{A}(A,B) \to \mathfrak{B}(FA,FB)$$

for all ordered pairs of objects A,B in $\mathfrak{A}$, 1-cells $\varphi_A : 1_{FA} \to F(1_A)$, and for all ordered triples of objects A,B,C in $\mathfrak{A}$, quasi-natural transformations $\varphi_{A,B,C}$ as illustrated

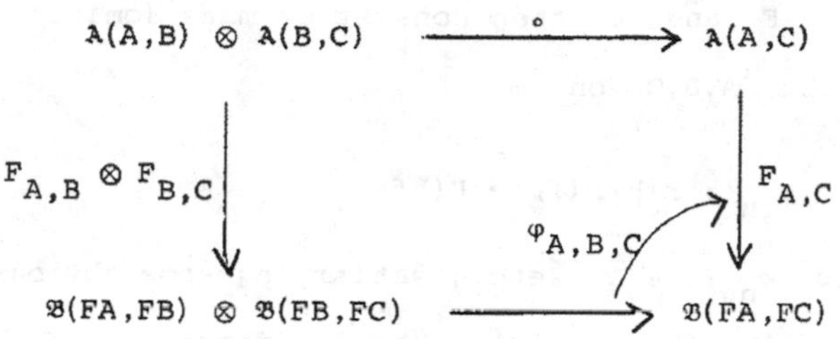

satisfying the analogous of PF4 and PF5 of I,3.2 in which $\times$ is replaced by $\otimes$. Note that a quasi-enriched functor between 2-categories regarded as 2-$\mathrm{Cat}_{\otimes}$-categories is the same as a pseudo-functor. There is an obvious extension of these notions to 2-$\mathrm{Cat}_{\otimes}$-bicategories.

Since the notion of quasi-enriched functor is rather complicated and since it will be used in complicated circumstances, we illustrate precisely what it means. Given two composable 2-cells in $\mathfrak{A}$

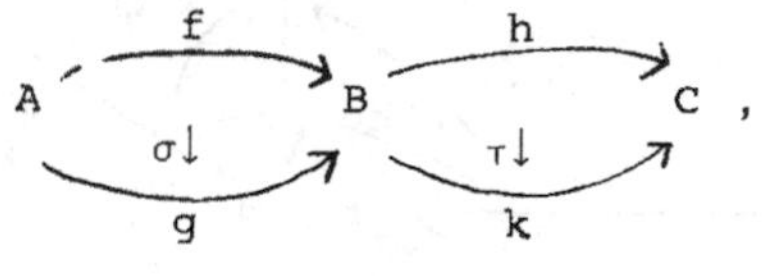

their "composition" is a diagram

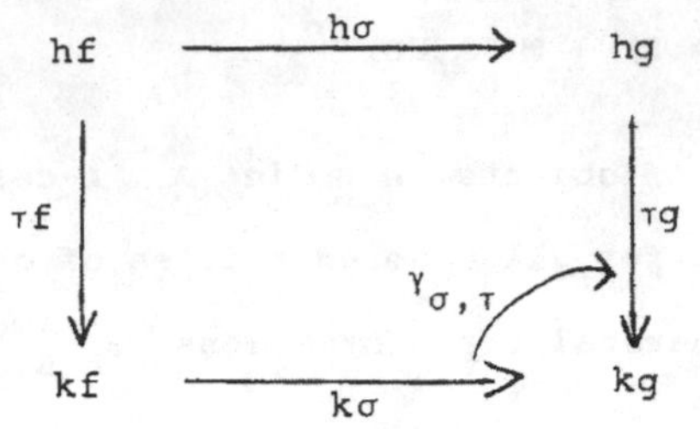

in $\mathbf{A}(A,C)$. F and φ then consist of maps (omitting the subscript A,B,C on φ)

$$\varphi_{fh} : F(h)F(f) \to F(hf)$$

and 2-cells $\varphi_{h\sigma}$, $\varphi_{\tau f}$, etc., satisfying some obvious properties, plus the property that the diagram

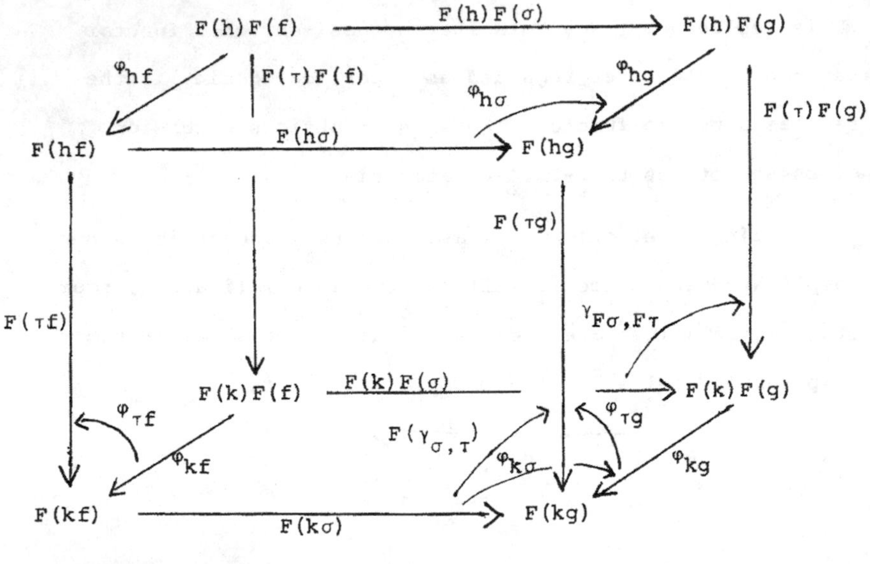

commutes. (Note that a $2\text{-Cat}_{\otimes}$-functor is the same sort of thing, with the φ's all identities.)

Among the various constructions for 2-categories that can also be carried out for $2\text{-Cat}_{\otimes}$-categories, the most important here is the quasi 3-comma category $[F_1,F_2]_\otimes$ constructed as in I,2.7 for 2-$Cat_\otimes$-functors F_1 and F_2 between 2-$Cat_\otimes$-categories. The cells are exactly as in I,2.7; the only difference being that the resulting category is a 2-$Cat_\otimes$-category rather than a 2-Cat-category. This does not generalize to quasi-enriched functors unless the φ's are isomorphisms.

There is an obvious dual situation given by the internal hom-functor $Fun_d(-,-)$ and its associated tensor product which we denote here by $\otimes d$. This monoidal closed category is written 2-$Cat_{\otimes d}$. A quasi-enriched functor is defined <u>exactly</u> as before. In both cases, one can also consider quasi$_d$-enriched functors in which the $\varphi_{A,B,C}$'s go the other way. Both types of duals will occur in I.7.

Duals of such categories can be found by reversing appropriate cells, as in I,2.6. We observe for future reference that 2-$Cat_\otimes^{op}$, ^{op}2-$Cat_\otimes$ and $_{op}$2-$Cat_\otimes$ are 2-$Cat_{\otimes d}$-categories while double duals are 2-$Cat_\otimes$-categories and $^{op}_{op}$2-$Cat_\otimes^{op}$ is a 2-$Cat_{\otimes d}$-category.

Finally, we shall want to consider quasi-natural transformations between quasi-functors between 2-$Cat_\otimes$-categories, $\mathfrak{A}$ and $\mathfrak{B}$. There are various ways to explain this using $[-,-]_\otimes$ or applying Fun(-) to the 2-category-hom objects of $\mathfrak{B}$. However, since there is only one example of this in I,7, we shall just give an elementary description here. Let (F,φ) and (G,ψ) be quasi-enriched functors between 2-$Cat_\otimes$-categories $\mathfrak{A}$ and $\mathfrak{B}$. A <u>quasi-natural transformation</u> $\mu : F \to G$ consists of the following data

(i) for each $A \in \mathbb{A}$, $\mu_A : FA \to GA$

(ii) for each $f : A \to B$ in $\mathbb{A}$, $\mu_f : \mu_B(Ff) \to (Gf)\mu_A$

(iii) for each $\sigma : f \to g$ in $\mathbb{A}$, there is a 3-cell as illustrated

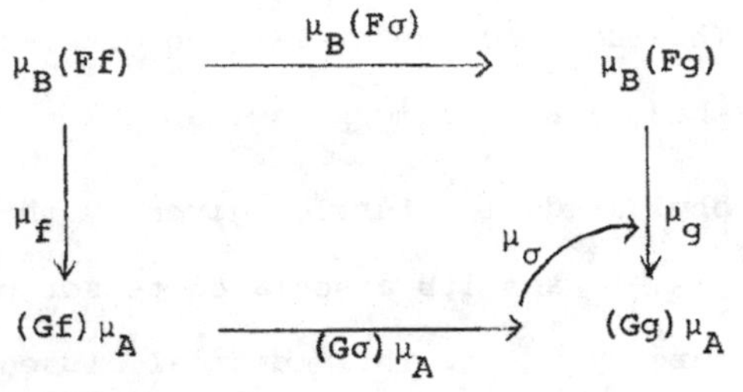

(iv) for each composition $A \xrightarrow{f} B \xrightarrow{h} C$, there is a 3-cell

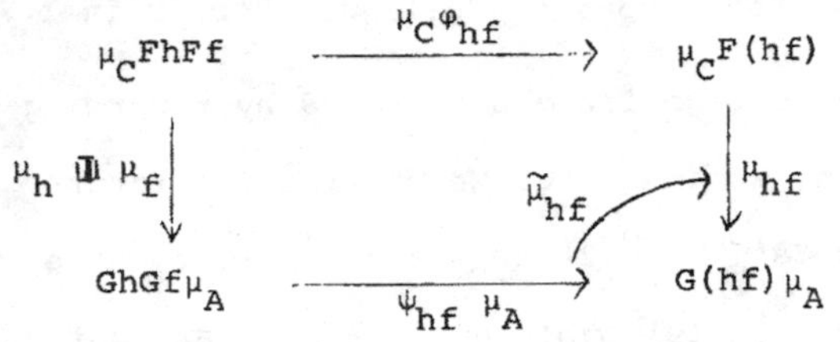

These data are required to be coherent in the sense that all diagrams formed from them commute. This can be described by a finite number of conditions which will not be described here.

I,5. <u>Properties of 2-comma categories</u>. As an application of the formulas in the preceeding section, we derive some properties of 2-comma categories (see I,2.5). These will be used in I,7. There are four types of properties discussed here.

i) $\mathrm{Fun}(\chi,-)$ commutes with 2-comma categories (I,5.1). Suitably interpreted (I,5.2), this allows one to adapt a suggestion of Lawvere's about ordinary comma categories to give a neat derivation of the <u>algebraic</u> properties of 2-comma categories (I,5.3).

ii) The fibration properties of 2-comma categories (I,5.6 and I,5.8)

iii) The measure of the failure of Godement's Fifth Rule [20], (I,5.7, ii)).

iv) The bimodule properties of 2-comma categories (I,5.8) and the characterization of quasi-homomorphic and homomorphic 2-functors between them (I,5.9).

The first two properties are exact analogues of the general properties of comma objects in representable 2-categories (Ch. III). The last two are different and account for the richness of the theory of quasi-adjointness (I,7).

If $F_i : \mathfrak{A}_i \to \mathfrak{B}$, $i = 1,2$, are 2-functors, then we take as a definition of $[F_1,F_2]$, the formula

$$(5.1)\quad [F_1,F_2] = \varprojlim(\mathfrak{A}_1 \xrightarrow{F_1} \mathfrak{B} \xleftarrow{\delta_o} \mathrm{Fun}\ \mathfrak{B} \xrightarrow{\delta_1} \mathfrak{B} \xleftarrow{F_2} \mathfrak{A}_2)$$

Here $\delta_i = \mathrm{Fun}_d(\partial_i,\mathfrak{B}) : \mathrm{Fun}_d(\underline{2},\mathfrak{B}) \to \mathrm{Fun}_d(\underline{1},\mathfrak{B}) \cong \mathfrak{B}$. We write $P_i : [F_1,F_2] \to \mathfrak{A}_i$ and $P_\theta : [F_1,F_2] \to \mathrm{Fun}\ \mathfrak{B}$ for the projections;

these satisfy $\delta_i P_\theta = P_i$.

A translation of diagrams

$$
(5.2) \qquad
\begin{array}{ccccccccc}
\mathfrak{A}_1 & \xrightarrow{F_1} & \mathfrak{B} & \longleftarrow & \operatorname{Fun}\mathfrak{B} & \longrightarrow & \mathfrak{B} & \xleftarrow{F_2} & \mathfrak{A}_2 \\
{\scriptstyle V_1}\big\downarrow & & \big\downarrow{\scriptstyle W} & & \big\downarrow{\scriptstyle \operatorname{Fun} W} & & \big\downarrow{\scriptstyle W} & & \big\downarrow{\scriptstyle V_2} \\
\mathfrak{A}'_1 & \xrightarrow[F'_1]{} & \mathfrak{B}' & \longleftarrow & \operatorname{Fun}\mathfrak{B}' & \longrightarrow & \mathfrak{B}' & \xleftarrow[F'_2]{} & \mathfrak{A}'_2
\end{array}
$$

induces a 2-functor

$$(5.3) \qquad (V_1, W, V_2) : [F_1, F_2] \to [F'_1, F'_2]$$

I,5.1. <u>Proposition</u>.

i) $[\operatorname{Fun}(\chi,F), \operatorname{Fun}(\chi,G)] = \operatorname{Fun}(\chi,[F,G])$

ii) Given an inverse system of diagrams

$$\{\mathfrak{A}_i \xrightarrow{F_i} \mathfrak{C}_i \xleftarrow{G_i} \mathfrak{B}_i\}$$

then

$$[\varprojlim F_i, \varprojlim G_i] = \varprojlim [F_i, G_i]$$

<u>Proof</u>. i) By I,4.4, $\operatorname{Fun}(\chi,-)$ preserves inverse limits. Thus the result follows by applying it to the inverse limit diagram defining $[F,G]$, and using I,4.15, ii).

ii) Since $\operatorname{Fun}\mathfrak{B} = \operatorname{Fun}_d(\underline{2},\mathfrak{B}) = {}^{op}\operatorname{Fun}(\underline{2},{}^{op}\mathfrak{B})$, it follows from I,4.4 that $\operatorname{Fun}(-)$ preserves inverse limits. Since inverse limits commute with each other, the result is immediate.

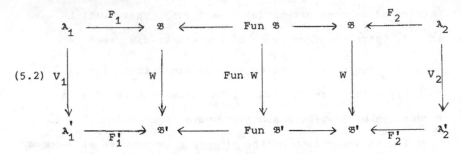

Although statement i) above and its proof are apparently simple, they say a great deal about the structure of quasi-natural transformations.

I,5.2. <u>Corollary</u>. i) There is a diagram

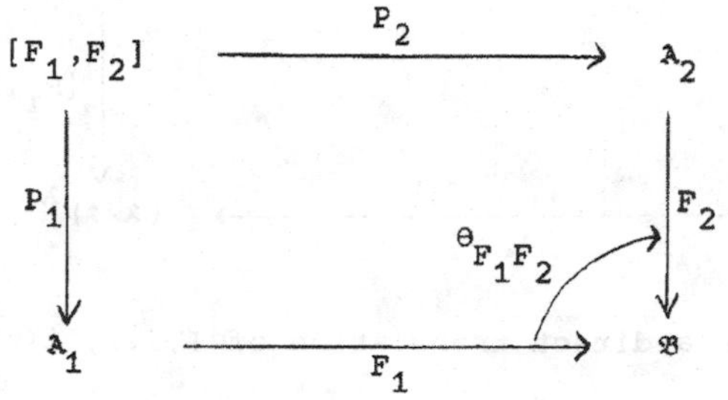

where $\theta_{F_1F_2}$ is a quasi-natural transformation, satisfying the following universal property, which characterizes $[F_1,F_2]$ up to an isomorphism: given any diagram

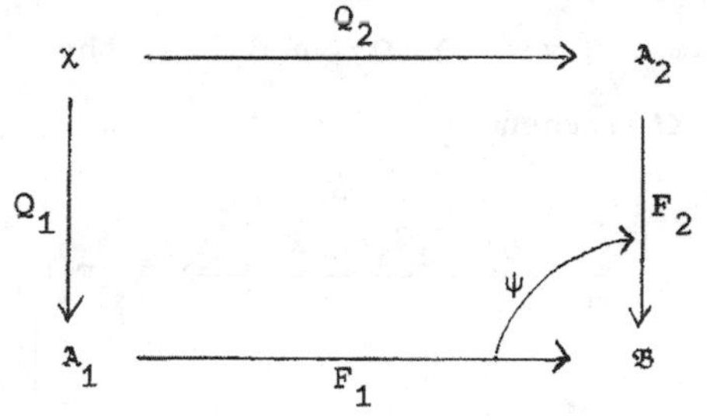

with ψ a quasi-natural transformation, then there is a unique 2-functor $Q : \chi \to [F_1,F_2]$ such that $P_iQ = Q_i$, $i = 1,2$, and $\theta_{F_1F_2}Q = \psi$.

ii) Furthermore, there is a one-to-one correspondence between modifications $s : \psi \to \psi'$ of quasi-natural transformations as in i) and <u>natural</u> transformations $\sigma : Q \to Q'$ such that $P_i\sigma = Q_i$, $i = 1,2$ and $\theta_{F_1F_2}\sigma = S$.

iii) In particular, if $F,G : \mathfrak{A} \to \mathfrak{B}$, then there is a pullback (n 2-Cat)

$$\begin{array}{ccc} \mathrm{Fun}(\mathfrak{A},\mathfrak{B})(F,G) & \xrightarrow{\overline{(-)}} & [F,G]^{\mathfrak{A}} \\ \downarrow & & \downarrow {\scriptstyle \{P_1,P_2\}^{\mathfrak{A}}} \\ \underline{1} & \xrightarrow{\ulcorner\Delta\urcorner} & (\mathfrak{A}\times\mathfrak{A})^{\mathfrak{A}} \end{array}$$

__Proof__. i) This is a direct translation of I,5.1, i) for objects, taking account of I,4.15, ii). In particular, $P_\theta : [F_1,F_2] \to \mathrm{Fun}\ \mathfrak{B}$ corresponds to a 2-functor $\underline{2} \to \mathrm{Fun}([F_1,F_2],\mathfrak{B})$ which gives the quasi-natural transformation $\theta_{F_1F_2}$. In terms of components, for an object $f : F_1A_1 \to F_2A_2$ of $[F_1,F_2]$, $(\theta_{F_1F_2})_f = f$ and for a morphism (h,φ,k) , $(\theta_{F_1F_2})_{(h,\varphi,k)} = \varphi$. Given Q_1,Q_2 and ψ , then $Q(X) = \psi_X$ and, if $k : X \to X'$, then

$$Q(k) = \begin{array}{ccc} F_1Q_1X & \xrightarrow{F_1Q_1k} & F_1Q_1X' \\ {\scriptstyle \psi_X}\downarrow & \overset{\psi_k}{\nearrow} & \downarrow{\scriptstyle \psi_{X'}} \\ F_2Q_2X & \xrightarrow[F_2Q_2k]{} & F_2Q_2X' \end{array}$$

ii) The effect of I,5.1, i) on morphisms can also be described; namely, if $Q,Q' : \chi \to [F_1,F_2]$, then quasi-natural transformations $\sigma : Q \to Q'$ correspond to pairs of quasi-natural transformations

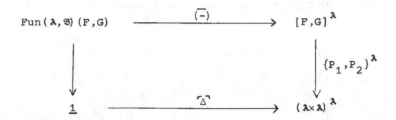

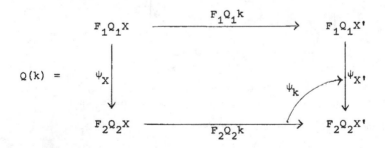

$$\sigma_1 : P_1Q \to P_1Q' \ , \ \sigma_2 : P_2Q \to P_2Q'$$

together with a morphism in $\mathrm{Fun}(\mathrm{Fun}(\chi,\mathfrak{B}))$

(5.4)

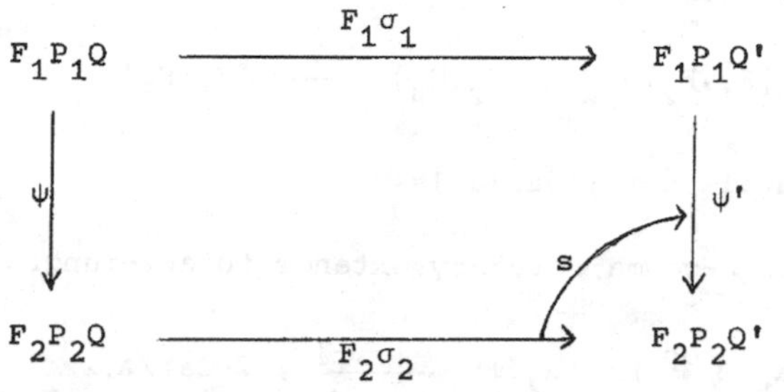

where $\psi = P_\theta Q$, $\psi' = P_\theta Q'$ and s is a modification.

Clearly, if σ_1 and σ_2 are identities, then only the modification $s : \psi \to \psi'$ is left. However, we claim more; that in these circumstances σ is natural. In general, one finds that if the above diagram is evaluated at a morphism $k : X \to X'$ in χ , then σ_k is the 2-cell in $\mathrm{Fun}(\mathrm{Fun}(\chi,\mathfrak{B}))$ represented by

(5.5) $$\langle F_1(\sigma_1)_k \ , \ F_2(\sigma_2)_k \rangle$$

which is an identity 2-cell if and only if σ_1 and σ_2 are natural; in particular, this is satisfied if they are identities.

iii) This is how $\mathrm{Fun}(\mathfrak{A},\mathfrak{B})$ is defined in [CCS]. Here it is the further specialization of ii) to the case where $Q_1 = Q_2 = \mathfrak{A}$. We shall frequently use the notation indicated here: a quasi-natural transformation $\varphi : F \to G$ corresponds to a 2-functor $\bar{\varphi} : \mathfrak{A} \to [F,G]$ with $P_1\bar{\varphi} = F$, $P_2\bar{\varphi} = G$ and a modification $s : \varphi \to \varphi'$ corresponds to a Cat-natural transformation $\bar{s} : \bar{\varphi} \to \bar{\varphi}'$ with $P_1\bar{s} = F$, $P_2\bar{s} = G$.

I,5.3. Theorem.

i) There is a "strictly associative and unitary" composition 2-functor

$$[F_1,F_2] \underset{\mathfrak{A}_2}{\times} [F_2,F_3] \xrightarrow{\ \circ\ } [F_1,F_3]$$

which is natural in all variables.

ii) The 2-comma category extends to a 2-functor

$$\mathrm{Fun}(\mathfrak{A}_1,\mathfrak{B})^{op}\times\mathrm{Fun}(\mathfrak{A}_2,\mathfrak{B}) \xrightarrow{[-,-]} 2\text{-}\mathrm{Cat}/\mathfrak{A}_1\times\mathfrak{A}_2$$

which is "compatible with the composition in i)" (the slash / means 2-categories over $\mathfrak{A}_1\times\mathfrak{A}_2$), and is natural in all variables.

Proof. i) This is a reflection of the fact that Fun $\mathfrak{B}$ is a double category (cf. [CCS] and I,2.8). Here we proceed as follows. Composition is the unique 1-cell determined by the composed quasi-natural transformation in the diagram

(5.6)

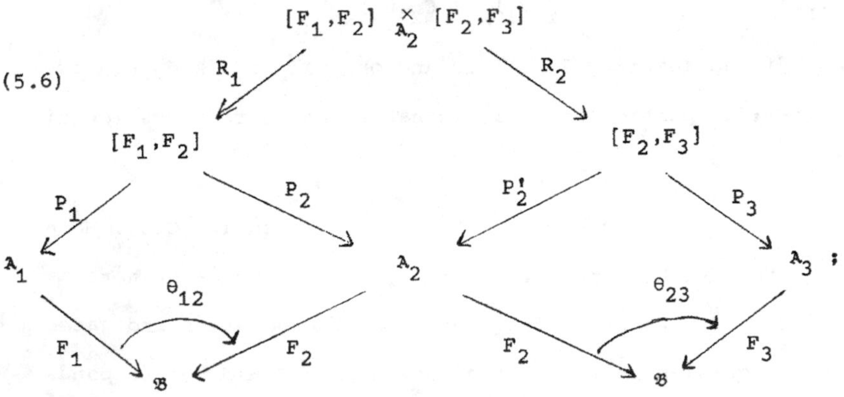

i.e., $\theta_{13}("\circ") = (\theta_{23}R_2)\cdot(\theta_{12}R_1)$.

For each $F : \mathfrak{A} \to \mathfrak{B}$, there is a unit $j_F : \mathfrak{A} \to [F,F]$ determined by the commutative diagram

$$
(5.7)\qquad
\begin{array}{ccc}
\mathfrak{A} & \xrightarrow{\ \mathfrak{A}\ } & \mathfrak{A} \\
{\scriptstyle \mathfrak{A}}\big\downarrow & & \big\downarrow{\scriptstyle F} \\
\mathfrak{A} & \xrightarrow[\ F\ ]{} & \mathfrak{B}
\end{array}
$$

It follows from uniqueness that this is associative and unitary in the obvious sense. Naturality means that given 2-functors and commutative diagrams

$$
\begin{array}{ccc}
\mathfrak{A}_i & \xrightarrow{\ F_i\ } & \mathfrak{B} \\
{\scriptstyle V_i}\big\downarrow & & \big\downarrow{\scriptstyle W} \\
\mathfrak{A}'_i & \xrightarrow[\ F'_i\ ]{} & \mathfrak{B}'
\end{array}
\qquad i = 1,2,3
$$

then the diagram

$$
(5.8)\qquad
\begin{array}{ccc}
[F_1,F_2] \underset{\mathfrak{A}_2}{\times} [F_2,F_3] & \xrightarrow{\ \circ\ } & [F_1,F_3] \\
\big\downarrow{\scriptstyle (V_1,W,V_2) \underset{V_2}{\times} (V_2,W,V_3)} & & \big\downarrow{\scriptstyle (V_1,W,V_3)} \\
[F'_1,F'_2] \underset{\mathfrak{A}'_2}{\times} [F'_2,F'_3] & \longrightarrow & [F'_1,F'_3]
\end{array}
$$

commutes.

ii) Let $\varphi : G_1 \to F_1$ and $\psi : F_2 \to G_2$ be quasi-natural transformations. Then

$$
(5.9)\qquad
\begin{array}{ccc}
[\varphi,\psi]:[F_1,F_2] & \longrightarrow & [G_1,G_2] \\
& {\scriptstyle \{P_1,P_2\}}\searrow \quad \swarrow{\scriptstyle \{P'_1,P'_2\}} & \\
& \mathfrak{A}_1\times\mathfrak{A}_2 &
\end{array}
$$

is the 2-functor determined by the composed quasi-natural transformation in the diagram

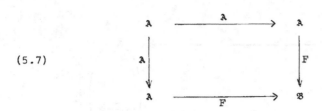

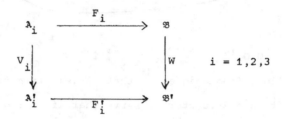

(5.10)

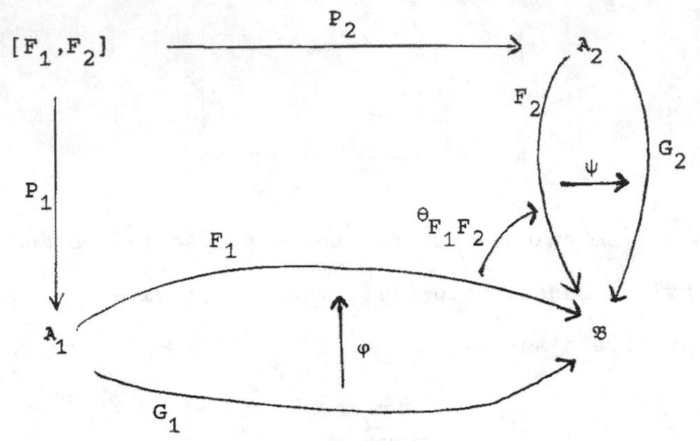

;

i.e., $\theta_{G_1G_2}[\varphi,\psi] = \psi P_2 \cdot \theta_{F_1F_2} \cdot \varphi P_1$.

It follows from uniqueness that $[-,-]$ is a functor. To see that it is a 2-functor, observe that if $u : \varphi \to \varphi'$ and $v : \psi \to \psi'$ are modifications, then

(5.11) $vP_2 \cdot \theta_{F_1F_2} \cdot uP_1 : \psi P_2 \cdot \theta_{F_1F_2} \cdot \varphi P_1 \to \psi' P_2 \cdot \theta_{F_1F_2} \cdot \varphi' P_2$

is a modification and hence corresponds to a natural transformation

(5.12) $$[u,v] : [\varphi,\psi] \to [\varphi',\psi']$$

Since this is natural, we get a 2-functor with codomain $2\text{-Cat}/\mathcal{A}_1 \times \mathcal{A}_2$.

Compatibility with composition means that, given $\varphi_i : F_i \to F_i'$, $i = 1,2,3$, the diagrams

$$\begin{array}{ccc}
[F_1',F_2] \underset{\mathfrak{A}_2}{\times} [F_2,F_3] & \xrightarrow{\quad\circ\quad} & [F_1',F_3] \\
\Big\downarrow {\scriptstyle [\varphi_1,F_2] \underset{\mathfrak{A}_2}{\times} [F_2,\varphi_3]} & & \Big\downarrow {\scriptstyle [\varphi_1,\varphi_3]} \\
[F_1,F_2] \underset{\mathfrak{A}_2}{\times} [F_2,F_3'] & \xrightarrow{\quad\circ\quad} & [F_1,F_3']
\end{array}$$

(5.13)

$$\begin{array}{ccc}
[F_1,F_2] \underset{\mathfrak{A}_2}{\times} [F_2',F_3] & \xrightarrow{[F_1,\varphi_2] \underset{\mathfrak{A}_2}{\times} [F_2',F_3]} & [F_1,F_2'] \underset{\mathfrak{A}_2}{\times} [F_2',F_3] \\
\Big\downarrow {\scriptstyle [F_1,F_2] \underset{\mathfrak{A}_2}{\times} [\varphi_2,F_3]} & & \Big\downarrow {\scriptstyle \circ} \\
[F_1,F_2] \underset{\mathfrak{A}_2}{\times} [F_2,F_3] & \xrightarrow{\quad\circ\quad} & [F_1,F_3]
\end{array}$$

commute. This follows by inserting 2-cells into the diagram defining composition in part i) in the same way that they are inserted into the basic diagram in defining $[\varphi,\psi]$.

Naturality in all variables means, given a commutative diagram of 2-functors, quasi-natural transformations and modifications

(5.14)

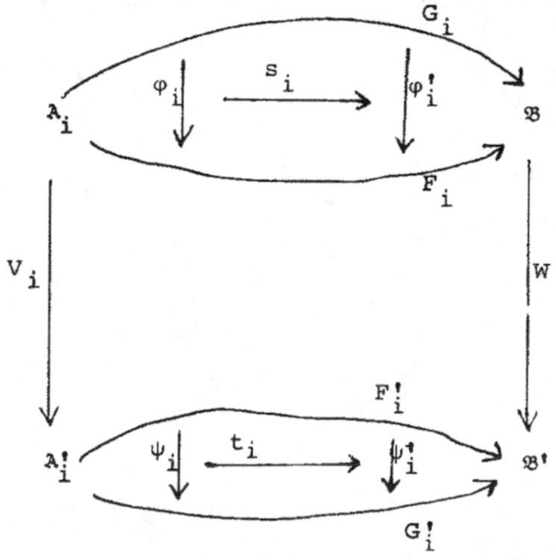

i=1,2,

then the diagram

(5.15)

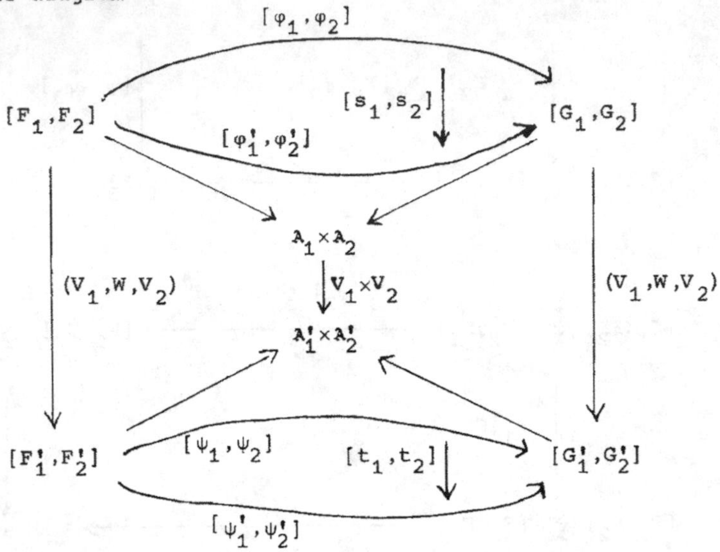

commutes. This is easily established.

<u>Remarks</u>. The construction in part ii) above can be generalized to give a 3-functor defined on 3-comma categories (cf. I,2.7),

$$^{op}[^{op}\widetilde{(2\text{-Cat})},\ulcorner\mathfrak{B}\urcorner]_3\times[\widetilde{(2\text{-Cat})},\ulcorner\mathfrak{B}\urcorner]_3 \to (\widetilde{2\text{-Cat}},P_1\times P_2)$$

where a pair of morphisms

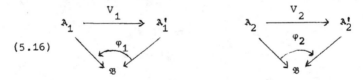

(5.16)

give rise to a diagram

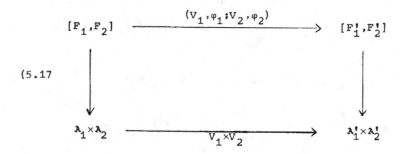

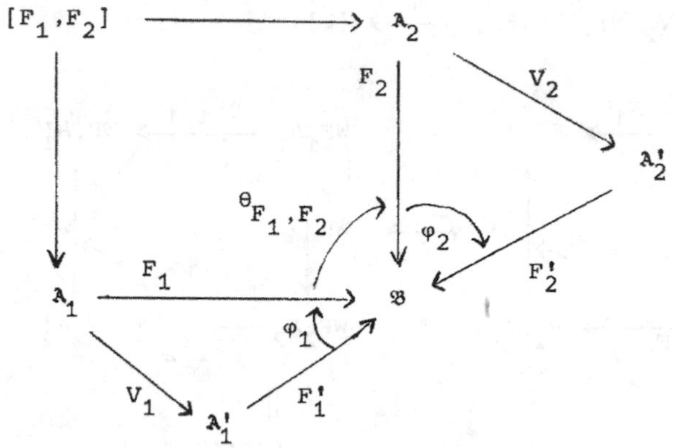

which induces a morphism (in the imprecisely described comma category)

(5.17

$$\begin{array}{ccc} [F_1,F_2] & \xrightarrow{(V_1,\varphi_1;V_2,\varphi_2)} & [F_1',F_2'] \\ \downarrow & & \downarrow \\ \mathbb{A}_1\times\mathbb{A}_2 & \xrightarrow[V_1\times V_2]{} & \mathbb{A}_1'\times\mathbb{A}_2' \end{array}$$

and 2-cells and 3-cells give rise to corresponding 2-cells and 3-cells "over" their components in the place of $V_1 \times V_2$.

I,5.4. <u>Explicit formulas</u>. Using the description in terms of components in the proof of I,5.2, these operations can be expressed as follows:

a) $(V_1,W,V_2) : [F_1,F_2] \longrightarrow [F'_1,F'_2]$

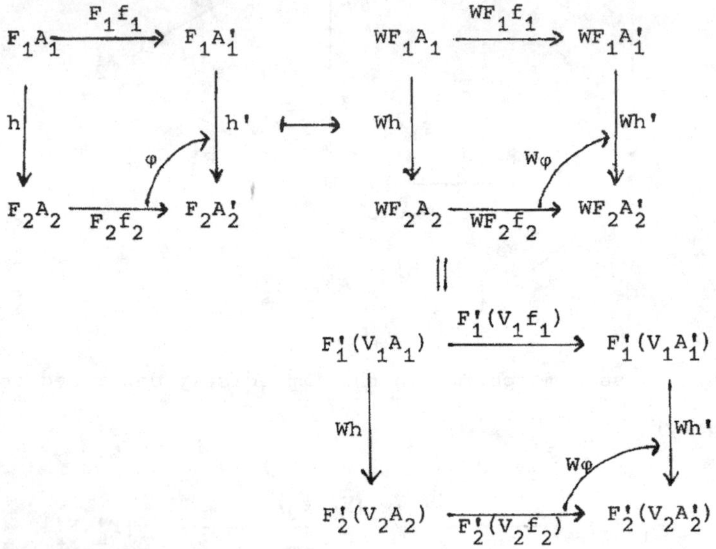

b) $[F_1,F_2] \underset{A_2}{\times} [F_2,F_3] \xrightarrow{\quad\circ\quad} [F_1,F_3]$

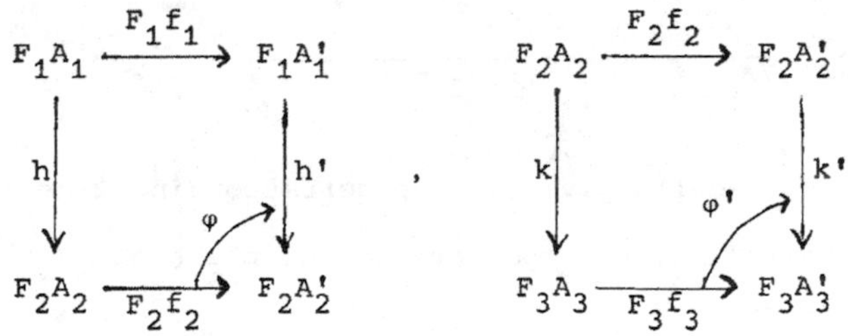

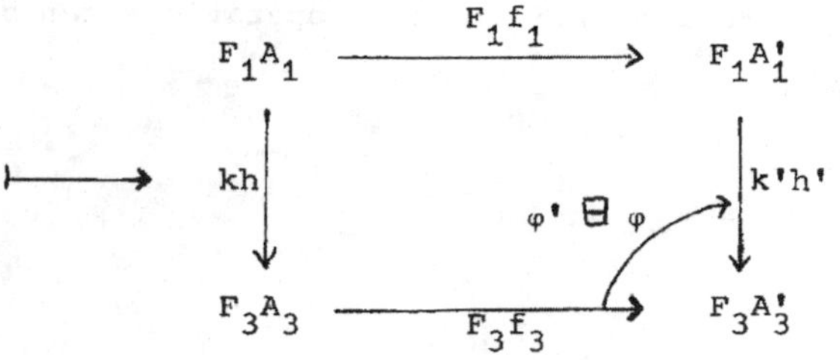

c) $[\varphi,\psi] : [F_1,F_2] \longrightarrow [G_1,G_2]$

A morphism in $[F_1,F_2]$ as in a) is taken into

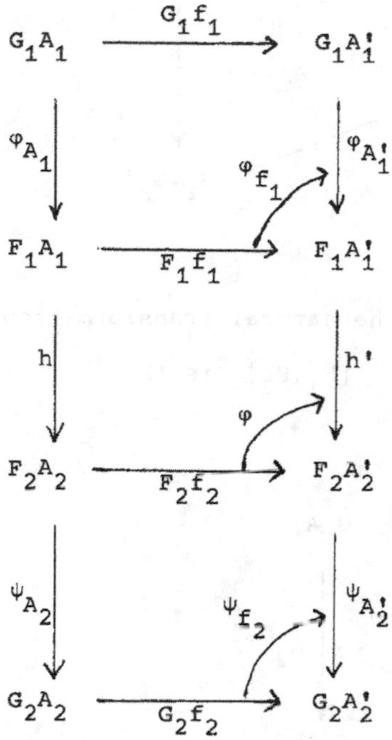

=

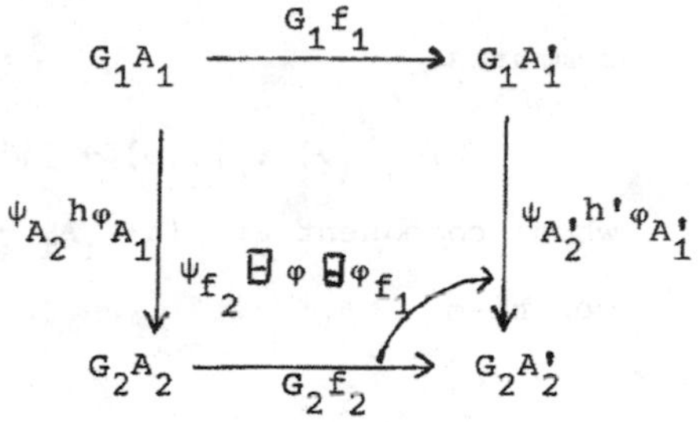

$[-,-]$ being a functor means that if

$[\varphi',\psi'] : [G_1,G_2] \to [H_1,H_2]$, then

$$[\varphi',\psi'][\varphi,\psi] = [\varphi\varphi',\psi'\psi] ,$$

which is obvious from the explicit formula. Note that there is another way to express $[\varphi,\psi]$; namely, $\varphi : G_1 \to F_1$ corresponds to a functor $\bar{\varphi} : \mathbb{A}_1 \to [G_1,F_1]$ and ψ to $\bar{\psi} : \mathbb{A}_2 \to [F_2,G_2]$. It follows, either from the universal mapping property or from the explicit formulas, that the diagram

$$
\begin{array}{ccc}
[F_1,F_2] \cong A_1 \underset{A_1}{\times} [F_1,F_2] \underset{A_2}{\times} A_2 & \xrightarrow{\bar{\varphi}\times 1\times\bar{\psi}} & [G_1,F_1] \underset{A_1}{\times} [F_1,F_2] \underset{A_2}{\times} [F_2,G_2] \\
& \searrow^{[\varphi,\psi]} & \downarrow \circ \\
& & [G_1,G_2]
\end{array}
$$

commutes.

d) $[u,v] : [\varphi,\psi] \to [\varphi',\psi']$ is the natural transformation whose component at $(h:F_1A_1 \to F_2A_2)$ in $[F_1,F_2]$ is the morphism

$$
\begin{array}{ccc}
G_1A_1 & \xrightarrow{\quad 1 \quad} & G_1A_1 \\
{\scriptstyle \psi_{A_2}h\varphi_{A_1}}\downarrow & vhu & \downarrow{\scriptstyle \psi'_{A_2}h\varphi'_{A_1}} \\
G_2A_2 & \xrightarrow[\quad 1 \quad]{} & G_2A_2
\end{array}
$$

in $[G_1,G_2]$. $[-,-]$ being a 2-functor means that

$$[u',v']\cdot[u,v] = [u'\cdot u,v'\cdot v]$$

$$[u',v']\ [u,v] = [uu',v'v]$$

when defined.

e) $(V_1,\varphi_1;V_2,\varphi_2) : [F_1,F_2] \longrightarrow [F_1',F_2']$

A morphism in $[F_1,F_2]$ as in a) is taken into

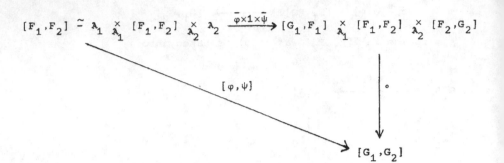

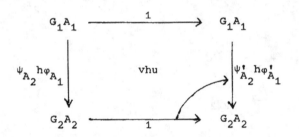

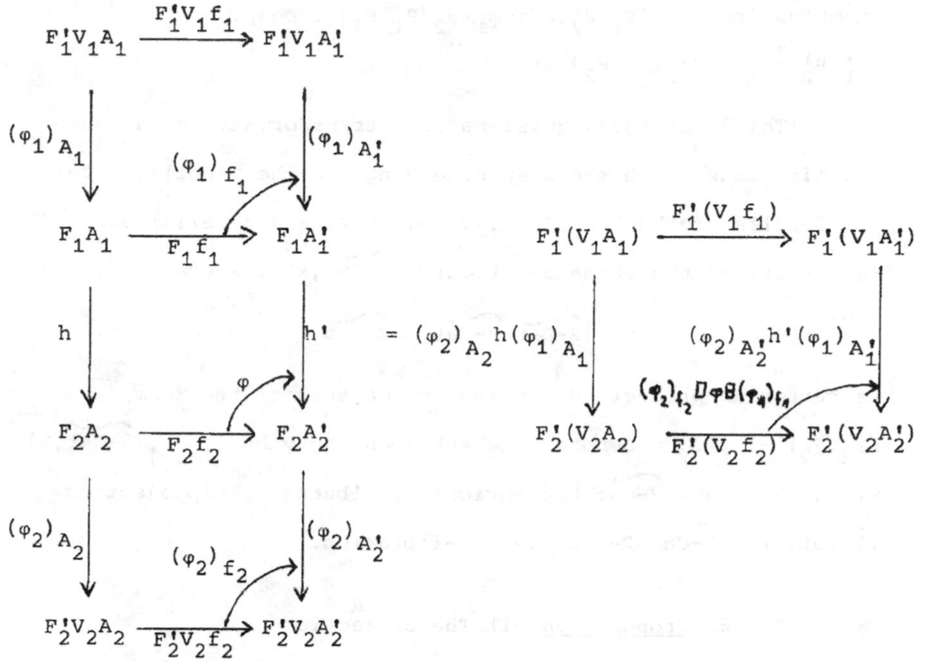

I,5.5. Definition. Given pairs of 2-functors

$$\mathfrak{A}_1 \xrightarrow{F_1} \mathfrak{B} \xleftarrow{F_2} \mathfrak{A}_2$$

$$\mathfrak{A}'_1 \xrightarrow{F'_1} \mathfrak{B}' \xleftarrow{F'_2} \mathfrak{A}'_2 \, ,$$

a 2-functor $T : [F_1, F_2] \to [F'_1, F'_2]$ (resp., quasi-natural transformation $\varphi : T \to T'$; resp., modification $s : \varphi \to \varphi'$) is over $V_1 \times V_2$ (resp., $\varphi_1 \times \varphi_2$; resp., $s_1 \times s_2$) if

$$\begin{array}{ccc}
[F_1, F_2] & \xrightarrow{T} & [F'_1, F'_2] \\
\Big\downarrow{\scriptstyle \{P_1, P_2\}} & & \Big\downarrow{\scriptstyle \{P'_1, P'_2\}} \\
\mathfrak{A}_1 \times \mathfrak{A}_2 & \xrightarrow[V_1 \times V_2]{} & \mathfrak{A}'_1 \times \mathfrak{A}'_2
\end{array}$$

commutes (resp., $\{P_1^!,P_2^!\}\varphi = \varphi_1\times\varphi_2\{P_1,P_2\}$; resp., $\{P_1^!,P_2^!\}s = s_1\times s_2\{P_1,P_2\}$.).

The 2-functors, quasi-natural transformations, and modifications which are over something are the 1-cells, 2-cells, and 3-cells of the "full" (i.e., all 1,2, and 3-cells) sub 3-category of the comma 3-category $(\widetilde{2\text{-Cat}},\times)$, where

$$\times : \widetilde{2\text{-Cat}}\times\widetilde{2\text{-Cat}} \to \widetilde{2\text{-Cat}}$$

is cartesian product, determined by objects of the form $[F_1,F_2] \to \mathbf{A}_1\times\mathbf{A}_2$. Denote this sub 3-category by ${}_{[-,-]}(\widetilde{2\text{-Cat}},\times)$. By I,2.9, since $\widetilde{2\text{-Cat}}$ has enriched pullbacks, the projection $(\widetilde{2\text{-Cat}},\times) \to \widetilde{2\text{-Cat}}\times\widetilde{2\text{-Cat}}$ is a 3-fibration.

I,5.6. Proposition. i) The projection

$${}_{[-,-]}(\widetilde{2\text{-Cat}},\times) \to \widetilde{2\text{-Cat}}\times\widetilde{2\text{-Cat}}$$

is a 3-fibration and the inclusion ${}_{[-,-]}(\widetilde{2\text{-Cat}},\times)$ in $(\widetilde{2\text{-Cat}},\times)$ is cleavage preserving.

ii) In particular, given $T : [F_1,F_2] \to [F_1^!,F_2^!]$ over $V_1\times V_2$, there is a unique $T' : [F_1,F_2] \to [F_1^!V_1,F_2^!V_2]$ over $\mathbf{A}_1\times\mathbf{A}_2$ such that the diagram

$$\begin{array}{ccc}
[F_1,F_2] & & \\
{\scriptstyle T'}\downarrow & \searrow{\scriptstyle T} & \\
[F_1^!V_1,F_2^!V_2] & \xrightarrow{(V_1,1,V_2)} & [F_1^!,F_2^!]
\end{array}$$

commutes.

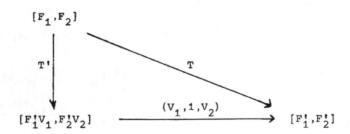

iii) Similarly, there is a bijective correspondence between natural transformations $\varphi : T \to T_1$ over $V_1 \times V_2$ and natural transformations $\varphi' : T' \to T_1'$ over $A_1 \times A_2$, where T, T_1, T', and T_1' are as in ii).

__Proof__. i) $(\widetilde{2\text{-Cat}}, \times)$ is a 3-fibration via pullbacks, so this follows from the observation (which is immediate from (5.1)) that

$$\begin{array}{ccc} [F_1'V_1, F_2'V_2] & \xrightarrow{(V_1,1,V_2)} & [F_1', F_2'] \\ \downarrow & & \downarrow \\ A_1 \times A_2 & \xrightarrow[V_1 \times V_2]{} & A_1' \times A_2' \end{array}$$

is a pullback in $\widetilde{2\text{-Cat}}$.

ii) and iii) are special cases of i). Explicitly, T' just reparenthesises T; i.e., if

$$T(F_1(A_1) \to F_2(A_2)) = F_1'(V_1(A_1)) \to F_2'(V_2(A_2))$$

then

$$T'(F_1(A_1) \to F_2(A_2)) = F_1'V_1(A_1) \to F_2'V_2(A_2)$$

We note that any quasi-natural transformation $\varphi : T \to T_1$ over $V_1 \times V_2$ is automatically natural and that a modification over $V_1 \times V_2$ is automatically the identity (cf., I,5.2, ii).)

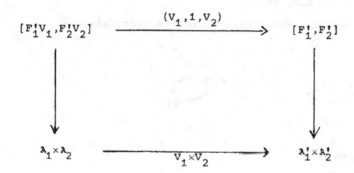

Either by identifying $(V_1,1,V_2) = (V_1,\mathrm{id};V_2,\mathrm{id})$ and using (2.18) or directly one can deduce what happens to a 2-functor $(V_1,1,V_2)$ when V_1 and V_2 are varied by quasi-natural transformations. The corresponding behavior in W differs in a crucial way from the analogous behavior of comma objects. This is, in fact, equivalent to the failure of composition of quasi-natural transformations to be 2-functorial. For brevity we write

(5.18) $$W_* = (1,W,1) : [F_1,F_2] \to [WF_1,WF_2]$$

I,5.7. <u>Proposition</u>.

Given 2-functors and quasi-natural transformations

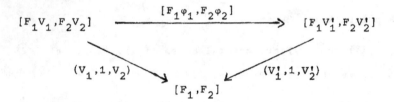

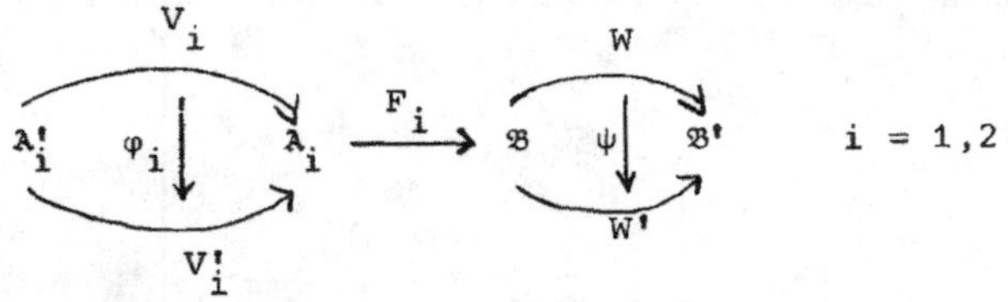

then

i) the diagram

$$\begin{array}{ccc} [F_1V_1,F_2V_2] & \xrightarrow{[F_1\varphi_1,F_2\varphi_2]} & [F_1V'_1,F_2V'_2] \\ & {\scriptstyle (V_1,1,V_2)} \searrow \quad \swarrow {\scriptstyle (V'_1,1,V'_2)} & \\ & [F_1,F_2] & \end{array}$$

commutes, and

ii) there is a diagram

$$\begin{array}{ccc} [F_1,F_2] & \xrightarrow{\quad W_* \quad} & [WF_1,WF_2] \\ {\scriptstyle W'_*}\downarrow & \overset{\psi_*}{\nearrow} & \downarrow{\scriptstyle [1,\psi F_2]} \\ [W'F_1,W'F_2] & \xrightarrow[\quad [\psi F_1,1] \quad]{} & [WF_1,W'F_2] \end{array}$$

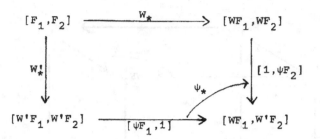

over $A_1 \times A_2$, where ψ_* is a Cat-natural transformation.

<u>Proof</u>. i) is clear. To prove ii) observe that W_* and W'_* are induced respectively by $W\theta_{F_1F_2}$ and $W'\theta_{F_1F_2}$ and this diagram reflects the composition $\psi\theta_{F_1F_2}$. However, it only seems possible to describe it in terms of components. Let $f : F_1A_1 \to F_2A_2$ be an object in $[F_1,F_2]$. Then, in the diagram

(5.19)
$$\begin{array}{ccc} WF_1A_1 & \xrightarrow{\quad Wf \quad} & WF_2A_2 \\ {\scriptstyle \psi_{F_1A_1}}\downarrow & \overset{\psi_f}{\nearrow} & \downarrow{\scriptstyle \psi_{F_2A_2}} \\ W'F_1A_1 & \xrightarrow[\quad W'f \quad]{} & W'F_2A_2 \end{array}$$

the clockwise composition is $[1,\psi F_2]W_*(f)$ while the counter-clockwise composition is $[\psi F_2,1]W'_*(f)$. We define $(\psi_*)_f$ to be the morphism in $[WF_1,W'F_2]$ represented by $(1,\psi_f,1)$. Explicit calculations show that this is Cat-natural. E.g., given a 1-cell $(h_1,\varphi,h_2) : (A_1,f,A_2) \to (A'_1,f',A'_2)$ in $[F_1,F_2]$, then the equation

$$(\psi_*)_{f'} \cdot [\psi F_2,1]W'_*(h_1,\varphi,h_2) = [1,\psi F_2]W_*(h_1,\varphi,h_2) \cdot (\psi_*)_f$$

is equivalent to the commutativity of the cube

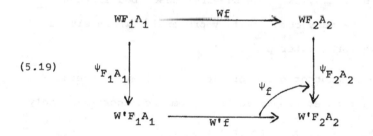

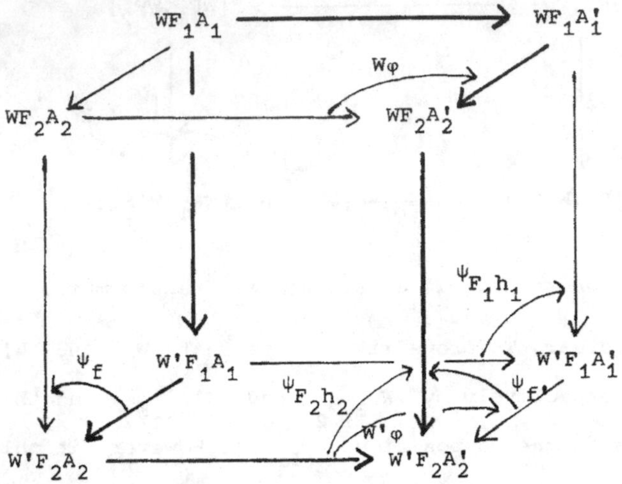

which follows from I,2.4, QN1 and QN3. Cat-naturality follows similarly from I,2.4, QN1. The diagrams described in this proposition have many compatibility properties whose elucidation we leave for a later paper.

There is another point of view which is very useful in describing further properties of 2-comma categories; namely,

$$\mathfrak{A}_1 \leftarrow [F_1,F_2] \rightarrow \mathfrak{A}_2$$

can be regarded as a 1-cell in Spans(2-Cat) (cf., I,3.4 In particular, if $F : \mathfrak{A} \rightarrow \mathfrak{B}$, then $[F,F] \rightarrow \mathfrak{A}\times\mathfrak{A}$ is an object in the multiplicative category [Spans(2-Cat)]$(\mathfrak{A},\mathfrak{A})$.

I,5.8. Proposition. Let $F_i : \mathfrak{A}_i \rightarrow \mathfrak{B}$, $i = 1,2$ be 2-functors

i) $\{P_1,P_2\} : [F_1,F_2] \rightarrow \mathfrak{A}_1\times\mathfrak{A}_2$ is a split-normal 2-fibration (cf., I,2.9).

ii) $[F_1,F_1] \rightrightarrows \mathfrak{A}_1$ is a monoid in Spans(2-Cat)$(\mathfrak{A}_1,\mathfrak{A}_1)$

and if $W : \mathfrak{B} \to \mathfrak{B}'$ then $W_* : [F_1,F_1] \to [MF_1,MF_1]$ is a monoid homomorphism.

iii) $\mathfrak{A}_1 \leftarrow [F_1,F_2] \to \mathfrak{A}_2$

is a left $[F_1,F_1]$-right $[F_2,F_2]$-bimodule in Spans(2-Cat)$(\mathfrak{A}_1,\mathfrak{A}_2)$ and the bimodule structure induced on $[F_1,F_2]$ by the change of monoids

$$(\tilde{F}_i)_* = (F_i)_*|\mathfrak{A}_i^2 : \mathfrak{A}_i^2 \to [F_i,F_i]$$

coincides with that given by the bifibration structure.

iv) If $\varphi_i : F_i \to F_i'$, $i = 1,2$ are quasi-natural transformations, then $[F_1,\varphi_2]$ and $[\varphi_1,F_2]$ are left and right module homomorphisms respectively, and hence cleavage and opcleavage preserving respectively. Similarly, $W_* : [F_1,F_2] \to [WF_1,WF_2]$ is a bimodule homomorphism.

Proof. i) In [CCS], §6, Example 1, it is pointed out that $[F_1,F_2]_o \to (\mathfrak{A}_1\times\mathfrak{A}_2)_o$ is the (1,o)-bifibration corresponding to the functor

$$\mathfrak{B}(F_1(-),F_2(-)) : \mathfrak{A}_1\times\mathfrak{A}_2 \to \mathrm{Cat}$$

and as a 2-category it is given by the fundamental construction of [CCS], §5. We claim here that it is a 2-bifibration in the sense of (I,2.9). The split cleavage and opcleavage are given by taking as cartesian and opcartesian morphisms in $[F_1,F_2]$ those morphisms of the forms

(5.20)
$$\begin{array}{ccccccc} F_1A_1 & \xrightarrow{F_1m} & F_1A_1' & & F_1A_1 & \xrightarrow{1} & F_1A_1 \\ {\scriptstyle h'(F_1m)}\downarrow & \text{cartesian} & \downarrow{\scriptstyle h'} & \text{and} & {\scriptstyle h}\downarrow & \text{opcartesian} & \downarrow{\scriptstyle (F_2n)h} \\ F_2A_2' & \xrightarrow[1]{} & F_2A_2' & & F_2A_2 & \xrightarrow[F_2n]{} & F_2A_2' \end{array}$$

respectively. Treating only the cartesian case, this defines an object function

$$|L| \;:\; |(A_1,P_1)| \to |([F_1,F_2])^{\underline{2}}|$$

taking the object (m,h') to the indicated cartesian morphism. It is easily verified that this extends to a 2-functor L satisfying $SL = id$, where

$$S = \{P_1^{\underline{2}}, [F_1,F_2]^{\partial_1}\} \;:\; [F_1,F_2]^{\underline{2}} \to (A_1,P_1) \quad ,$$

and one must only verify that $S \underset{Cat}{\longrightarrow\!\!\!\!\!\!\!\!\;\;\dashv} L$.

Consider the diagram

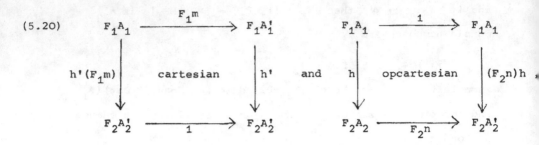

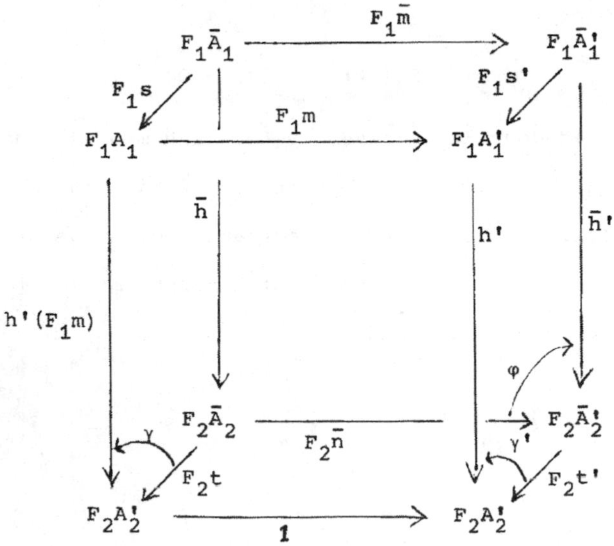

where $(\bar{n},\varphi,\bar{m}) : \bar{h} \to \bar{h}'$ is regarded as an object in $[F_1,F_2]^{\underline{2}}$ mapped into $L(m,h')$ by the maps (t,γ,s) and (t',γ',s'). Commutativity of the cube shows that $t = t'\bar{n}$ and $\gamma = \gamma' \square \varphi$ which shows that there is a bijection between such 1-cells and 1-cells

$$<(s,s') : \bar{m} \to m, (t',\gamma',s') : \bar{h}' \to h'>$$

from $S(\bar{n},\varphi,\bar{m})$ to (m,h') in $(\mathcal{A}_1,P_1)$. This is easily extended to 2-cells, showing that one has a Cat-adjunction.

ii) The monoid structure is given by composition as in I,5.3 and it follows from naturality that W_* is a monoid homomorphism.

iii) Similarly, I,5.3 implies that $[F_1,F_2]$ is a bimodule. The second statement means that, for instance,

$$\begin{array}{ccc} \mathcal{A}_1^{\underline{2}} \underset{\mathcal{A}_1}{\times} [F_1,F_2] & \xrightarrow{(\tilde{F}_1)_* \underset{\mathcal{A}_1}{\times} 1} & [F_1,F_1] \underset{\mathcal{A}_1}{\times} [F_1,F_2] \\ & \searrow^{M} & \downarrow{\circ} \\ & & [F_1,F_2] \end{array}$$

commutes, where M is derived from L as in I,3.5 by setting $M = [F_1,F_2]^{\partial_0}L$. But, by the definition of L, this gives $M(m,h') = h'(F,m)$, as desired.

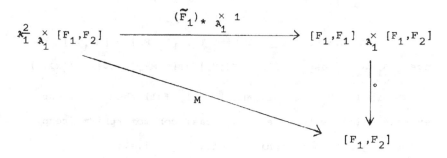

iv) It follows from (5.13) that $[F_1,\varphi_2]$ is a homomorphism for the $[F_1,F_1]$-module structure and hence, by iii) above, for the $\mathcal{A}_1^{\underline{2}}$-module structure, which shows that it is cleavage preserving. This is also evident from the explicit formula for $[F_1,\varphi_2]$ in I,5.4 c) and (5.20) above. The correspondence in I,3.5 between morphisms of fibrations and quasi-homomorphisms of the fibrations regarded as modules does not apply directly to $[F_1,F_2]$ regarded as a module over $[F_1,F_1]$, without further conditions. (We do not know the situation with regard to the change of monoids $\tilde{F}_*$ and the possible existence of adjoints with respect to quasi-homomorphisms). The further conditions can be stated in various ways. Here we consider the following (non-commutative) collection of 2-functors

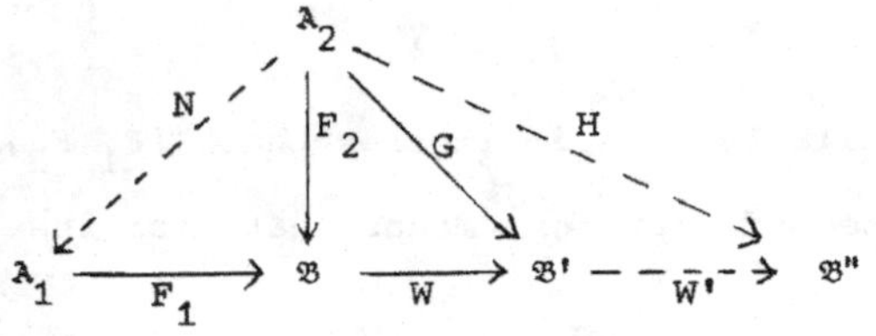

I,5.9. <u>Theorem</u>: Let $T : [F_1,F_2] \to [MF_1,G]$ be over $\mathfrak{A}_1\times\mathfrak{A}_2$ (resp., $T' : [F_2,F_1] \to [G,MF_1]$ over $\mathfrak{A}_2\times\mathfrak{A}_1$)

i) If F_1 is full and locally full then T (resp., T') extends uniquely to a left W_*-quasi-cohomomorphism (resp., right W_*-quasi-homomorphism) (cf., I,8); i.e.,

a) there exists a natural transformation $\gamma_{W,T}$ over $\mathfrak{A}_1\times\mathfrak{A}_2$ (resp., $\gamma'_{W,T'}$ over $\mathfrak{A}_2\times\mathfrak{A}_1$)

$$\begin{array}{ccc} [F_1,F_1] \underset{\mathbb{A}_1}{\times} [F_1,F_2] & \xrightarrow{\circ} & [F_1,F_2] \\ \Big\downarrow {\scriptstyle W_* \underset{\mathbb{A}_1}{\times} T} & \gamma_{W,T} & \Big\downarrow {\scriptstyle T} \\ [WF_1,WF_1] \underset{\mathbb{A}_1}{\times} [WF_1,G] & \xrightarrow[\circ]{} & [WF_1,G] \end{array} \quad \text{resp.,} \quad \begin{array}{ccc} [F_2,F_1] \underset{\mathbb{A}_1}{\times} [F_1,F_1] & \xrightarrow{\circ} & [F_2,F_1] \\ \Big\downarrow {\scriptstyle T' \underset{\mathbb{A}_1}{\times} W_*} & \gamma'_{W,T'} & \Big\downarrow {\scriptstyle T'} \\ [G,WF_1] \underset{\mathbb{A}_1}{\times} [WF_1,WF_1] & \xrightarrow[\circ]{} & [G,WF_1] \end{array}$$

b) $\gamma_{W,T}$ (resp., $\gamma'_{W,T'}$) is compatible with units and associativity.

c) If, furthermore, $T_1 : [WF_1,G] \to [W'WF_1,H]$ is over $\mathbb{A}_1 \times \mathbb{A}_2$ (resp., $T_1' : [G,WF_1] \to [H_2,W'WF_1]$) then

$$\gamma_{W'W,T_1T} = \gamma_{W',T_1} \boxminus \gamma_{W,T}$$

$$(\text{resp.}, \gamma_{W'W,T_1'T'} = \gamma'_{W',T_1'} \boxminus \gamma_{W,T'}) .$$

ii) a) T (resp., T') is a homomorphism (i.e., $\gamma_{W,T} = \mathrm{id}$ or resp., $\gamma'_{W,T'} = \mathrm{id}$) if and only if T is cleavage (resp., opcleavage) preserving.

b) If there exists an N with $F_2 = F_1N$ then T (resp., T') is a homomorphism if and only if there is a quasi-natural transformation $\varphi : WF_2 \to G$ such that $T = [WF_1,\varphi]\cdot W_*$ (resp., $\varphi : G \to WF_2$ such that $T' = [\varphi,WF_1]\cdot W_*$).

c) If further, F_1 is fully faithful (resp., a full embedding) then φ is unique up to an isomorphism (resp., unique).

<u>Note</u>: The special case of functors

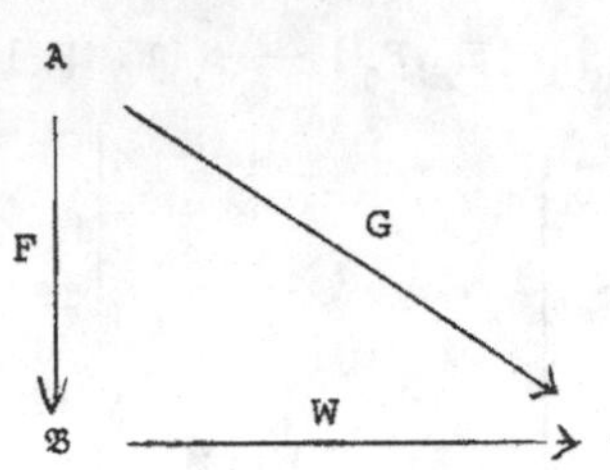

and

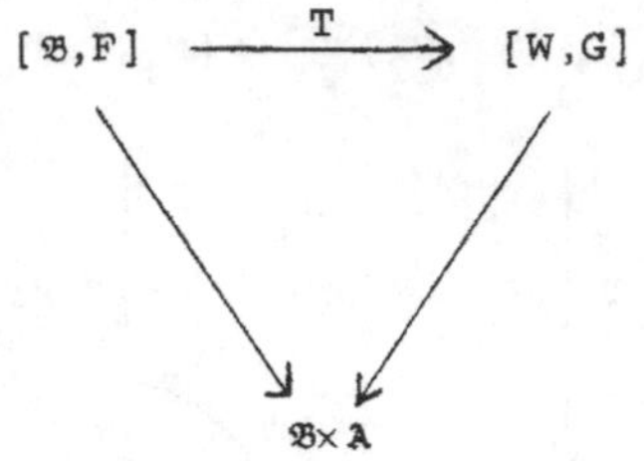

(5.21)

leads to T being a homomorphism if and only if it is of the form $T = [W,\varphi]\cdot W_*$ where $\varphi : WF \to G$. This (and its dual) will be specialized further in I,5.10 to the cases where either W or G is the identity.

<u>Proof</u>: i) Consider the diagram

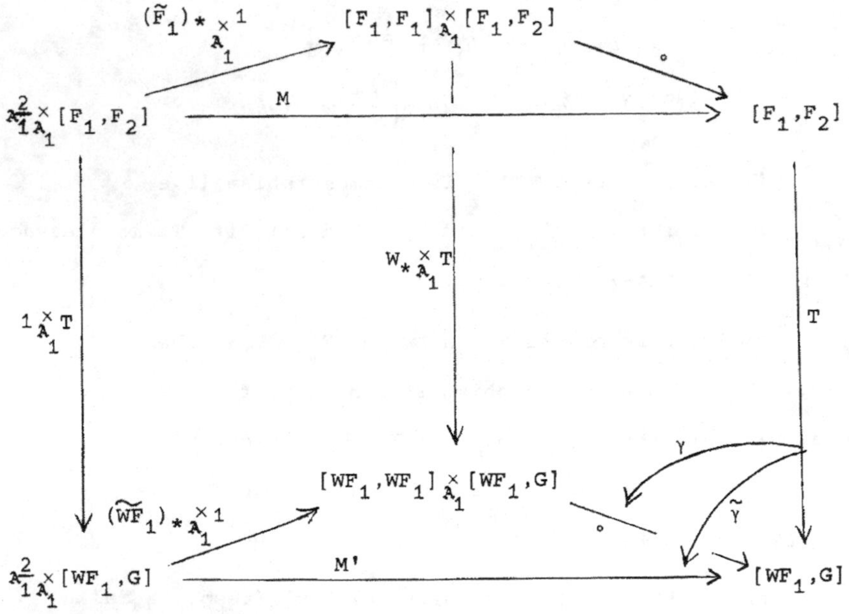

where M and M' are given as in the proof of I,5.8 iii) and $\tilde{\gamma}$ is the natural transformation for T as in I,3.5. We want to factor this through γ as indicated. The hypotheses

are needed to make this work. The procedure in I,3.5 suggests the following:
given an object

$$((F_1A_1' \xrightarrow{m} F_1A_1),(F_1A_1 \xrightarrow{h} F_2A_2)) \in [F_1,F_1]\underset{A_1}{\times}[F_1,F_2] \ ,$$

since F_1 is full we can write $m = F_1 n$ for some n and then apply T to the cartesian morphism for (n,h) ; i.e.,

$$T\left(\begin{array}{ccc} F_1A_1' & \xrightarrow[F_1 n]{m \atop \|} & F_1A_1 \\ \Big\downarrow{\scriptstyle hm} & & \Big\downarrow{\scriptstyle h} \\ F_2A_2 & \xrightarrow{1} & F_2A_2 \end{array}\right) = \left(\begin{array}{ccc} WF_1A_1' & \xrightarrow[WF_1 n]{Wm \atop \|} & WF_1A_1 \\ \Big\downarrow{\scriptstyle T(hm)} & \overset{(\gamma_{W,T})_{m,h}}{\nearrow} & \Big\downarrow{\scriptstyle T(h)} \\ GA_2 & \longrightarrow & GA_2 \end{array}\right) \qquad (5.22)$$

The result is not a cartesian morphism in general. The domain μ of the component of the comparison natural transformation from it to the cartesian morphism for $(n,T(h))$ is the morphism

$$\begin{array}{ccc} WF_1A_1' & \xrightarrow{1} & WF_1A_1' \\ \Big\downarrow{\scriptstyle T(hm)} & \overset{(\gamma_{W,T})_{m,h}}{\nearrow} & \Big\downarrow{\scriptstyle T(h)W(m)} \\ GA_2 & \longrightarrow & GA_2 \end{array} \qquad (5.23)$$

in $[WF_1,G]$ from $T(\text{"}\circ\text{"})(m,h)$ to $(\text{"}\circ\text{"})(W_*\underset{A_1}{\times}T)(m,h)$.
This does not depend on the choice of n since if n' also satisfies $F_1(n') = F_1(n) = m$, then, since F is locally full, there is a 2-cell $\nu : n \to n'$ with $F_1(\nu) = id$.

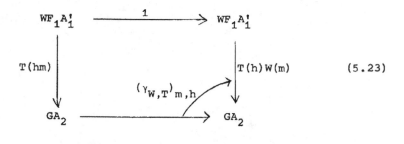

The result follows by putting this 2-cell in (5.22) and using I,2.5 and the fact that T is a 2-functor over $\mathbb{A}_1 \times \mathbb{A}_2$.

Naturality means that given a pair of morphisms

$$(g', \tau, g) : (A_1', m, A_1) \to (\tilde{A}_1', \tilde{m}, \tilde{A}_1) \quad \text{in} \quad [F_1, F_1]$$

$$(g, \sigma, f) : (A_1, h, A_2) \to (\tilde{A}_1, \tilde{h}, \tilde{A}_2) \quad \text{in} \quad [F_1, F_2]$$

then

$$(\gamma_{M,T})_{\tilde{m},\tilde{n}} \boxbar T(\sigma \boxminus \tau) = [T(\sigma) \boxminus M(\tau)] \boxbar (\gamma_{M,T})_{m,n} \tag{5.24}$$

where we have written $T(\sigma)$ for the 2-cell in $T(g,\sigma,f)$, etc. This follows by expanding (5.22) into a cube with $\tau = F(\tau')$ (since F is locally full) as a 2-cell in the top face and using I,2.5 again.

i) b) Compatibility with units means that

$$(\gamma_{W,T})(j_{F_1} \overset{\times}{\mathbb{A}_1} [F_1, F_2]) = \text{id} . \tag{5.25}$$

This follows since T is a functor, so $(\gamma_{W,T})_{F_1 A_1, h} = \text{id}$. Compatibility with associativity means that the diagram

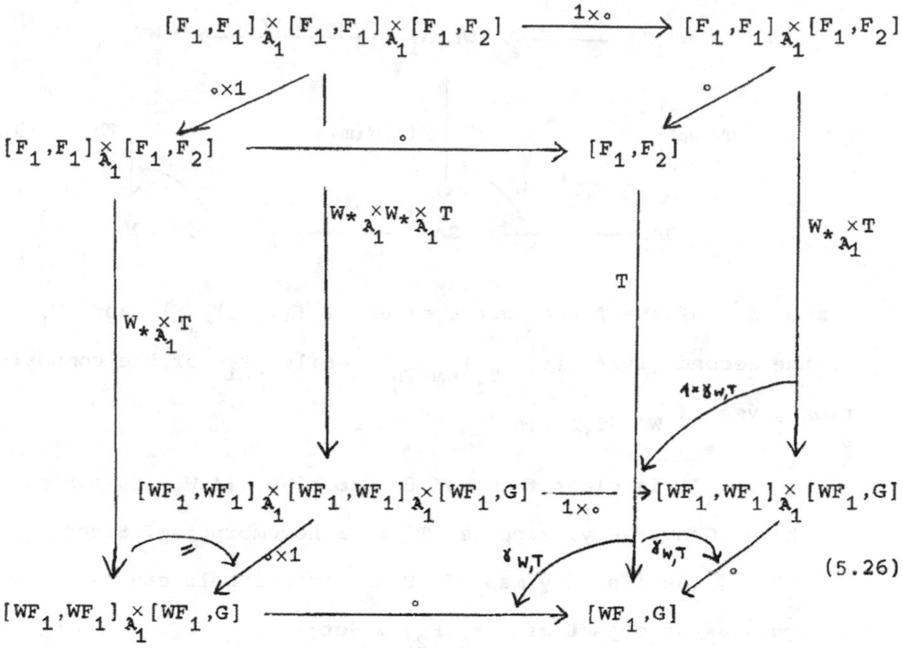

commutes. This follows since applying T to a composition

$$\begin{array}{ccccc} F_1A_1'' & \xrightarrow{m'} & F_1A_1' & \xrightarrow{m} & F_1A_1 \\ \downarrow{\scriptstyle hmm'} & & \downarrow{\scriptstyle hm} & & \downarrow{\scriptstyle h} \\ F_2A_2 & \xrightarrow{1} & F_2A_2 & \xrightarrow{1} & F_2A_2 \end{array} \tag{5.27}$$

as in (5.22), shows that

$$\gamma_{mm',h} = \gamma_{m,n} \boxed{\;} \gamma_{m',hm} \;;$$

i.e.,
$$\gamma_{mm',h} = [\gamma_{m,h}(Wm')] \cdot \gamma_{m',hm} \,,$$

which is what is needed when γ is interpreted as in (5.23).

i) c) This follows by applying T_1 to the composition

$$\begin{array}{ccccc} WF_1A_1' & \xrightarrow{\;1\;} & WF_1A_1' & \xrightarrow{\;W(m)\;} & WF_1A_1 \\ {\scriptstyle T(hm)}\downarrow & \overset{(\gamma_{W,T})_{m,h}}{\nearrow} & \downarrow{\scriptstyle T(h)W(m)} & \overset{=}{\nearrow} & \downarrow{\scriptstyle Th} \\ GA_2 & \longrightarrow & GA_2 & \longrightarrow & GA_2 \end{array} \qquad (5.29)$$

since T_1 of the first square gives $T_1((\gamma_{W,T})_{m,h})$ and T_1 of the second gives $(\gamma_{W',T_1})_{Wm,Th}$, while T_1 of the composition gives $(\gamma_{W'W,T_1T})_{m,h}$.

ii) It is clear from I,5.8 that $[WF_1,\varphi]\cdot W_*$ is a homomorphism. Conversely, suppose T is a homomorphism. Since $F_2 = F_1N$, the identity map of F_2A_2 into itself can be regarded as an object of $[F_1,F_2]$. Set

$$\varphi_{A_2} = T(\mathrm{id}_{F_2A_2} : F_1NA_2 \to F_2A_2) : WF_2A_2 \to GA_2 \qquad (5.30)$$

and for $f : A_2 \to A_2'$, φ_f is the 2-cell in the diagram

$$(5.31) \qquad T\left(\begin{array}{ccc} F_1NA_2 & \xrightarrow{\;F_1Nf\;} & F_1NA_2' \\ {\scriptstyle 1}\downarrow & & \downarrow{\scriptstyle 1} \\ F_2A_2 & \xrightarrow[\;F_2f\;]{} & F_2A_2' \end{array}\right) = \begin{array}{ccc} WF_2A_2 & \xrightarrow{\;WF_2f\;} & WF_2A_2' \\ {\scriptstyle \varphi_{A_2}}\downarrow & \overset{\varphi_f}{\nearrow} & \downarrow{\scriptstyle \varphi_{A_2'}} \\ GA_2 & \xrightarrow[\;Gf\;]{} & GA_2 \end{array}$$

Since T is a 2-functor, φ is a quasi-natural transformation. Alternatively, let $\bar{T}$ be the unique 2-functor over $A_2\times A_2$ (by I,5.5) making the diagram

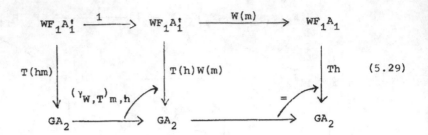

$$\begin{array}{ccc} [F_2,F_2]=[F_1N,F_2] & \xrightarrow{(N,1,1)} & [F_1,F_2] \\ {\scriptstyle \bar{T}}\downarrow & & \downarrow{\scriptstyle T} \\ [WF_2,G]=[WF_1N,G] & \xrightarrow{(N,1,1)} & [WF_1,G] \end{array} \tag{5.32}$$

commute and then φ is the quasi-natural transformation corresponding to

$$\bar{\varphi} = A_2 \xrightarrow{j_{F_2}} [F_2,F_2] \xrightarrow{\bar{T}} [WF_2,G] \tag{5.33}$$

This formula always defines a $\varphi : WF_2 \to G$, whether or not T is a homomorphism. When T is a homomorphism, then

$$T = [WF_1,\varphi] \circ W_*$$

follows for objects by considering

$(h:F_1A_1 \to F_2A_2) \in [F_1,F_2]$ as the first component in

$(h:F_1A_1 \to F_1NA_2, id_{F_2A_2}:F_1NA_2 \to F_2A_2) \in [F_1,F_1] \underset{A_1}{\times} [F_1,F_2]$.

Since $(\gamma_{W,T})_{h,id_{F_2A_2}} = id$, we get that

$$T(h) = T(id_{F_2A_2})W(h) = \varphi_{A_2}W(h) = ([WF_1,\varphi]\circ W_*)(h) .$$

The equation for morphisms follows from (5.24).

Alternatively, the equation follows from the commutativity of (5.34), where the regions labeled I commute by I,5.3 i), II commutes by I,5.4 c), III commutes by the definition of $\bar{T}$, IV commutes by the definition of $\bar{\varphi}$, and (the only thing which does not hold in general) V commutes because T is a homomorphism.

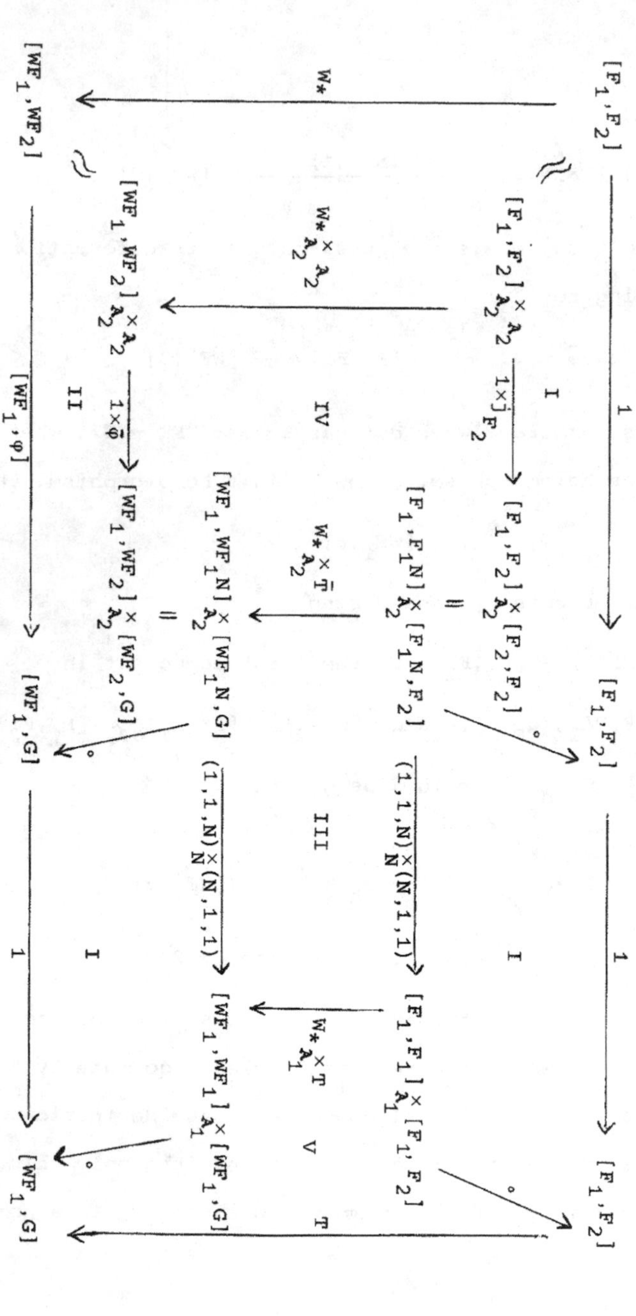
[F1,F2]
1
[F1,F2]
1
[F1,F2]
W*
[F1,F2]×λ2 λ2
1×jF2
[F1,F2]×λ2 [F2,F2]
[F1,F1N]×λ2 [F1N,F2]
(1,1,N)×N(N,1,1)
[F1,F1]×λ1 [F1,F2]
I
I
IV
III
V
T
W*×λ2 λ2
W*×λ2 T̄
W*×λ1 T
[WF1,WF2]×λ2 λ2
1×φ̄
[WF1,WF2]×λ2 [WF2,G]
[WF1,WF1N]×λ2 [WF1N,G]
(1,1,N)×N(N,1,1)
[WF1,WF1]×λ1 [WF1,G]
II
I
[WF1,WF2]
[WF1,φ]
[WF1,G]
1
[WF1,G]

(5.34)

It is clear that φ depends only on N ; i.e., on the way that F_2A_2 is represented as F_1A_1' . If F_1 is fully faithful and if $F_1A_1' = F_1A_1'' = F_2A_2$, then the identity map of F_2A_2 is F_1 applied to a unique isomorphism $m : A_1' \to A_1''$ Then $T(m,\mathrm{id},1) = (\varphi_{A_2}' , s_{A_2} , \varphi_{A_2}'')$ provides the components for an isomorphic modification $s : \varphi' \to \varphi''$ between the quasi-natural transformations constructed with $NA_2 = A_1'$ or A_1'' respectively (since $WF_1(m) = \mathrm{id}$.) If F_1 is an embedding, this ambiguity cannot occur, so φ is unique.

There are four special cases of this theorem which will be used in I,7. Other cases follow from these together with I,5.7. We omit consideration of part c) here since it will be dealt with more carefully there.

I,5.10. **Corollary:**

i) Let $F,G : A \to \mathfrak{B}$ and let $T : [\mathfrak{B},F] \to [\mathfrak{B},G]$ be over $\mathfrak{B} \times A$ (resp., $T' : [F,\mathfrak{B}] \to [G,\mathfrak{B}]$ over $A \times \mathfrak{B}$). Then there is a diagram

$$[\mathfrak{B},\mathfrak{B}] \underset{\mathfrak{B}}{\times} [\mathfrak{B},F] \xrightarrow{\quad\circ\quad} [\mathfrak{B},F] \qquad\qquad [F,\mathfrak{B}] \underset{\mathfrak{B}}{\times} [\mathfrak{B},\mathfrak{B}] \xrightarrow{\quad\circ\quad} [F,\mathfrak{B}]$$

$$1 \underset{\mathfrak{B}}{\times} T \downarrow \qquad \gamma_T \searrow \qquad \downarrow T \qquad \text{resp.,} \qquad T' \underset{\mathfrak{B}}{\times} 1 \downarrow \qquad \gamma_{T'}' \nearrow \qquad \downarrow T'$$

$$[\mathfrak{B},\mathfrak{B}] \underset{\mathfrak{B}}{\times} [\mathfrak{B},G] \xrightarrow[\circ]{\quad} [\mathfrak{B},G] \qquad\qquad [G,\mathfrak{B}] \underset{\mathfrak{B}}{\times} [\mathfrak{B},\mathfrak{B}] \xrightarrow[\circ]{\quad} [G,\mathfrak{B}]$$

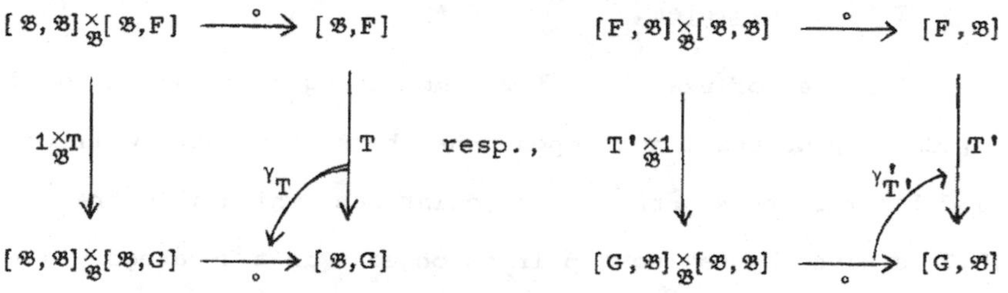

and γ_T (resp., $\gamma_{T'}'$) is the identity if and only if there is a unique quasi-natural transformation $\varphi : \mathbb{E} \to G$ with $T = [\mathfrak{B},\varphi]$ (resp., $\varphi' : G \to F$ with $T' = [\varphi',\mathfrak{B}]$).

ii) Let $F : \mathfrak{A} \to \mathfrak{B}$ and $U : \mathfrak{B} \to \mathfrak{A}$ and let $T : [\mathfrak{A},U] \to [F,\mathfrak{B}]$ be over $\mathfrak{A}\times\mathfrak{B}$ (resp., $T' : [F,\mathfrak{B}] \to [\mathfrak{A},U]$ over $\mathfrak{A}\times\mathfrak{B}$). Then there is a diagram

$$\begin{array}{ccc} [\mathfrak{A},\mathfrak{A}]\times_{\mathfrak{A}}[\mathfrak{A},U] & \xrightarrow{\circ} & [\mathfrak{A},U] \\ {\scriptstyle F_*\times_{\mathfrak{A}} T}\downarrow & \swarrow{\scriptstyle \gamma_{F,T}} & \downarrow{\scriptstyle T} \\ [F,F]\times_{\mathfrak{A}}[F,\mathfrak{B}] & \xrightarrow[\circ]{} & [F,\mathfrak{B}] \end{array} \quad \text{resp.,} \quad \begin{array}{ccc} [F,\mathfrak{B}]\times_{\mathfrak{B}}[\mathfrak{B},\mathfrak{B}] & \xrightarrow{\circ} & [F,\mathfrak{B}] \\ {\scriptstyle T'\times_{\mathfrak{B}} U_*}\downarrow & \nearrow{\scriptstyle \gamma'_{U,T'}} & \downarrow{\scriptstyle T'} \\ [\mathfrak{A},U]\times_{\mathfrak{B}}[U,U] & \xrightarrow[\circ]{} & [\mathfrak{A},U] \end{array}$$

and $\gamma_{F,T}$ (resp., $\gamma'_{U,T'}$) is the identity if and only if there is a unique quasi-natural transformation $\varepsilon : FU \to \mathfrak{B}$ with $T = [F,\varepsilon]\circ F_*$ (resp., $\eta : \mathfrak{A} \to UF$ with $T' = [\eta,U]\circ U_*$).

<u>Proof</u>: i) Let $F_1 = \mathfrak{B}$, which is certainly full, locally full and a full embedding, $F_2 = F$, $G = G$, $W = \mathfrak{B}$ and $N = F$.

ii) For the first part, let $F_1 = \mathfrak{A}$, $F_2 = U$, $G = \mathfrak{B}$, $W = F$ and $N = U$. For the second part, let $F_1 = \mathfrak{B}$, $F_2 = F$, $G = A$, $W = U$ and $N = F$.

I,5.11. <u>Examples</u>.

A number of examples of 2-comma categories can be found in [CCS] and we shall not repeat them here. Many more will be found later in this work. A particular case which illustrates I,5.7 is given by taking a pair of objects in a 2-category $\mathfrak{B}$. Denote the corresponding 2-functors from $\underline{1}$ by

$$\underline{1} \xrightarrow{\ulcorner B \urcorner} \mathfrak{B} \xleftarrow{\ulcorner C \urcorner} \underline{1}$$

It is easily seen that $[\ulcorner B \urcorner, \ulcorner C \urcorner] = \mathfrak{B}(B,C)$ and, if $\sigma : F \to G$ is a quasi-natural transformation between 2-functors from $\mathfrak{B}$

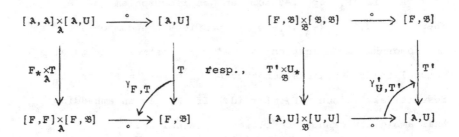

to $\mathfrak{B}'$, then the σ_* in I,5.7 is the same as the σ_{BC} in I,2.4.

I,6. Adjoint morphisms in 2-categories.

In this section we discuss properties of adjoint 1-cells in a 2-category and, briefly, in a bicategory. The usual equational description of adjoints is taken as the definition. It is impossible to state everything that follows from this definition, but we have tried to organize one class of phenomena around the notion of the category of adjoint squares in a 2-category $\mathbb{A}$, and using this, another class around the notion of Kan extensions along a 1-cell in a 2-category. In this section we ignore questions of existence of Kan extensions (since nothing can be said about this in a general 2-category) and concentrate on describing the relations they satisfy when they exist. It should be emphasized that any such thing is "purely formal". What is interesting is how much actually fits that description. As applications, we discuss preservation of Kan extensions and, in particular, of colimits, including the "formal criterion for the existence of an adjoint" of Dubuc [9] as well as the usual interchange 2-cells between limits and between limits and colimits. We also briefly discuss "final" morphisms and "dual" Kan extensions in Cat.

I,6.1 <u>Definition</u>:

i) Let $\mathbb{A}$ be a 2-category. A pair of morphisms

$$A \underset{u}{\overset{f}{\rightleftarrows}} B$$

is called <u>adjoint</u>, written $f \dashv u$, if there exist 2-cells

$$\varepsilon\colon fu \longrightarrow B, \quad \eta\colon A \longrightarrow uf$$

such that

$$\varepsilon f \cdot f\eta = f \qquad (6.1)$$

$$u\varepsilon \cdot \eta u = u$$

ii) Let $\mathfrak{B}$ be a bicategory. A pair of morphisms (f,u) as above is called <u>adjoint</u> if there exist 2-cells

$$\varepsilon\colon fu \longrightarrow I_B$$

$$\eta\colon I_A \longrightarrow uf$$

such that the compositions

$$f \xrightarrow{r^{-1}} fI_A \xrightarrow{f\eta} f(uf) \xrightarrow{\alpha^{-1}} (f\eta)f \xrightarrow{\varepsilon f} I_B f \xrightarrow{\ell} f \qquad (6.2)$$

$$u \xrightarrow{\ell^{-1}} I_A u \xrightarrow{\eta u} (uf)u \xrightarrow{\alpha} u(fu) \xrightarrow{u\varepsilon} uI_B \xrightarrow{r} u$$

are identity 2-cells

I,6.2 <u>Examples</u>

1) Adjoint morphisms in a 2-category dualize in a very simple fashion. If f is a morphism in $\mathfrak{A}$ and if ${}^{op}f$, f^{op} and ${}^{op}f^{op}$ denote the corresponding morphisms in ${}^{op}\mathfrak{A}$, $\mathfrak{A}^{op}$ and ${}^{op}\mathfrak{A}^{op}$, respectively, then the following are equivalent:

a) $f \dashv u$

b) ${}^{op}u \dashv {}^{op}f$

c) $u^{op} \dashv f^{op}$

d) ${}^{op}f^{op} \dashv {}^{op}u^{op}$.

2) In a 2-category $\mathfrak{A}^{\mathfrak{B}}$, Cat-natural transformations $\varphi\colon F \longrightarrow G$ and $\psi\colon G \longrightarrow F$ are adjoint, $\varphi \dashv \psi$, if and only if there are modifications $\varepsilon\colon \varphi\psi \longrightarrow G$ and $\eta\colon F \longrightarrow \psi\varphi$ satisfying the required equations. This is equivalent to

requiring that for all B, $\varphi_B \dashv \psi_B$ in $\mathfrak{A}$ and that the 2-cells expressing these adjunctions, ε_B and η_B, can be chosen "naturally"; i.e., as components of modifications.

3) In a 2-category $\mathrm{Fun}(\mathfrak{B},\mathfrak{A})$, quasi-natural transformations $\varphi\colon F \longrightarrow G$ and $\psi\colon G \longrightarrow F$ are adjoint, $\varphi \dashv \psi$, if and only if there are modifications $\varepsilon\colon \varphi\psi \longrightarrow G$ and $\eta\colon F \longrightarrow \psi\varphi$ satisfying the required equations. In this case, this is equivalent to requiring that for all B, $\varphi_B \dashv \psi_B$ in $\mathfrak{A}$ and that the 2-cells expressing these adjunctions, ε_B and η_B can be chosen so that, given a 2-cell

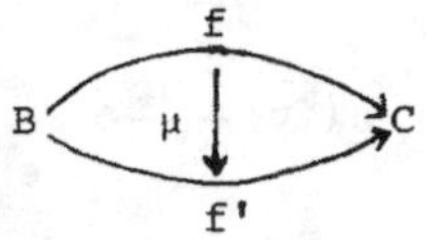

in $\mathfrak{B}$, they satisfy

$$(\psi_C\varphi_{f'} \cdot \psi_{f'}\varphi_B)\cdot[(F\mu)\eta_B] = \eta_C(F\mu)$$

$$[\varepsilon_C(G\mu)]\cdot(\varphi_C\psi_f \cdot \varphi_f\psi_B) = (G\mu)\varepsilon_B \qquad (6.3)$$

(See I,2.4, MQN.)

Similar formulas hold for adjunctions in a bicategory $\mathrm{Bifun}(\mathfrak{B},\mathfrak{B}')$.

4) In the 2-category 2-Cat (see I,2.3), a pair of 2-functors $F\colon \mathfrak{A} \longrightarrow \mathfrak{B}$, $U\colon \mathfrak{B} \longrightarrow \mathfrak{A}$ are adjoint (also called Cat-<u>adjoint</u>, and written $F \underset{\mathrm{Cat}}{\dashv} U$ if there are Cat-natural transformations $\varepsilon\colon FU \longrightarrow \mathfrak{B}$ and $\eta\colon \mathfrak{A} \longrightarrow UF$ satisfying $\varepsilon F \cdot F\eta = F$ and $U\varepsilon \cdot \eta U = U$. As usual, this is equivalent to the existence of a Cat-natural isomorphism

$$\mathfrak{B}(F(-),-) \cong \mathfrak{A}(-,U(-)) : \mathfrak{A}^{op} \times \mathfrak{B} \longrightarrow \text{Cat}$$

5) In particular, Cat-adjoints to a constant imbedding $\Delta : \mathfrak{A} \longrightarrow \mathfrak{A}^{\underline{D}}$, where $\underline{D}$ is a small category, are called Cat-<u>limits</u> and Cat-<u>colimits</u>, written

$$\text{Cat-}\varinjlim_{\underline{D}} \underset{\text{Cat}}{\dashv} \Delta \underset{\text{Cat}}{\dashv} \text{Cat-}\varprojlim_{\underline{D}}$$

An ordinary limit is a Cat-limit if and only if it is preserved by the Cat-representable functors (See, e.g., Dubuc [9].) More generally, Cat-adjoints to a constant embedding $\Delta : \mathfrak{A} \longrightarrow \mathfrak{A}^{\mathfrak{D}}$, where $\mathfrak{D}$ is a small 2-category, are called Cat_2-<u>limits</u> and Cat_2-<u>colimits</u>, written

$$\text{Cat}_2\text{-}\varinjlim_{\mathfrak{D}} \underset{\text{Cat}}{\dashv} \Delta \underset{\text{Cat}}{\dashv} \text{Cat-}\varprojlim_{\mathfrak{D}}$$

For instance, the Cat_2-limit of a diagram of the form

$$\underline{A} \underset{G}{\overset{F}{\Downarrow \varphi}} \underline{B} \qquad (6.4)$$

in Cat is the subcategory of $\underline{A}$ consisting of the objects A such that $\varphi_A = \text{id}$ and the morphisms f such that $F(f) = G(f)$; i.e., on the equalizer $\underline{A}_o$ of F and G, one equalizes the two functors

$$\underline{A}_o \underset{\ulcorner \varphi | \underline{A}_o \urcorner}{\overset{\text{id}_F}{\rightrightarrows}} \underline{B}^{\underline{2}} .$$

I,6.3 <u>Proposition</u>. In a 2-category or a bicategory:

i) Adjoints are unique, up to an isomorphism,

ii) Given

$$A \underset{u}{\overset{f}{\rightleftarrows}} B \underset{u'}{\overset{f'}{\rightleftarrows}} C$$

with $f \dashv u$ and $f' \dashv u'$, then $f'f \dashv uu'$.

iii) Adjoints are preserved by 2-functors.

<u>Proof</u>. i) Let $\mathbb{A}$ be a 2-category. Suppose $f \dashv u$ and $f \dashv u'$ with adjunction 2-cells $\varepsilon\colon fu \longrightarrow B$, $\eta\colon A \longrightarrow uf$ and $\varepsilon'\colon fu' \longrightarrow B$, $\eta'\colon A \longrightarrow u'f$ respectively. Then

$$u\varepsilon' \cdot \eta u\colon u' \longrightarrow u \qquad (6.5)$$

is an isomorphism with $u'\varepsilon \cdot \eta' u$ as its inverse. The proof follows from the commutativity of the diagram

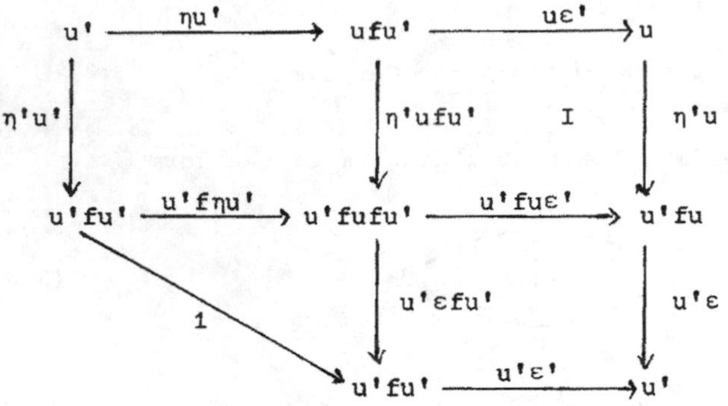

The square labled I commutes, for instance,because composition is a functor of two variables. This fact is applied to the composition

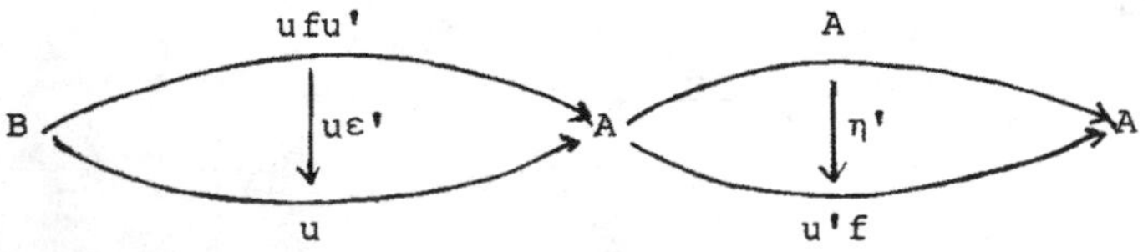

In a bicategory, the isomorphism $u' \longrightarrow u$ is given by the composition

$$u' \xrightarrow{\ell^{-1}} I_A u' \xrightarrow{\eta u'} (uf)u' \xrightarrow{\alpha} u(fu') \xrightarrow{u\varepsilon'} uI_B \xrightarrow{r} u \qquad (6.6)$$

with an analogous formula for the inverse.

The same diagram as above, considerably expanded to take care of the associativities and units, provides the proof via the same argument -that composition is a functor of two variables.

ii) Let $\mathbb{A}$ be a 2-category. If the adjunction 2-cells are

$$\varepsilon\colon fu \longrightarrow B \qquad \eta\colon A \longrightarrow uf$$

$$\varepsilon'\colon f'u' \longrightarrow C \qquad \eta'\colon B \longrightarrow u'f'$$

then define ε'' and η'' to be the compositions

$$\begin{aligned} \varepsilon'' &= (f'fuu' \xrightarrow{f'\varepsilon u'} f'u' \xrightarrow{\varepsilon'} C) \\ \eta'' &= (A \xrightarrow{\eta} uf \xrightarrow{u\eta' f} uu'f'f) \end{aligned} \qquad (6.7)$$

These describe an adjunction $f'f \dashv uu'$. The equation $(uu'\varepsilon'').(\eta''uu') = id$ holds, for instance, since the diagram

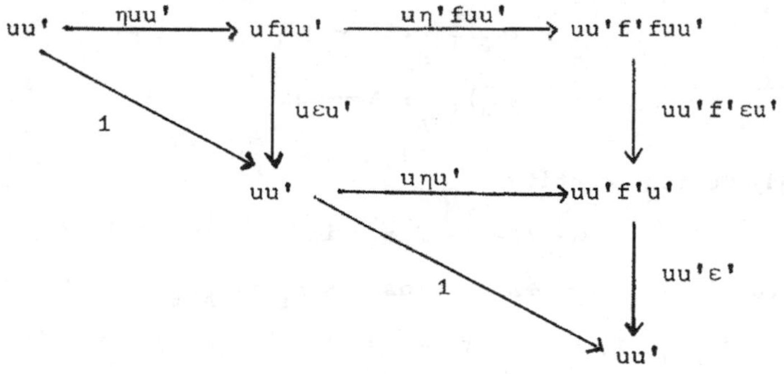

commutes. Again the crucial ingredient is that composition is a functor of two variables. In the bicategory case, appropriate instances of the isomorphisms α, ℓ, and r must be inserted, as in i).

As these results suggest, almost any equational property of adjoints carries over automatically to 2-categories, and with sufficient care to bicategories. However, this is not true for results which use the Yoneda embedding in an essential way, as is the case in the following proposition, and its consequences.

I,6.4 <u>Proposition</u>. The following are equivalent, for a 2-category $\mathcal{A}$:

i) $f \dashv u$ in $\mathcal{A}$,

ii) $\mathcal{A}(-,f) \dashv \mathcal{A}(-,u)$ in $\mathrm{Cat}^{\mathcal{A}^{op}}$,

iii) $\mathcal{A}(u,-) \dashv \mathcal{A}(f,-)$ in $\mathrm{Cat}^{\mathcal{A}}$.

<u>Proof</u>. The implications i) $\rightarrow$ ii) and iii) follow immediately from I,6.3,iii). Conversely, given ii), for instance, then since Yoneda is full, the modifications

$$\tilde{\varepsilon}\colon \mathcal{A}(-,f) \circ \mathcal{A}(-,u) \longrightarrow \mathcal{A}(-,B)$$
$$\tilde{\eta}\colon \mathcal{A}(-,A) \longrightarrow \mathcal{A}(-,u) \circ \mathcal{A}(-,f)$$

are induced by 2-cells

$$\varepsilon = (\tilde{\varepsilon}_B)_{id_B} : fu \longrightarrow B$$

$$\eta = (\tilde{\eta}_A)_{id_A} : A \longrightarrow uf$$

The adjunction identity

$$\tilde{\varepsilon}\mathcal{A}(-,f) \cdot \mathcal{A}(-,f)\tilde{\eta} = id$$

applied to $id_A\colon A \longrightarrow A$ yields $(\tilde{\varepsilon}_A)_f \cdot f(\tilde{\eta}_A)_{id_A} = id$. By (2.8), $(\tilde{\varepsilon}_A)_f = (\tilde{\varepsilon}_A)_{id_A} f$, so $\varepsilon f \cdot f\eta = id$. The other identity is derived analogously.

For a bicategory, Yoneda is not locally full and this argument cannot be carried out. We have not attempted to find out what is true in this case, except that, clearly, i) implies ii) and iii).

I,6.5 <u>Remarks</u>: Condition ii) means (by I,6.2,(1)) that for all C, the functors

$$\mathbf{A}(C,A) \underset{\mathbf{A}(C,u)}{\overset{\mathbf{A}(C,f)}{\rightleftarrows}} \mathbf{A}(C,B)$$

are adjoint in the usual sense and that the adjunction natural transformations are natural in C. This, in turn, means that given $h\colon C \longrightarrow A$, $k\colon C \longrightarrow B$, there are bijections

$$\mathbf{A}(C,B)(fh,k) \simeq \mathbf{A}(C,A)(h,uk) \tag{6.8}$$

which are natural with respect to varying h and k by 2-cells and by composition with $c\colon C' \longrightarrow C$. The dual properties hold for condition iii) which gives bijections

$$\mathbf{A}(A,C)(hu,k) \simeq \mathbf{A}(B,C)(h,kf) \tag{6.9}$$

for $h\colon A \longrightarrow C$, $k\colon B \longrightarrow C$.

I,6.6 <u>Corollary</u>. (Cf., Palmquist, [37]) Given adjoint morphisms $f \dashv u$ and $f' \dashv u'$ and morphisms h and k as indicated

$$\begin{array}{ccc} B & \xrightarrow{\quad k \quad} & B' \\ f \uparrow\downarrow u & & f' \uparrow\downarrow u' \\ A & \xrightarrow[\quad h \quad]{} & A' \end{array} \tag{6.10}$$

there is a commutative square of bijections

$$\begin{array}{ccc} \mathbf{A}(A,B')(f'h,kf) & \simeq & \mathbf{A}(B,B')(f'hu,k) \\ \wr & & \wr \\ \mathbf{A}(A,A')(h,u'kf) & \simeq & \mathbf{A}(B,A')(hu,u'k) \end{array} \tag{6.11}$$

<u>Proof</u>. The bijections are given by the preceeding remark. The horizontal ones can be decomposed by standard arguments by

inserting a column

$$\mathbb{A}(B,B')(f'hu,kfu) \cong \mathbb{A}(B,A')(hu,u'kfu)$$

in the middle . The left side maps to it by $\mathbb{A}(u,B')$ and $\mathbb{A}(u,A')$ and a commutative square results by naturality with respect to composition with the map $u: B \longrightarrow A$. It maps to the right side by $\mathbb{A}(B,B')(f'hu,k\varepsilon)$ and $\mathbb{A}(B,A')(hu,u'k\varepsilon)$ and a commutative square results by naturality with respect to varying k by the 2-cell $k\varepsilon: kfu \longrightarrow k$. Hence the original square commutes.

I,6.7. <u>Definition</u>. An <u>adjoint square</u> in $\mathbb{A}$ consists of a pair of adjoint morphisms $f \longrightarrow\!\!\dashv u$ and $f' \longrightarrow\!\!\dashv u'$ and morphisms h and k as in I,6.6, together with a matrix

$$\begin{pmatrix} \varphi_{11} & \varphi_{12} \\ \varphi_{21} & \varphi_{22} \end{pmatrix}$$

of compatible 2-cells arranged in the pattern of I,6.6. (These have also been studied by Palmquist [37] and Maranda [36].) Thus we have

$$\begin{pmatrix} \varphi_{11}: f'h \longrightarrow kf & \varphi_{12}: f'hu \longrightarrow k \\ \varphi_{21}: h \longrightarrow u'kf & \varphi_{22}: hu \longrightarrow u'k \end{pmatrix}$$

and these are related by the equations

$$\begin{aligned} \varphi_{11} &= (\varphi_{12}f)\cdot(f'h\eta) = (\varepsilon'kf)\cdot(f'\varphi_{21}) \\ &= (\varepsilon'kf)\cdot(f'\varphi_{22}f)\cdot(f'h\eta) \end{aligned} \tag{6.12}$$

$$\begin{aligned} \varphi_{12} &= (k\varepsilon)\cdot(\varphi_{11}u) = (\varepsilon'k\varepsilon)\cdot(f'\varphi_{21}u) \\ &= (\varepsilon'k)\cdot(f'\varphi_{22}) \end{aligned} \tag{6.13}$$

$$\varphi_{21} = (u'\varphi_{11})\cdot(\eta' h) = (u'\varphi_{12} f)\cdot(\eta' h\eta) = (\varphi_{22} f)\cdot(h\eta) \qquad (6.14)$$

$$\varphi_{22} = (u'k\varepsilon)\cdot(u'\varphi_{11} u)\cdot(\eta' hu) = (u'\varphi_{12})\cdot(\eta' hu) \qquad (6.15)$$
$$= (u'k\varepsilon)\cdot(\varphi_{21} u)$$

We shall call such 2-cells <u>transposes</u> of each other.

It follows directly from these equations that if one has two adjoint squares

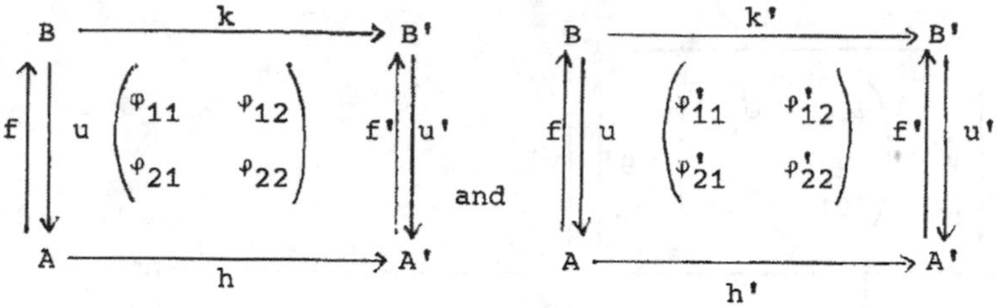

and a pair of 2-cells $\mu: h \longrightarrow h'$ and $\nu = k \longrightarrow k'$ then the equations

$$\begin{aligned} \nu f\cdot\varphi_{11} &= \varphi'_{11}\cdot f'\mu \\ \nu\cdot\varphi_{12} &= \varphi'_{12}\cdot f'\mu u \\ u'\nu\cdot\varphi_{22} &= \varphi'_{22}\cdot\mu u \\ u'\nu f\cdot\varphi_{21} &= \varphi'_{21}\cdot\mu \end{aligned} \qquad (6.16)$$

are equivalent. We call μ and ν a <u>compatible pair</u> of 2-cells if these equations hold.

Adjoint squares are a very handy bookeeping devise for treating relations between adjoint morphisms. I first discussed their properties in a lecture at Oberwolfach in 1966. Similar things were treated previously by Maranda [36] and, presumably, Benabou [unpublished], and subsequently by Dubuc [9] and Palmquist [37]. The following theorem is the basis for the calculus of adjoint squares.

I,6.8 <u>Theorem-Definition</u>. Let λ be a 2-category. The class
of adjoint squares in λ forms a double category, denoted by Ad-
Fun(λ). It is the underlying double category of a category object
in double categories (triple category, cf.I,2.8) $\widetilde{\text{Ad-Fun}}(\lambda)$.

<u>Proof</u>. Consider the following adjoint squares

$$
\begin{array}{ccccc}
C & \xrightarrow{\quad \ell \quad} & C' & & \\
\uparrow g \;\; v\downarrow & \begin{pmatrix} \psi_{11} & \psi_{12} \\ \psi_{21} & \psi_{22} \end{pmatrix} & g'\uparrow \;\; v'\downarrow & & \\
B & \xrightarrow{\quad k \quad} & B' & \xrightarrow{\quad k' \quad} & B'' \\
\uparrow f \;\; u\downarrow & \begin{pmatrix} \varphi_{11} & \varphi_{12} \\ \varphi_{21} & \varphi_{22} \end{pmatrix} & f'\uparrow \;\; u'\downarrow & \begin{pmatrix} \varphi'_{11} & \varphi'_{12} \\ \varphi'_{21} & \varphi'_{22} \end{pmatrix} & f''\uparrow \;\; u''\downarrow \\
A & \xrightarrow{\quad h \quad} & A' & \xrightarrow{\quad h' \quad} & A''
\end{array}
$$

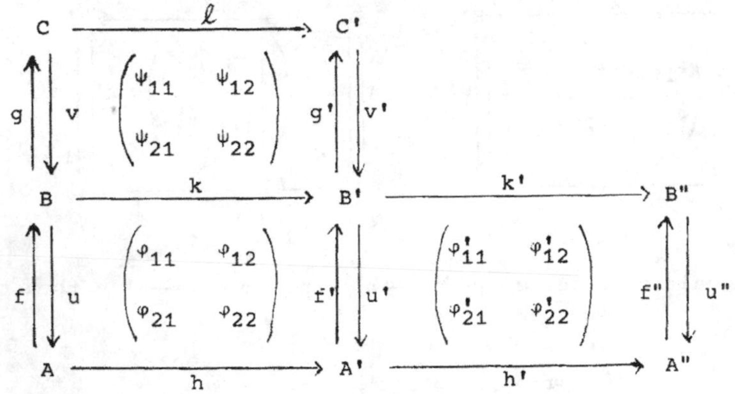

The **horizontal** domain and codomain of an adjoint square are the
left and right sides respectively, while the vertical domain
and codomain are the top and bottom. Horizontal and vertical
identities are represented by the following adjoint squares:

$$
\begin{array}{ccc}
B \xrightarrow{\;\; B \;\;} B & & A \xrightarrow{\;\; h \;\;} A' \\
f\uparrow \; u\downarrow \begin{pmatrix} f & \varepsilon \\ \eta & u \end{pmatrix} f\uparrow \; u\downarrow & & A\uparrow \; A\downarrow \begin{pmatrix} h & h \\ h & h \end{pmatrix} A'\uparrow \; A'\downarrow \\
A \xrightarrow{\;\; A \;\;} A & & A \xrightarrow{\;\; h \;\;} A'
\end{array}
$$

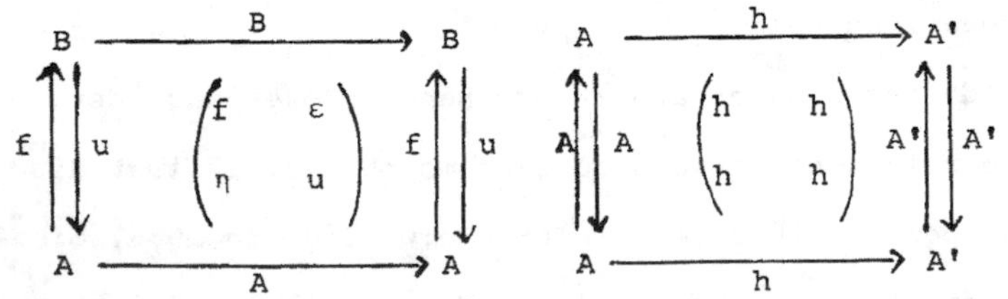

Horizontal composition is given by

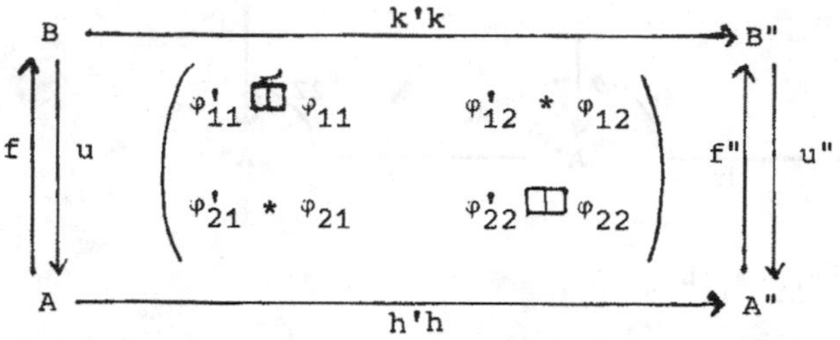

where $\varphi_{12}' * \varphi_{12} = (\varphi_{12}'\varphi_{12}) \cdot (f''h'\eta'hu)$

(6.17)

$\varphi_{21}' * \varphi_{21} = (u''k'\varepsilon'kf) \cdot (\varphi_{21}'\varphi_{21})$

and ▥ is defined in I,2.1, equations (2.3) and (2.3)'. Vertical
composition is given by

(6.18)

$$
\begin{CD}
C @>\ell>> C'
\end{CD}
$$

gf uv $\begin{pmatrix} \psi_{11} \boxminus \varphi_{11} & \psi_{12} \cdot (g'\varphi_{12}v) \\ (u'\psi_{21}f) \cdot \varphi_{21} & \varphi_{22} \boxminus \psi_{22} \end{pmatrix}$ $g'f'$ $u'v'$

$$
\begin{CD}
A @>h>> A'
\end{CD}
$$

It is obvious from (6.1) that the identities are adjoint squares
and it is easily seen that they serve as identities for the
compositions. What is not immediate is that the compositions of
adjoint squares are again adjoint squares. By (2.3) and (2.3')
applied to the diagrams

$$
\begin{CD}
B @>k>> B' @>k'>> B'' \\
@AfAA @Af'AA @Af''AA \\
A @>h>> A' @>h'>> A''
\end{CD}
$$

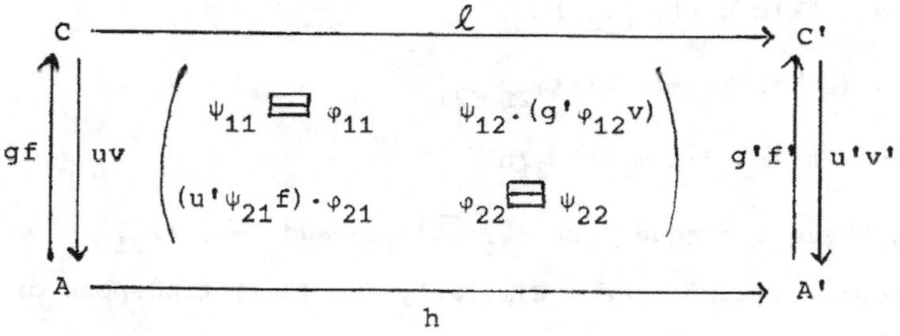

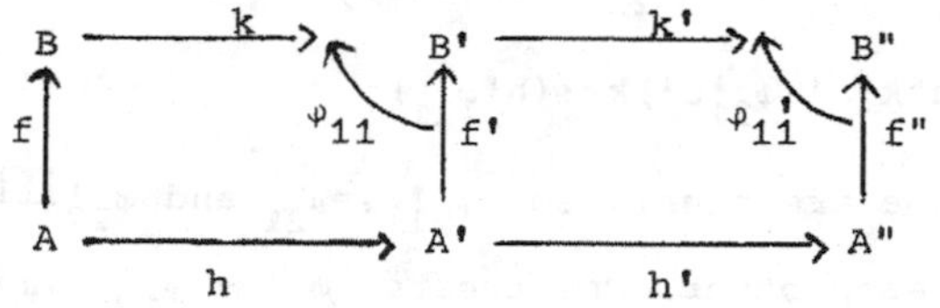

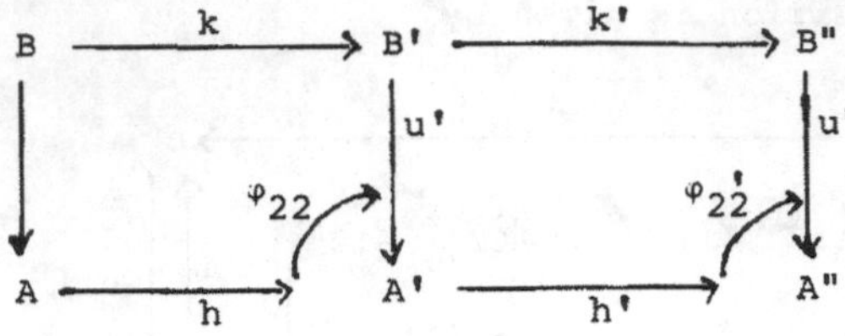

$$\varphi_{11}' \boxminus \varphi_{11} = k'\varphi_{11} \cdot \varphi_{11}'h$$

$$\varphi_{22}' \boxminus \varphi_{22} = \varphi_{22}'k \cdot h'\varphi_{22}.$$

By (6.14), the (2,1) transpose of $\varphi_{11}' \boxminus \varphi_{11}$ is

$$u''(\varphi_{11}' \boxminus \varphi_{11}) \cdot \eta''h'h = u''k'\varphi_{11} \cdot u''\varphi_{11}'h \cdot \eta''h'h$$

$$= u''k'\varphi_{11} \cdot \varphi_{21}'h$$

By definition,

$$\varphi_{21}' * \varphi_{21} = (u''k'\varepsilon'kf)(\varphi_{21}'\varphi_{21})$$

$$= (u''k'\varepsilon'kf) \cdot (u''k'f'\varphi_{21} \cdot \varphi_{21}'h)$$

$$= u''k'(\varepsilon'kf \cdot f'\varphi_{21}) \cdot \varphi_{21}'h$$

By (6.12), these are equal, so $\varphi_{11}' \boxminus \varphi_{11}$ and $\varphi_{21}' * \varphi_{21}$ are transposes of each other. Similarly the (2,1) transpose of $\varphi_{22}' \boxminus \varphi_{22}$ is

$$(\varphi_{22}' \boxminus \varphi_{22})f \cdot h'h\eta = \varphi_{22}'kf \cdot h'\varphi_{22}f \cdot h'h\eta = \varphi_{22}'kf \cdot h'\varphi_{21}$$

But we can also write

$$\varphi_{21}' * \varphi_{21} = (u''k'\varepsilon'kf)(\varphi_{21}'\varphi_{21})$$

$$= (u''k'\varepsilon'kf) \cdot (\varphi_{21}'u'kf) \cdot (h'\varphi_{21})$$

$$= (u''k'\varepsilon' \cdot \varphi_{21}'u')kf \cdot (h'\varphi_{21})$$

By (6.15), these are equal, so $\varphi_{21}' * \varphi_{21}$ and $\varphi_{22}' \boxminus \varphi_{22}$ are transposes of each other. One treats $\varphi_{12}' * \varphi_{12}$ similarly.

Therefore, the horizontal composition of adjoint squares is an adjoint square. (The original proof of this in Oberwolfach was a large diagram. See also, Maranda [36]). The calculations for vertical composition are much simpler and are left to the reader. That both compositions are associative and satisfy an interchange law, follows since the (1,1) components have this property. Hence Ad-Fun($\mathbb{A}$) is a double category.

In order to obtain the triple category structure $\widetilde{\text{Ad-Fun}(\mathbb{A})}$, consider the special case of an adjoint square of the form

$$\begin{array}{ccc} B & \xrightarrow{\quad B \quad} & B \\ f\uparrow\downarrow u & \begin{pmatrix}\varphi_{11} & \varphi_{12}\\ \varphi_{21} & \varphi_{22}\end{pmatrix} & f'\uparrow\downarrow u' \\ A & \xrightarrow[\quad A \quad]{} & A \end{array}$$

The 2-cells $\varphi_{11}: f' \longrightarrow f$ and $\varphi_{22}: u \longrightarrow u'$ are often called <u>conjugates.</u> They determine not only each other but also the other two transposes, $\varphi_{12}: f'u \longrightarrow B$ and $\varphi_{21}: A \longrightarrow u'f$. Vertical composition shows that conjugation is compatible with composition of adjoints, while horizontal composition shows that it is "functorial". The proof of I,6.3 is a special case.

Now define a 3-cell of $\widetilde{\text{Ad-Fun}(\mathbb{A})}$ between a pair of adjoint squares of the form

$$\begin{array}{ccc} B & \xrightarrow{\quad k \quad} & B' \\ f\uparrow\downarrow u & \begin{pmatrix}\varphi_{11} & \varphi_{12}\\ \varphi_{21} & \varphi_{22}\end{pmatrix} & f'\uparrow\downarrow u' \\ A & \xrightarrow[\quad h \quad]{} & A' \end{array} \qquad \begin{array}{ccc} B & \xrightarrow{\quad k' \quad} & B' \\ g\uparrow\downarrow v & \begin{pmatrix}\varphi'_{11} & \varphi'_{12}\\ \varphi'_{21} & \varphi'_{22}\end{pmatrix} & g'\uparrow\downarrow v' \\ A & \xrightarrow[\quad h' \quad]{} & A' \end{array}$$

(note that the four objects are the same) to consist of

a) conjugate 2-cells

$$\psi_{11}: g \longrightarrow f, \quad \psi_{22}: u \longrightarrow v$$

$$\psi_{11}': g' \longrightarrow f', \quad \psi_{22}': u' \longrightarrow v'$$

b) 2-cells $\mu: h \longrightarrow h'$, $\nu: k \longrightarrow k'$ such that μ and ν are compatible with the composed adjoint squares

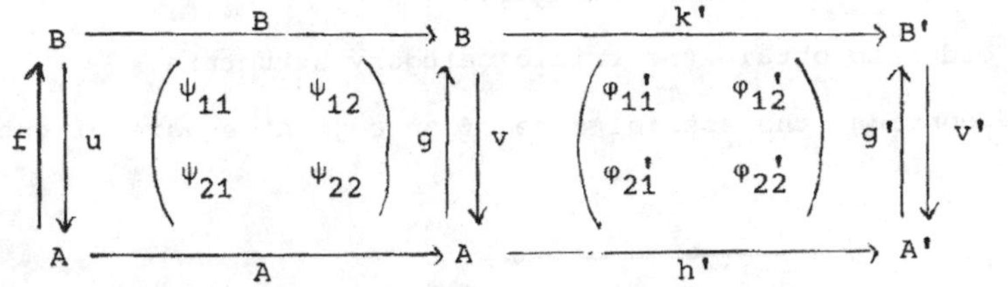

$$
\begin{array}{ccccccc}
B & \xrightarrow{\quad B \quad} & B & \xrightarrow{\quad k' \quad} & B' \\
f \uparrow \;\; u \downarrow & \begin{pmatrix} \psi_{11} & \psi_{12} \\ \psi_{21} & \psi_{22} \end{pmatrix} & g \downarrow \;\; v \downarrow & \begin{pmatrix} \varphi_{11}' & \varphi_{12}' \\ \varphi_{21}' & \varphi_{22}' \end{pmatrix} & g' \uparrow \;\; v' \downarrow \\
A & \xrightarrow{\quad A \quad} & A & \xrightarrow{\quad h' \quad} & A'
\end{array}
$$

and

$$
\begin{array}{ccccccc}
B & \xrightarrow{\quad k \quad} & B' & \xrightarrow{\quad B' \quad} & B' \\
f \uparrow \;\; u \downarrow & \begin{pmatrix} \varphi_{11} & \varphi_{12} \\ \varphi_{21} & \varphi_{22} \end{pmatrix} & f' \uparrow \;\; u' \downarrow & \begin{pmatrix} \psi_{11}' & \psi_{12}' \\ \psi_{21}' & \psi_{22}' \end{pmatrix} & g' \uparrow \;\; v' \downarrow \\
A & \xrightarrow{\quad h \quad} & A' & \xrightarrow{\quad A' \quad} & A'
\end{array}
$$

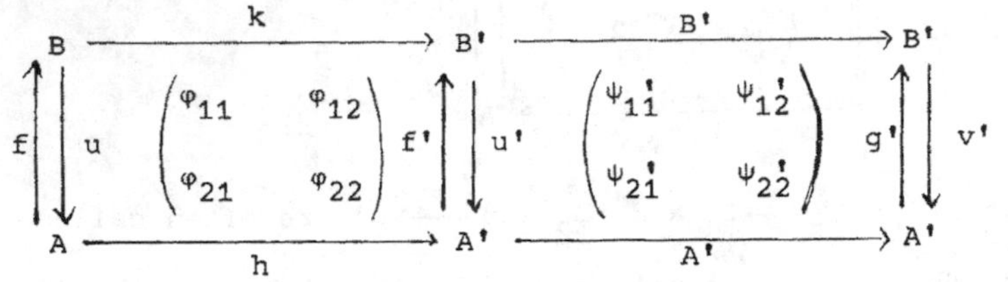

If one concentrates just on the $(2,2)$ component, then compatibility says that a cube like the one in I,4.1, $QF_2 3$ commutes. From this it is clear that the third composition, besides the obvious horizontal and vertical ones, given by horizontal composition of conjugates and weak composition of 2-cells makes Ad-Fun $\mathcal{A}$ into a triple category and that this third structure restricted to Ad-Fun $\mathcal{A}$ is discrete.

I,6.9 <u>Proposition</u>. The following are equivalent for an adjoint square as in I,6.7

i) The square is an isomorphism in Ad-Fun(A) with respect to horizontal composition.

ii) h,k and φ_{11} are isomorphisms.

iii) h,k and φ_{22} are isomorphisms.

In particular, if h and k are isomorphisms, then φ_{11} is an isomorphism if and only if φ_{22} is an isomorphism.

<u>Proof</u>. It is clear from the definition of composition that in the diagram

$$
\begin{array}{ccccc}
B & \xrightarrow{\quad k \quad} & B' & \xrightarrow{\quad k^{-1} \quad} & B \\
f\Big\uparrow\Big\downarrow u & \begin{pmatrix} \varphi_{11} & \varphi_{12} \\ \varphi_{21} & \varphi_{22} \end{pmatrix} & f'\Big\uparrow\Big\downarrow u' & \begin{pmatrix} \psi_{11} & \psi_{12} \\ \psi_{21} & \psi_{22} \end{pmatrix} & f\Big\uparrow\Big\downarrow u \\
A & \xrightarrow[\quad h \quad]{} & A' & \xrightarrow[\quad h^{-1} \quad]{} & A
\end{array}
$$

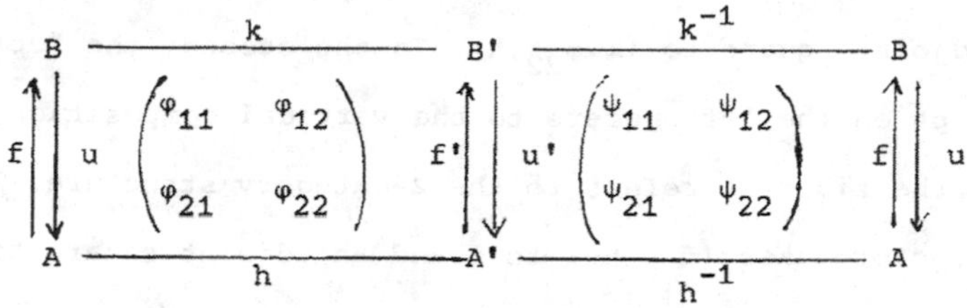

a candidate for the inverse to the first square must have the form of the second square. Suppose the second square is the inverse of the first. Then

$$k^{-1}\varphi_{11}\cdot\psi_{11}h = f \quad\text{and}\quad k\psi_{11}\cdot\varphi_{11}h^{-1} = f' \tag{6.19}$$

which implies that

$$\varphi_{11}\cdot k\psi_{11}h = kf \quad\text{and}\quad k\psi_{11}h\cdot\varphi_{11} = f'h;$$

i.e. that $k\psi_{11}h = \varphi_{11}^{-1}$. Hence i) $\Longrightarrow$ ii). Similarly, i) $\Longrightarrow$ iii). Conversely, if φ_{11}^{-1} exists, then $\psi_{11} = k^{-1}\varphi_{11}^{-1}h^{-1}$ satisfies (6.19), so the composed squares have the forms

$$\begin{pmatrix} f & ? \\ ? & ? \end{pmatrix} \quad\text{and}\quad \begin{pmatrix} f' & ? \\ ? & ? \end{pmatrix}$$

Since transposes are unique, these must be the corresponding

identity squares. Hence ii) and, similarly, iii) imply i).

I,6.10 <u>Proposition</u>. There are "forgetful" double functors

i)   Ad-Fun $\lambda \longrightarrow$ (Fun λ)$_o$

ii) ^{op}(Ad-Fun λ) $\longrightarrow [^{op}$Fun $^{op}\lambda]_o$

which, in fact, extend to triple functors.

<u>Proof</u>: The first double functor takes an object $(f \dashv u)$ to u,

and an adjoint square to (k, φ_{22}, h). In the second, the "op"

superscript on the left refers to the vertical composition,

while on the right it refers to the 2-category structure. The

double functor takes $(f \dashv u)$ to f and an adjoint square to

(h, φ_{11}, k).

I,6.11. <u>Examples</u>.

1. The bijection (6.8) in I,6.5 corresponds to adjoint

squares of the form

$$
\begin{array}{ccc}
C & \xrightarrow{\quad k \quad} & B \\
C \uparrow\downarrow C & \begin{pmatrix} \varphi_{11} & \varphi_{11} \\ \varphi_{22} & \varphi_{22} \end{pmatrix} & f \uparrow\downarrow u \\
C & \xrightarrow{\quad h \quad} & A
\end{array}
$$

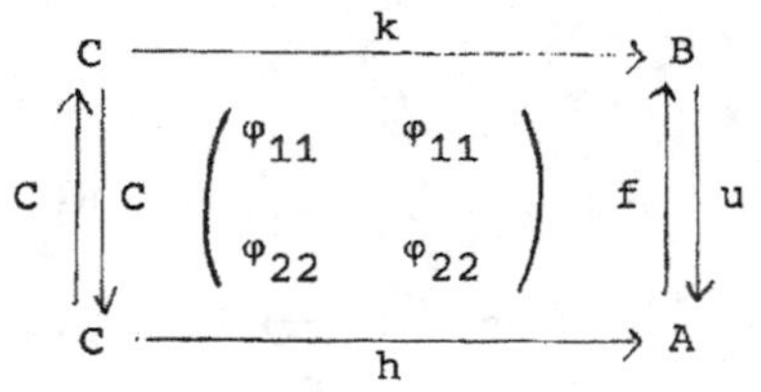

while the bijection (6.9) is given by taking the right side to

consist of identities.

2. In studying fibrations later, we will need horizontal

composition **to** show that the composition of cartesian morphisms

is cartesian. There is no other proof for cofibrations in Cat;
i.e., fibrations in Cat^{op}. This was overlooked in [FCC].

3. A square of the form

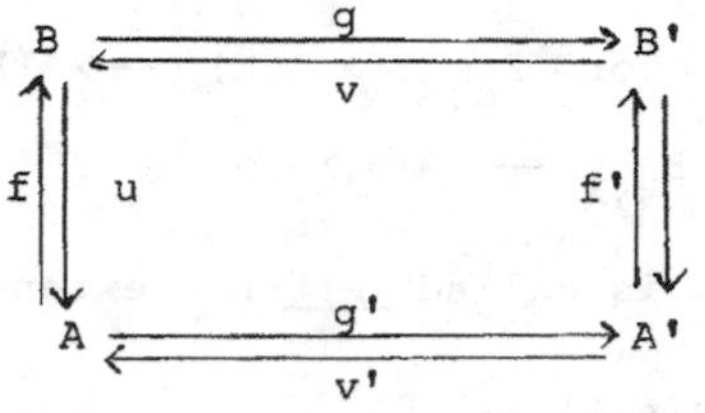

with the f's left adjoint to the u's and the g's left adjoint
to the v's contains four different adjoint squares which lead
to various transposes. If we denote the adjunction 2-cells by

$$\varepsilon\colon fu \longrightarrow 1, \qquad \eta\colon 1 \longrightarrow uf$$
$$\tilde{\varepsilon}\colon gv \longrightarrow 1, \qquad \tilde{\eta}\colon 1 \longrightarrow vg$$

with primes for the others, then a 2-cell $\theta\colon v'u' \longrightarrow uv$ has
successive transposes

$$\tilde{\theta} = (\varepsilon vf')\cdot(f\theta f')\cdot(fv'\eta')\colon fv' \longrightarrow vf'$$
$$\tilde{\tilde{\theta}} = (\tilde{\varepsilon}f'g')\cdot(g\tilde{\theta}g')\cdot(gf\tilde{\eta}')\colon gf \longrightarrow f'g' \tag{6.20}$$

while a 2-cell $\psi\colon uv \longrightarrow v'u'$ has transposes

$$\tilde{\psi} = (\tilde{\varepsilon}'u'g)\cdot(g'\psi g)\cdot(g'u\tilde{\eta})\colon g'u \longrightarrow u'g$$
$$\tilde{\tilde{\psi}} = (\varepsilon'gf)\cdot(f'\tilde{\psi}f)\cdot(f'g'\eta)\colon f'g' \longrightarrow gf \tag{6.21}$$

It can be checked that if $\psi = \theta^{-1}$ then $\tilde{\tilde{\psi}} = \tilde{\tilde{\theta}}^{-1}$. For instance,
if $g' = v' = A$ and $u' = uv$ then there is a transpose isomorphism
$f' \simeq gf$ which coincides with the isomorphism given by combining
i) and ii) of I,6.3.

4. Many of the more interesting examples involve limits, or more generally, Kan extensions, to which we now turn.

I.6.12 <u>Definition</u>. Let $s: A \longrightarrow B$ be a morphism in a 2-category $\mathbb{A}$, and let X be an object in $\mathbb{A}$. The <u>right Kan extension</u> along s for X is a functor $E_{s,X}: \mathbb{A}(A,X) \longrightarrow \mathbb{A}(B,X)$ such that

$$E_{s,X} \dashv \mathbb{A}(s,X)$$

The weakly dual notion is called <u>left</u> Kan extension, i.e.,

$$\mathbb{A}(s,X) \dashv E^{s,X}$$

The two strong duals have no names since they have never been considered in Cat. (see I,6.14(5).)

<u>Note</u>: Given $h: A \longrightarrow X$, $k: B \longrightarrow X$, this says there is an isomorphism

$$\mathbb{A}(B,X)[E_{s,X}h,k] \cong \mathbb{A}(A,X)(h,ks)$$

and that these isomorphisms are natural with respect to varying h and k by 2-cells. For a given s and X, $E_{s,X}$ need not exist at all, or may be defined only for certain values of h. In what follows $\mathbb{C} \subset \mathrm{Fun}(\mathbb{A}^{op}) \times \mathbb{A}$ denotes the full (triple) subcategory determined by objects (s,X) such that $E_{s,X}$ exists. We assume it is chosen for all such (s,X). Many particular cases of what follows are, of course, valid if $E_{s,X}$ is only partially defined.

I,6.13 <u>Theorem</u>. There is an operation

$$E_{-}^{-} : \mathbb{C} \longrightarrow \widetilde{\text{Ad-Fun}}(\text{Cat})$$

which is a 2-functor with respect to horizontal composition and a homomorphic pseudo-functor (I,3.2) with respect to vertical composition.

<u>Proof</u>: Consider a morphism

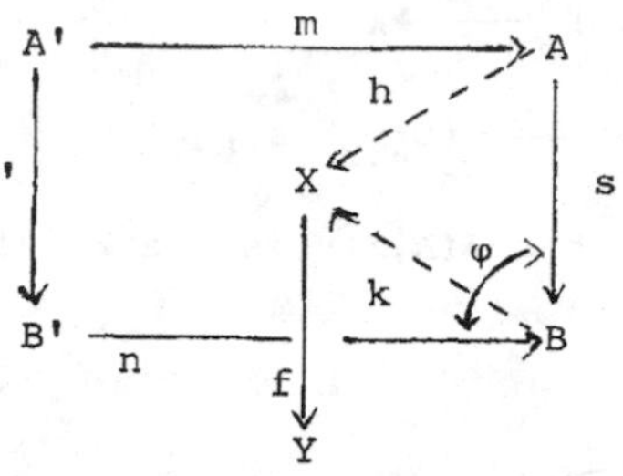

in $\mathfrak{C}$ (i.e., $E_{s,X}$ and $E_{s',Y}$) are defined. Then $\varphi\colon sm \longrightarrow ns'$ induces a natural transformation

$$\varphi_*\colon \mathbb{A}(m,f)\circ\mathbb{A}(s,X) \longrightarrow \mathbb{A}(s',Y)\circ\mathbb{A}(n,f)$$

whose component at $k \in \mathbb{A}(B,X)$ is

$$(\varphi_*)_k = fk\varphi\colon fksm \longrightarrow fkns'$$

We define the value of E_-^- on this morphism to be the adjoint square

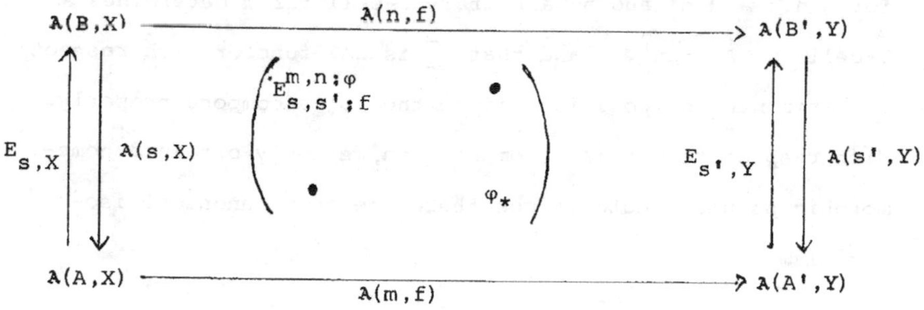

Thus

$$E^{m,n;\varphi}_{s,s';f} : E_{s',Y}\circ\mathbb{A}(m,f) \longrightarrow \mathbb{A}(n,f)\circ E_{s,X}$$

is the transpose natural transformation to φ_*.

If the adjunction natural transformations are denoted by

$$\varepsilon_{s,X}\colon \mathfrak{C}_{s,X}\circ\mathbb{A}(s,X) \longrightarrow \mathbb{A}(B,X)$$

$$\eta_{s,X}\colon \mathbb{A}(A,X) \longrightarrow \mathbb{A}(s,X)\circ E_{s,X}$$

with components

$$(\varepsilon_{s,X})_k \colon E_{s,X}(ks) \longrightarrow k$$

$$(\eta_{s,X})_h \colon h \longrightarrow (E_{s,X}h)s \ ,$$

then its component at $h \in \mathbb{A}(A,X)$ is the 2-cell given by the composition

$$\begin{array}{lcl}
E_{s',Y}(fhm) & & \\
\Big\downarrow E_{s',Y}[f(\eta_{s,X})_h m] & & \\
E_{s',Y}[f(E_{s,X}h)sm] & & \\
\Big\downarrow E_{s',Y}[(\varphi_*)_{E_{s,X}h}] & & \\
E_{s',Y}[f(E_{s,X}h)ns'] & \xrightarrow{(\varepsilon_{s',Y})_{f(E_{s,X}h)n}} & f(E_{s,X}h)n
\end{array}
\qquad E_{s',Y}(fhm) \xrightarrow{\left(E^{m,n;\varphi}_{s,s';f}\right)_h} f(E_{s,X}h)n \tag{6.22}$$

It is clear that a 3-cell in $\mathbb{C}$ given by taking compatible 2-cells for m,n,s and s' and an arbitrary 2-cell for f determines a 3-cell in $\widetilde{\text{Ad-Fun}}|\mathbb{A}|$ and that E_-^- is a 2-functor with respect to horizontal composition, since the φ_*'s compose properly. With respect to vertical composition, we only obtain a homomorphic pseudo-functor since there are only canonical isomorphisms

$$E_{t,X} \circ E_{s,X} \cong E_{ts,X}$$

rather than identities.

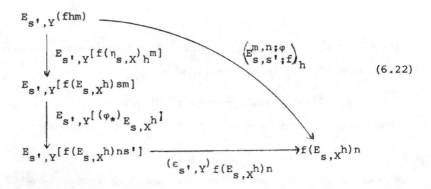

I,6.14 <u>Examples</u>. In the following special cases any constituent of E_-^- which is an identity is omitted from the notation. The purpose of these examples is to illustrate the contention that transposes are the general case of a "canonical induced morphism " and the point of I,6.13 is that, therefore, these canonical induced morphisms have all possible naturality properties. We have not investigated the corresponding situation for bicategories, but there does not appear to be any obstacle, other than finiteness of available publication space, to doing so.

1). If $f \dashv u$, then for all X, $E_{f,X}$ and $E^{u,X}$ exist and we may always assume they are chosen to be

$$E_{f,X} = \mathcal{A}(u,X)\,, \qquad E^{u,X} = \mathcal{A}(f,X)$$

by I,6.4.

2) <u>Definition</u>. $f: X \longrightarrow Y$ <u>preserves</u> $E_{s,-}$ if

$$\begin{array}{ccc} E_{s;f} \overset{=}{\text{def}} E_{s,s;f}: E_{s,Y} \cdot \mathcal{A}(A,f) & \rightarrow & \mathcal{A}(B,f) \circ E_{s,X} \\ \| & & \| \\ E_{s,Y}(f(-)) & \longrightarrow & fE_{s,X}(-) \end{array} \qquad (6.23)$$

is an isomorphism. The following are immediate from I,6.13. (Cf., Dubuc [9] and Gabriel-Ulmer [17].)

a) If f and g preserve $E_{s,-}$, then so does gf and $E_{s;qf} = E_{s;q} \boxed{\,|\,} E_{s;f}$

b) If f preserves $E_{s,-}$ and $E_{t,-}$, then it preserves $E_{ts,-}$ and $E_{ts;f} = E_{t;f} \boxminus E_{s;f}$

c) If f has a right adjoint u, then f preserves $E_{s,-}$ for all s. This follows by applying I,6.11,3) to the diagram

$$\begin{array}{ccc}
\mathcal{A}(B,X) & \underset{\mathcal{A}(B,u)}{\overset{\mathcal{A}(B,f)}{\rightleftarrows}} & \mathcal{A}(B,Y) \\
E_{s,X}\uparrow\downarrow \mathcal{A}(s,X) & & E_{s,Y}\uparrow\downarrow \mathcal{A}(s,Y) \\
\mathcal{A}(A,X) & \underset{\mathcal{A}(A,u)}{\overset{\mathcal{A}(A,f)}{\rightleftarrows}} & \mathcal{A}(A,Y)
\end{array}$$

where $\mathcal{A}(A,u)\circ\mathcal{A}(s,Y) = \mathcal{A}(s,X)\circ\mathcal{A}(B,u)$.

d) <u>Theorem</u> (Dubuc [9]) $f: A \longrightarrow B$ has a right adjoint $u: B \longrightarrow A$ iff $E_{f,A}(A)$ exists and is preserved by f.

<u>Proof</u>: (Dubuc-Street) The second condition means that $E_{f,B}(f)$ exists and that

$$(E_{f;f})_A : E_{f,B}(fA) \longrightarrow f\circ E_{f,A}(A)$$

is an invertible 2-cell. Since $E_{f,B}(f)$ is determined only up to an isomorphism, we can and will assume that $(E_{f,f})_A$ is the identity.

Suppose first that $u: B \longrightarrow A$ exists with $f \dashv u$. Then $E_{f,X} = \mathcal{A}(u,X)$ exists for all X. In particular $E_{f,A}(A) = u$ and

$$E_{f,B}(f) = \mathcal{A}(u,X)(f) = fu = f\circ E_{f,A}(A).$$

Conversely, suppose $E_{f,A}(A)$ exists and is preserved by f. Let $u = E_{f,A}(A)$. Then there are isomorphisms for all k and ℓ, natural in k and ℓ, and a commutative diagram

$$\begin{array}{ccc}
\mathcal{A}(B,A)(u,k) & \cong & \mathcal{A}(A,A)(A,kf) \\
\downarrow \mathcal{A}(B,f) & & \downarrow \mathcal{A}(A,f) \\
\mathcal{A}(B,B)(fu,fk) & \cong & \mathcal{A}(A,B)(f,fkf) \\
\mathcal{A}(B,B)(fu,\ell) & \cong & \mathcal{A}(A,B)(f,\ell f)
\end{array}$$

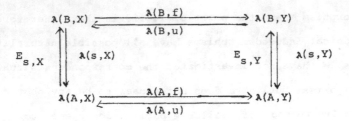

Let $k = u$. Then 1_u corresponds to $\eta_A: A \longrightarrow uf$, while 1_{fu} corresponds to $\eta_f: f \longrightarrow fuf$ and $\eta_f = f\eta_A$. Similarly, for $\ell = B$, 1_f corresponds to $\varepsilon_B: fu \longrightarrow B$. If ℓ is replaced by the 2-cell ε_B, one gets the corresponding commutative triangles

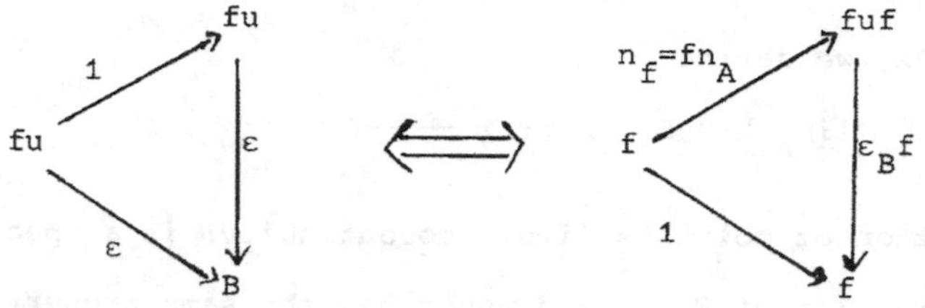

in $\mathbf{A}(B,B)(fu,-)$ and $\mathbf{A}(A,B)(f,-)$ which shows that ε and η satisfy one of the adjunction equations. For the other, observe that the diagram

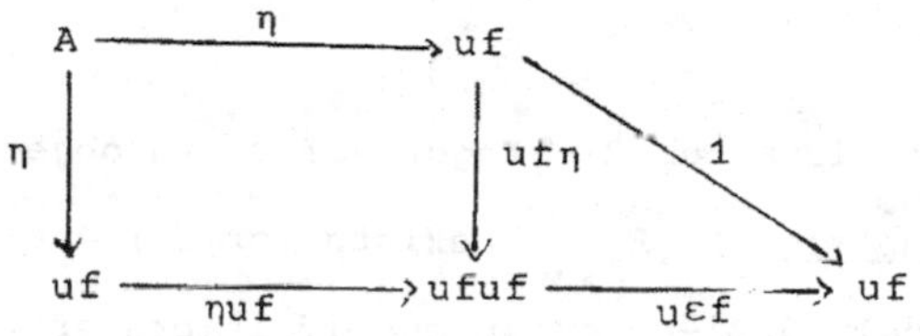

commutes, the square by the functoriality of composition and the triangle by composing the previous triangle with u. Hence, for the indicated 2-cells in place of k, one has corresponding commutative diagrams

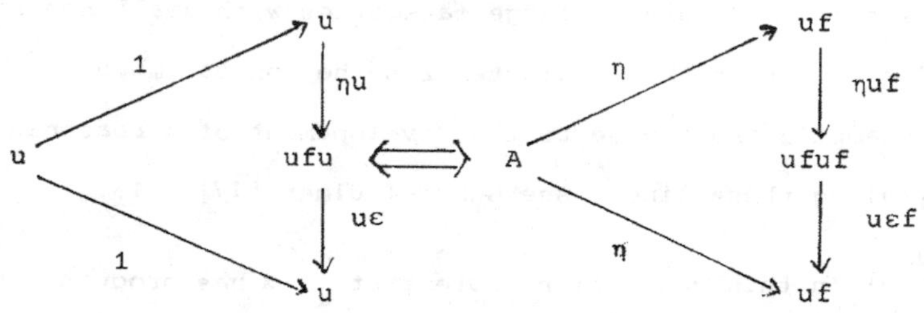

in $\mathbf{A}(B,A)(u,-)$ and $\mathbf{A}(A,A)(A,-f)$.

Hence the other adjunction equation is satisfied.

e) If A has a strict terminal object 1 (i.e., for each A there is exactly one 2-cell from A to 1 which is therefore the identity 2-cell of a unique 1-cell $\tau_A\colon A \longrightarrow 1$) then, for $h\colon A \longrightarrow X$, we define

$$\varinjlim_A h = E_{\tau_A, X}\, h\colon 1 \longrightarrow X$$

Preservation of colimits (i.e., cocontinuity) is a special case of preservation of $E_{s,-}$, and hence has the same properties as above. Limits are defined analoguously in terms of $E^{\tau_A, X}$. It is only under special assumptions in a representable 2-category that the preservation of colimits implies the preservation of $E_{s,-}$.

f) If $\mathbb{A}'$ is a sub 2-category of $\mathbb{A}$, an object $X \in \mathbb{A}$ is called $\mathbb{A}'$-<u>cocomplete</u> if $E_{\tau_A, X}$ exists for all $A \in \mathbb{A}'$. A morphism $f\colon X \longrightarrow Y$ is called $\mathbb{A}'$-<u>cocontinuous</u> if it preserves $E_{\tau_A,-}$ for all $A \in \mathbb{A}'$. Given an arbitrary X, one can ask for a morphism $h\colon X \longrightarrow \hat{X}$ where $\hat{X}$ is $\mathbb{A}'$-cocomplete such that given any $k\colon X \longrightarrow Y$ with Y $\mathbb{A}'$-complete there exists an $\mathbb{A}'$-cocontinuous $k'\colon \hat{X} \longrightarrow Y$, unique up to an isomorphism, such that $k'h = k$. If $\mathbb{A}$ is the 2-category of large categories with small Hom sets and $\mathbb{A}'$ = Cat, then this characterizes the Yoneda imbedding. For other examples and a clue to the developement of a reasonable theory along these lines, see Gabriel-Ulmer [17], §15.

g) In this connection, note that if $\mathbb{A}$ has products, then a diagram

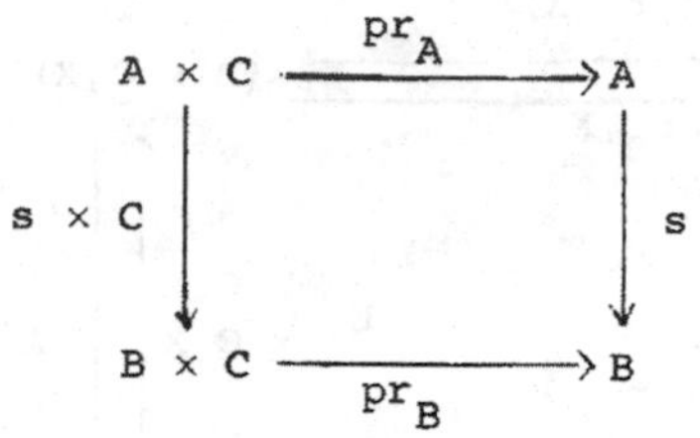

gives rise to a diagram

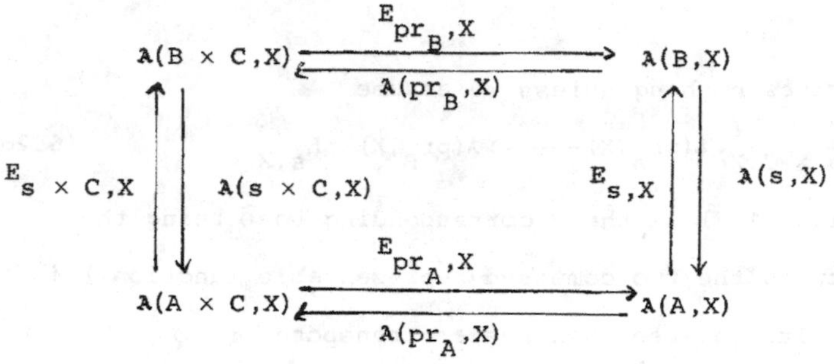

in which $A(pr_A,X) \circ A(s,X) = A(s \times C,X) \circ A(pr_B,X)$ and hence by I,6.11,3), there is an isomorphism

$$E_{s,X} \circ E_{pr_A,X} \cong E_{pr_B,X} \circ E_{s \times C,X} \tag{6.24}$$

with all possible naturality properties. Specializing to B = 1 so $pr_B = \tau_C$, $s = \tau_A$, and $s \times C = pr_C$, gives

$$\varinjlim_A \circ E_{pr_A,X} \cong \varinjlim_C \circ E_{pr_C,X} \tag{6.25}$$

which is the usual interchange of limits.

On the other hand, the diagram

$$\begin{array}{ccc}
\mathcal{A}(B,X) & \underset{E^{pr_B,X}}{\overset{\mathcal{A}(pr_B,X)}{\rightleftarrows}} & \mathcal{A}(B \times C,X) \\
E_{s,X} \big\uparrow \big\downarrow \mathcal{A}(s,X) & & E_{s \times C,X} \big\uparrow \big\downarrow \mathcal{A}(s \times C,X) \\
\mathcal{A}(A,X) & \underset{E^{pr_A,X}}{\overset{\mathcal{A}(pr_A,X)}{\rightleftarrows}} & \mathcal{A}(A \times C,X)
\end{array}$$

in general gives nothing unless we assume

$$E_{s \times C,X} \circ \mathcal{A}(pr_A,X) \longrightarrow \mathcal{A}(pr_B,X) \circ E_{s,X} \tag{6.26}$$

(which, in I,6.11.3) is the $\tilde{\tilde{\theta}}$ corresponding to $\tilde{\theta}$ being the identity between the two composed representable function) is an isomorphism. Its inverse then has as transpose

$$E_{s,X} \circ E^{pr_A,X} \longrightarrow E^{pr_B,X} \circ E_{s \times C,X} \tag{6.27}$$

Taking B = 1 as before, this specializes to

$$\varinjlim_A \circ E^{pr_A,X} \longrightarrow \varprojlim_C \circ E_{pr_C,X} \tag{6.28}$$

which is the usual non-isomorphic interchange of colimits with limits.

In Cat, (6.26) holds, using the cartesian closed structure to write $\mathcal{A}(A \times C,X) \cong \mathcal{A}(C,\bar{\mathcal{A}}(A,X))$ and the enriched nature of Kan extensions to write $E_{s \times C,X} \cong \mathcal{A}(C,\bar{E}_{s,X})$. This argument can, of course, be generalized to a suitable monoidal closed 2-category, but other than this, we do not know the range of validity of assuming that (6.26) is an isomorphism.

h) One can, of course, equally well talk about f reflecting

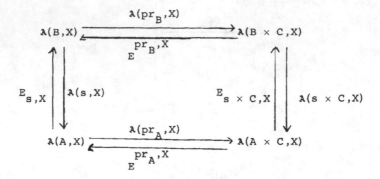

or creating limits or colimits. We shall use this in Chapter IV.

3) <u>Definition</u>. $m: A' \longrightarrow A$ is <u>final with respect to</u> $s: A \longrightarrow B$ if for all X, $E_{sm,X}$ exists iff $E_{s,X}$ exists and

$$E^m_{s,sm;X} : E_{sm,X} \circ \mathbb{A}(m,X) \longrightarrow E_{s,X}$$

is an isomorphism; i.e., there is a commutative diagram of isomorphisms

$$\begin{array}{ccc} \mathbb{A}(B,X)(E_{s,X}h,k) & \cong & \mathbb{A}(A,X)(h,ks) \\ \Big\| & & \Big\| \, \mathbb{A}(m,X) \\ \mathbb{A}(B,X)(E_{sm,X}hm,k) & \cong & \mathbb{A}(A',X)(hm,ksm) \end{array}$$

for all $h: A \longrightarrow X$, $k: B \longrightarrow X$.

a) In particular, m is <u>final</u> if it is final with respect to $\tau_A: A \longrightarrow 1$; i.e., there is an isomorphism

$$\varinjlim_{A'}[(-)m] \cong \varinjlim_A .$$

Equivalently, for all $x : 1 \longrightarrow X$, there is an isomorphism

$$\mathbb{A}(A,X)(h, x\tau_A) \cong \mathbb{A}(A',X)(hm, x\tau_{A'})$$

b) Given $n: A'' \longrightarrow A'$, if n is final with respect to sm and m is final with respect to s, then mn is final with respect to s.

c) If m has a left adjoint, then m is final with respect to s for all s.

d) As in 2), this can be used to describe various notions in a 2-category. For instance, if $\mathbb{A}$ has products, then $D \in \mathbb{A}$

is called <u>directed</u> if the diagonal $D \longrightarrow D \times D$ is final.

4) Another possibility is given by $s: A \longrightarrow B$ and $n: B \longrightarrow B'$ where one can require that

$$E^{n}_{s,ns;X}: E_{s,X} \longrightarrow \mathbb{A}(n,X) \circ E_{ns,X}$$

be an isomorphism. We do not know what this means.

5) The preceeding consideration hold in any 2-category. In particular, they hold in Cat^{op}. We are not aware that anyone has studied what we choose to call <u>dual Kan extensions</u>; i.e. Kan extensions in Cat^{op}. For instance, 2)d) holds in this situation. An example is given by taking $\underline{X} = \underline{1}$. Given $S: \underline{A} \longrightarrow \underline{B}$, $\tilde{E}_{S,\underline{1}}$ (resp., $\tilde{E}^{S,\underline{1}}$) exists if and only if S has a left (resp., right) adjoint (since $\underline{A}^{\underline{1}} = \underline{A}$, etc.). We have not been able to discover examples that are not variations of this one.

6) Kan extensions provide a typical example of adjoint quasi-natural transformations as in I,6.2 (3), providing we restrict attention to the locally full subcategory $\mathfrak{C}'$ of $\mathfrak{C}$ determined by morphisms (as in the beginning of the proof of I,6.13) in which φ is an identity 2-cell; i.e., $sm = ns'$. Taking the bottom row and top row respectively of the value of E^{-}_{-} on such a square provides two 2-functors

$$F,G: \mathfrak{C}' \longrightarrow \mathrm{Cat};$$

i.e., $F(m,n;f) = \mathbb{A}(m,f)$ and $G(m,n;f) = \mathbb{A}(n,f)$ The sides can be viewed as the components of quasi -natural transformations between F and G (i.e., morphisms in ${}^{op}\mathrm{Fun}({}^{op}\mathfrak{C}', {}^{op}\mathrm{Cat})$), where

$$\begin{cases} \varphi_{s,X} = E_{s,X}: \mathbb{A}(A,X) \longrightarrow \mathbb{A}(B,X) : F(s,X) \longrightarrow G(s,X) \\ \varphi_{m,n;f} = E^{m,n}_{s,s';f} \end{cases}$$

$$\begin{cases} \psi_{s,X} = \mathbb{A}(s,X) : \mathbb{A}(B,X) \longrightarrow \mathbb{A}(A,X) : G(s,X) \longrightarrow F(s,X) \\ \psi_{m,n;f} = \mathrm{id} \end{cases}$$

The adjunction natural transformations between $E_{s,X}$ and $\mathbb{A}(s,X)$ provide the modifications ε and η. The first equation in (6.3) follows from (6.14) and the second from (6.13); in both cases $\varphi_{11} = \mathrm{id}$.

I,7 Quasi-adjointness.

Basically, a quasi-adjunction between 2-functors $F: \mathcal{A} \longrightarrow \mathcal{B}$ and $U: \mathcal{B} \longrightarrow \mathcal{A}$ is a pair of quasi-natural transformations $\varepsilon: FU \longrightarrow \mathcal{B}$ and $\eta: \mathcal{A} \longrightarrow UF$ satisfying the usual equations. However, it turns out that this direct generalization of the usual notion is both overly and insufficiently general. Some (but by no means all) useful formulations are described in I,7.1. The analogues of I,6.3 and I,6.4 (which fail to hold in general) are discussed in I,7.3 and I,7.4. The general failure of quasi-adjoints to be "functorial" in the first variable seems to indicate that the technique of adjoint squares is not appropriate here. In I,7.5, a characterization in terms of 2-comma categories is given for the type of quasi-adjointness which is of particular interest here. This suggests another kind of quasi-adjointness, called transendental, which is defined in I,7.6. Its relations to the previous notions are described in I,7.7. In I,7.8, the connections of these notions with universal mapping properties are discussed. The rest of the chapter consists of examples, as follows:

I,7.9 Some general principles.
I,7.10 Some finite quasi-limits.
I,7.11 Quasi-colimits in Cat.
I,7.12 Quasi-limits in Cat.
I,7.13 Quasi-fibrations.
I,7.14 Quasi-Kan extensions in Cat.
I,7.15 The categorical comprehension scheme.
I,7.16 The global quasi-Yoneda lemma.
I,7.17 Globalized adjunction morphisms.

To begin, let $F: \mathcal{A} \longrightarrow \mathcal{B}$ and $U: \mathcal{B} \longrightarrow \mathcal{A}$ be 2-functors between 2-categories, let $\varepsilon: FU \longrightarrow \mathcal{B}$ and $\eta: \mathcal{A} \longrightarrow UF$ be quasi natural transformations (I,2.4) and let s and t be modifications (I,2.4, MQN) as indicated:

One can form the composed modifications

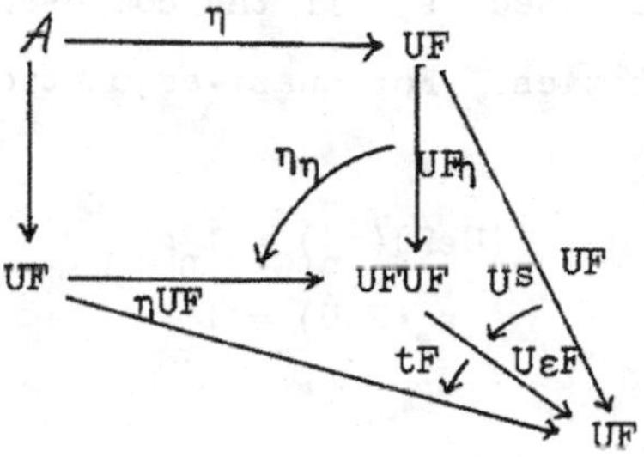

(7.1)

and

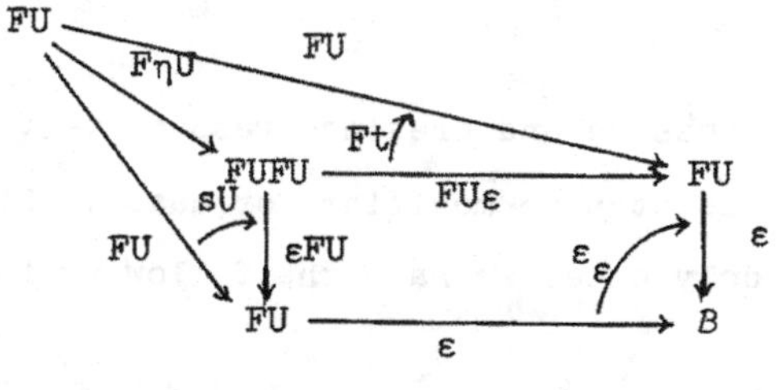

(7.1')

Here, for instance, η_η has components $\eta_{(\eta_A)}$.

I.7.1 <u>Definition</u>.

i) The four-tuple $(\varepsilon,\eta;s,t)$ is called a <u>weak quasi-adjunction</u>

ii) If s and t are isomorphic modifications then it is called i-<u>weak</u>.

iii) If ε and η are iso-quasi-natural (I,4.24), then it is called i-<u>quasi</u>.

iv) If s and t are identities; i.e., if $\varepsilon F \cdot F\eta = F$ and $U\varepsilon \cdot \eta U = U$, then (ε,η) is called a <u>quasi-adjunction</u>.

v) In any of the preceeding situations, the adjunction is called <u>strict</u> (abbreviated s) if the composed modifications in (7.1) and (7.1') are identities. For quasi-adjunctions, these reduce to

$$\begin{aligned}(U\varepsilon F)(\eta_\eta) &= 1_\eta \\ \varepsilon_\varepsilon(F\eta U) &= 1_\varepsilon\end{aligned} \quad . \tag{7.2}$$

We can thus speak of x-weak y-quasi-adjoints where

x = -, +, i, s, si

y = n, i, s, si.

Here -, + denote the absence and presence respectively of the adjective weak and n denotes no modifier for quasi. Since s appears as a modifier only once, we have the following twelve possibilities

y \ x	-	+	i	s	si
n	•	•	•	•	•
i	•	•	*	•	•
s	*				
si	•				

(7.3)

The *'s indicate the two cases which seem at the moment to be the principle ones; namely, i-weak i-quasi-adjoints and s-quasi-adjoints.

I,7.2 <u>Definition</u>. A pair of morphisms $F: \mathcal{B} \longrightarrow \mathcal{B}'$ and $U: \mathcal{B}' \longrightarrow \mathcal{B}$ between bicategories are called <u>quasi-adjoint</u> if there are quasi-natural transformations (I,3.3) $\varepsilon: FU \longrightarrow \mathcal{B}$ and $\eta: \mathcal{A} \longrightarrow UF$ such that $F\eta$ and $U\varepsilon$ are defined (I.4.20) and $\varepsilon F \cdot F\eta = F$, $U\varepsilon \cdot \eta U = U$. As with 2-functors, there are various other possibilities, but we do not treat them here.

For the analogue of I,6.3, consider 2-functors

$$\mathcal{A} \; \underset{\overline{U}}{\overset{\overset{F}{\longrightarrow}}{\underset{\longleftarrow}{\overset{U}{\longleftarrow}}}} \; \mathcal{B} \; \underset{U'}{\overset{F'}{\underset{\longleftarrow}{\longrightarrow}}} \; \mathcal{M}$$

with weak quasi-adjunctions

$$(\varepsilon,\eta;s,t): F \dashv U$$
$$(\overline{\varepsilon},\overline{\eta};\overline{s},\overline{t}): F \dashv \overline{U}$$
$$(\varepsilon',\eta';s',t'): F' \dashv U'$$

I,7.3. <u>Proposition</u>

i) a) There exist quasi-natural transformations $\varphi: U \longrightarrow \overline{U}$ and $\overline{\varphi}: \overline{U} \longrightarrow U$.

b) If t and $\bar{t}$ are isomorphic modifications, then there exist modifications $u: U \longrightarrow \bar{\varphi}\varphi$ and $\bar{u}: \bar{U} \longrightarrow \varphi\bar{\varphi}$

c) If, in addition, the composed modifications in (7.4) and (7.4') below are isomorphisms, then u and $\bar{u}$ are isomorphisms In particular, this holds for i-weak i-quasi-adjoints.

d) If U and $\bar{U}$ are i-quasi-adjoints and the compositions in (7.4) and (7.4') below are identities, then U and $\bar{U}$ are quasi-isomorphic.

ii) a) $F'F$ is weak quasi-adjoint to UU' .

b) If $(\varepsilon,\eta;s,t)$ and $(\varepsilon',\eta';s',t')$ are both i-weak i-quasi-adjunctions, then $F'F$ is i-weak i-quasi-adjoint to UU'.

c) If both are quasi-adjoints and if for all $C \in \mathcal{M}$, $A \in \mathcal{A}$,

$$(U\eta')_{\varepsilon U'_C} = \mathrm{id} \quad , \quad (F'\varepsilon)_{\eta' F_A} = \mathrm{id}$$

then $F'F$ and UU' are quasi-adjoint.

iii) If $\Gamma: 2\text{-Cat}_o \longrightarrow 2\text{-Cat}_o$ is a functor which is enriched with respect to the closed structure given by $\mathrm{Fun}(-,-)$, then Γ preserves x-weak y-quasi-adjunctions.

<u>Proof</u>: i) Consider the diagrams

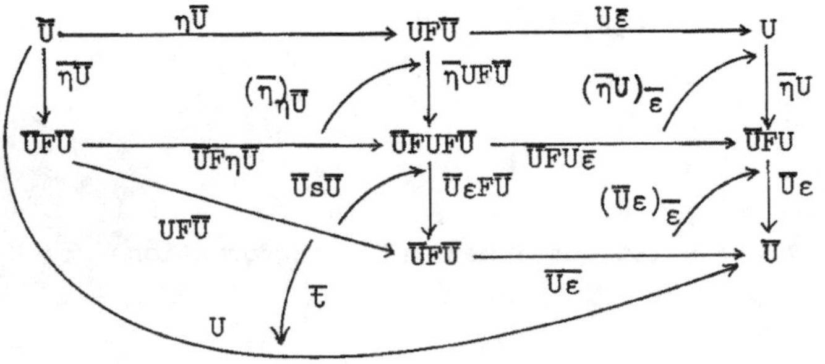

(7.4)

and (7.4') in which the roles of U and $\overline{U}$ are interchanged. Define

$$\varphi = (\overline{U}\varepsilon)(\overline{\eta}U) \colon U \longrightarrow \overline{U}$$
$$\overline{\varphi} = (U\overline{\varepsilon})(\eta\overline{U}) \colon \overline{U} \longrightarrow U .$$

The results then are immediate.

ii) These follow directly from the diagrams

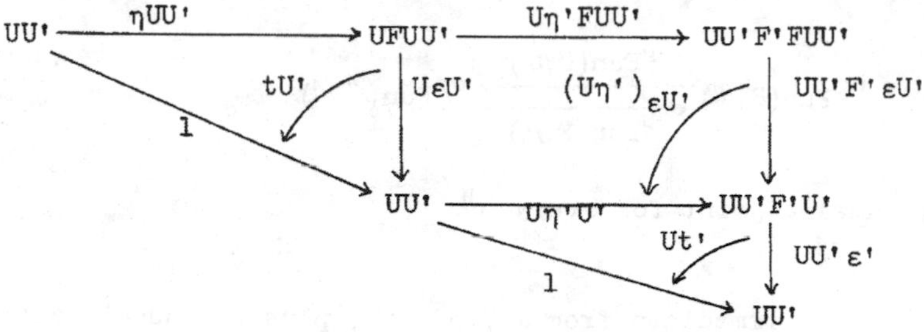

(7.5)

and

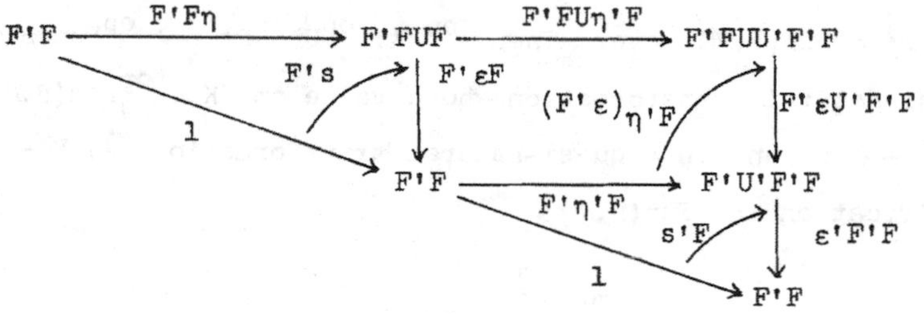

(7.5')

iii) This is immediate.

<u>Remark</u>: Parts i) and ii) clearly admit many special cases other than those specifically mentioned.

I,7.4. <u>Proposition</u>. Let $F\colon \mathcal{A} \longrightarrow \mathcal{B}$ and $U\colon \mathcal{B} \longrightarrow \mathcal{A}$ be x-weak y-quasi adjoint. Then

i) The pair of 2-functors

$$\mathrm{Fun}(\mathcal{M},\mathcal{A}) \underset{\mathrm{Fun}(\mathcal{M},U)}{\overset{\mathrm{Fun}(\mathcal{M},F)}{\rightleftarrows}} \mathrm{Fun}(\mathcal{M},\mathcal{B})$$

is x-weak y-quasi adjoint for every $\mathcal{M}$.

ii) The pair of 2-functors

$$^{op}\mathrm{Fun}(\mathcal{B},\mathcal{M}) \underset{^{op}\mathrm{Fun}(F,\mathcal{M})}{\overset{^{op}\mathrm{Fun}(U,\mathcal{M})}{\rightleftarrows}} {}^{op}\mathrm{Fun}(\mathcal{A},\mathcal{M})$$

is x-weak y-quasi adjoint for every $\mathcal{M}$.

<u>Proof</u>: Part i) is immediate from I.7.3, iii) plus the usual self-enrichment of the covariant hom-functor. To check part ii), let $(\varepsilon,\eta,s,t)\colon F \dashv U$ be the x-weak y-quasi adjunction from F to U and write $F^* = \mathrm{Fun}(F,\mathcal{M})$, etc. Then $^{op}\varepsilon^*\colon {}^{op}U^*\,{}^{op}F^* \longrightarrow {}^{op}\mathrm{Fun}(\mathcal{B},\mathcal{M})$ is the quasi-natural transformation whose value on $K \in {}^{op}\mathrm{Fun}(\mathcal{B},\mathcal{M})$ is $K\varepsilon\colon KFU \longrightarrow K$ and on a quasi-natural transformation $\psi\colon K \longrightarrow K'$ is the modification in $\mathrm{Fun}(\mathcal{B},\mathcal{M})$

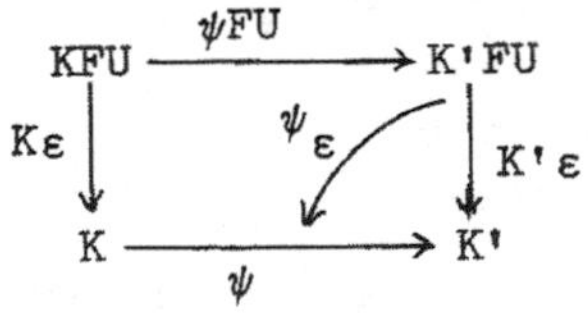

Thus in $^{op}\mathrm{Fun}(\mathcal{B},\mathcal{M})$, ψ_ε is a modification going in the proper direction. $^{op}\eta^*$ operates similarly. s^* is the modification in $\mathrm{Fun}(\mathcal{A},\mathcal{M})$ with components

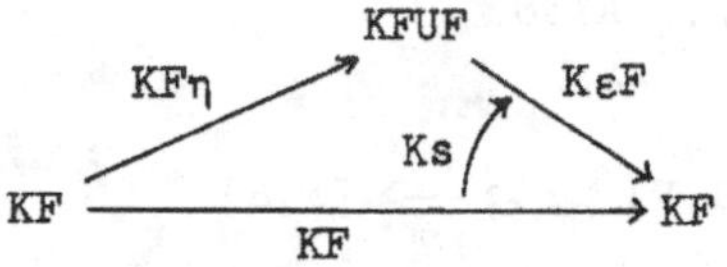

so in ${}^{op}\mathrm{Fun}(\mathcal{A},\mathcal{M})$, ${}^{op}s_*$ goes the other way. One checks easily then that

$$({}^{op}\varepsilon_*, {}^{op}\eta_*, {}^{op}t_*, {}^{op}s_*) : {}^{op}U_* \longrightarrow\!\mid {}^{op}F_*$$

is the desired x-weak y-quasi adjunction.

<u>Remark</u>: The occurrance of the weak dualization in part ii) has the effect that the covariant and contravariant instances of induced quasi-adjunctions cannot be combined into the study of quasi-adjoint squares, as is the case with ordinary adjunctions (cf., I.6.6 ff).

It would be nice to have a characterization of various types of quasi-adjointness in terms of properties of Cat-valued hom-functors. However, it is easily checked that, except in the case of i-weak i-quasi-adjunctions, studying hom-functors leads to a more complicated situation than that of quasi-adjunctions.

One of the purposes of the introduction of 2-comma categories is that they enable one to reduce the study of quasi-adjunctions to the study of ordinary adjunctions. In the following theorem, we have 2-functors and Cat-natural transformations over $\mathcal{A} \times \mathcal{B}$ (Cf., I.5.5)

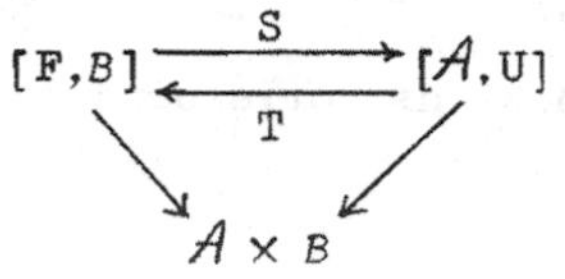

(7.6)

φ: ST —> id, ψ: id —> TS. Also.

$$\tilde{j}_F = (1.1,F)\, j_F\colon \mathcal{A} \longrightarrow [F,\mathcal{B}]$$
$$\tilde{j}_U = (U,1.1) j_U\colon \mathcal{B} \longrightarrow [\mathcal{A},U] \qquad (7.6')$$

(Cf., (5.3) and (5.7).) (Note that if φ and ψ are assumed to be either quasi-natural or natural then it follows that they are Cat-natural.)

It will always be assumed that S is a right U_*-homomorphism (i.e., opcleavage preserving) and T is a left F_*-homomorphism (i.e., cleavage preserving) so that

$$S = [\eta, U] \circ U_*$$
$$T = [F, \varepsilon] \circ F_*$$

for unique quasi-natural transformations

$$\eta\colon \mathcal{A} \longrightarrow UF \quad \text{and} \quad \varepsilon\colon FU \longrightarrow \mathcal{B}$$

(See I,5.10)

I,7.5 Theorem.

i) There is a bijection between four-tuples (S,T,ψ,φ) as above and weak quasi-adjunctions $(\varepsilon,\eta;s,t)$ between F and U.

ii) i-weak adjunctions correspond to four-tuples with $\varphi\tilde{j}_U$ and $\psi\tilde{j}_F$ isomorphisms.

iii) quasi-adjunctions correspond to four tuples with $\varphi\tilde{j}_U$ and $\psi\tilde{j}_F$ identities

iv) Strictness corresponds to a four-tuple in which φ and ψ define an ordinary (Cat-enriched) adjunction $S \dashv T$.

Remark: Thus strict quasi-adjunctions are equivalent to "homomorphic" adjoint functors $S \dashv T$ where the adjunction morphisms φ and ψ are over $\mathcal{A} \times \mathcal{B}$ and satisfy $\varphi\tilde{j}_U = id = \psi\tilde{j}_F$.

Proof: i) By assumption S and η determine each other as do T and ε. To determine the relation between ψ and s, let $(h\colon FA \longrightarrow \mathcal{B}) \in [F,\mathcal{B}]$. Then

$$TS(h) = \varepsilon_{\mathcal{B}}(FUh)(F\eta_A)\colon FA \longrightarrow \mathcal{B}.$$

A natural transformation ψ over $\mathcal{A} \times \mathcal{B}$ has components

$$\psi_h = (1,\overline{\psi}_h,1)\colon h \longrightarrow TS(h)$$

where $\overline{\psi}_h$ is a 2-cell from h to $TS(h)$. Define

$$s_A = \overline{\psi}_{FA} : FA \longrightarrow (\varepsilon F_A) \cdot (F\eta_A).$$

Computing ψ on the morphism

$$\begin{array}{ccc} FA & \xrightarrow{FA} & FA \\ {\scriptstyle FA}\downarrow & & \downarrow{\scriptstyle h} \\ FA & \xrightarrow[h]{} & \mathcal{B} \end{array}$$

in $[F,\mathcal{B}]$ shows that

$$\bar{\psi}_h = (\varepsilon_h(F\eta_A)) \cdot (hs_A) \tag{7.7}$$

One calculates directly that this formula describes a bijection between natural transformations ψ: id $\longrightarrow$ TS and modifications η: F $\longrightarrow$ $(\varepsilon F) \cdot (F\eta)$. Similarly, φ and ε determine each other.

ii) By construction $(\psi j_F)_A = (1, s_A, 1)$, and it is immediate that this is an isomorphism in $[F, \mathcal{B}]$ if and only if s_A is an isomorphic 2-cell.

iii) This follows as in ii).

iv) Let $(h: FA \longrightarrow \mathcal{B}) \in [F, \mathcal{B}]$ and consider the diagram

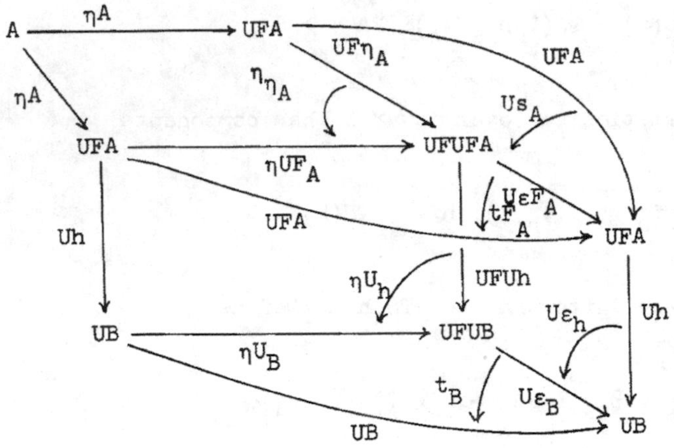

(7.8)

The clockwise and counterclockwise outer compositions from A to UB are

$$S(h) = (UB)(Uh)(\eta_A) = (Uh)(UFA)(\eta_A)$$

while in the middle there is the composition

$$STS(h) = (U\varepsilon_B)(UFUh)(UF\eta_A)(\eta_A) .$$

Furthermore

$$S\psi_h = [((U\varepsilon_h)(UF\eta_A)) \cdot ((Uh)(Us_A))]\eta_A$$
$$: S(h) \longrightarrow STS(h)$$

while

$$\varphi S_h = (t_B(Uh)\eta_A) \cdot ((U\varepsilon_B)(\eta U_h \boxplus \eta_{\eta_A})).$$

Thus, if the quasi-adjunction is strict. then since t is a modification, it follows from (7.1) that $\varphi S_h \cdot S\psi_h = \text{id}$. The other adjunction equation follows similarly from (7.1'). Conversely, if $S \dashv T$, then (7.1) follows from the particular case of the above large diagram in which $h = \text{id}: FA \longrightarrow FA$.

Part iv) of the preceeding theorem suggests the following definition.

I.7.6. <u>Definition</u>. A <u>transendental</u> <u>quasi-adjunction</u> between 2-functors $F: \mathcal{A} \longrightarrow \mathcal{B}$ and $U: \mathcal{B} \longrightarrow \mathcal{A}$ consists of a pair of 2-functors

$$[F,\mathcal{B}] \underset{T}{\overset{S}{\rightleftarrows}} [\mathcal{A},U]$$

over $\mathcal{A} \times \mathcal{B}$ such that $S \dashv T$ via Cat-natural transformations $\varphi: ST \longrightarrow \text{id}$ and $\psi: \text{id} \longrightarrow TS$ over $\mathcal{A} \times \mathcal{B}$.

The difference from the preceeding notions is that S and T need not be homomorphisms of any kind and hence need not be given by composition with appropriate quasi-natural transformations. However, as in I.5(5.33), S and T always determine quasi-natural transformations via

$$\overline{\varepsilon} = (\mathcal{B} \xrightarrow{j_U} [U,U] \xrightarrow{\overline{T}} [FU,\mathcal{B}])$$
$$\overline{\eta} = (\mathcal{A} \xrightarrow{j_F} [F,F] \xrightarrow{\overline{S}} [\mathcal{A},UF]) \quad .$$

One can ask when this ε and η are part of a suitable quasi-adjunction. Some of the results of I,7.5 hold in this situation.

I,7.7 Theorem.

i) Functors S and T and natural transformations φ and ψ (no conditions) determine a weak quasi-adjunction $(\varepsilon,\eta;s,t)$.

ii) If $\varphi\tilde{j}_U = id = \psi\tilde{j}_F$ and if S and T are partial homomorphisms (see proof) then (ε,η) is a quasi-adjunction.

iii) If $(S,T;\varphi,\psi)$ is a transendental quasi-adjunction then $(\varepsilon,\eta;s,t)$ is a strict weak quasi-adjunction.

Remark: Conditions ii) and iii) together thus describe when (ε,η) is a strict quasi-adjunction.

Proof. i) To define s and t, let

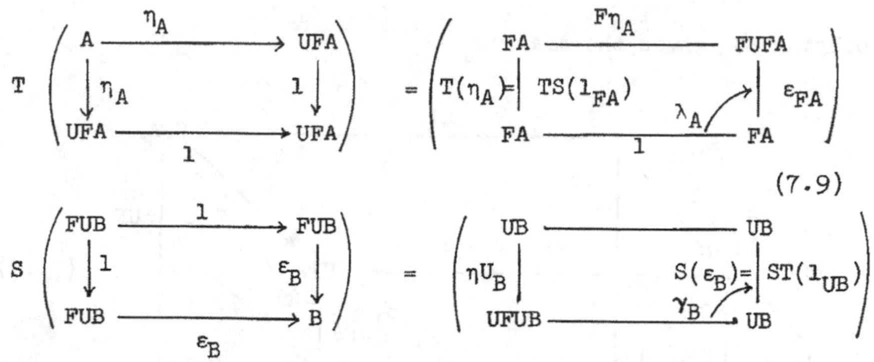

$$
T\begin{pmatrix} A \xrightarrow{\ \eta_A\ } UFA \\ \ \ \downarrow \eta_A \qquad\quad 1\downarrow \\ UFA \xrightarrow[\ 1\ ]{} UFA \end{pmatrix} = \begin{pmatrix} FA \xrightarrow{\ F\eta_A\ } FUFA \\ T(\eta_A)\!\!\mid \ TS(1_{FA}) \qquad\quad \Big| \ \varepsilon_{FA} \\ FA \xrightarrow[\ 1\ ]{} \overset{\lambda_A\nearrow}{} FA \end{pmatrix}
$$

$$(7.9)$$

$$
S\begin{pmatrix} FUB \xrightarrow{\ 1\ } FUB \\ \ \ \downarrow 1 \qquad\quad \varepsilon_B\downarrow \\ FUB \xrightarrow[\ \varepsilon_B\ ]{} B \end{pmatrix} = \begin{pmatrix} UB \xrightarrow{\qquad} UB \\ \eta U_B\Big| \qquad\quad S(\varepsilon_B)=\!\!\mid ST(1_{UB}) \\ UFUB \xrightarrow[\ \gamma_B\nearrow\ ]{} UB \end{pmatrix}
$$

Set $s_A = \lambda_A \cdot \psi_{1_{FA}}$ and $t_B = \varphi_{1_{UB}} \cdot \gamma_B$. Then $(\varepsilon, \eta; s, t)$ is a weak quasi-adjunction.

    ii) Consider the functors

$$\langle j_F, \overline{\varepsilon}\rangle\colon B \xrightarrow{\qquad\qquad} [F,F] \underset{B}{\times} [F,B]$$
$$\langle \eta, j_U\rangle\colon A \xrightarrow{\qquad\qquad} [A,U] \underset{A}{\times} [U,U]$$

The natural transformations of I,5.9 and I.5.10 which express the failure of S and T to be homomorphisms, when restricted to the images of these functors have values which are the λ_A's and γ_B's above. Thus we call S and T partial homomorphisms if all the λ_A's and γ_B's are identities. Part ii) is now immediate.

    iii) The situation here is a bit more complicated. Consider the diagram in $[F,B]$.

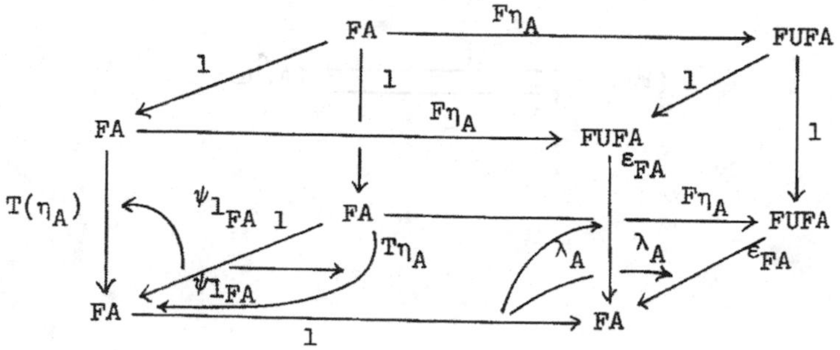

Applying S gives the diagram

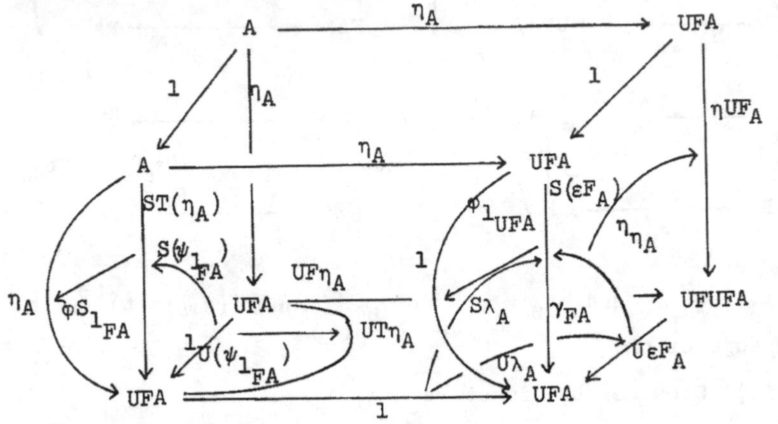

(7.10)

A careful examination of this yields the result. Note that the adjunction equations give $\varphi S_{1_{FA}} \cdot S\psi_{1_{FA}} = id$ while naturality of φ gives $(\varphi_{1_{UFA}}\eta_A) \cdot S(\lambda_A) = \varphi S_{1_{FA}}$. The bottom of the cube is $U(s_A)$ while the right side is t_{FA}. Finally, the cube commutes since it is S applied to a commutative cube.

I,7.8. <u>Universal mapping properties</u>

1) Let (S,T,φ,ψ) be a transendental quasi-adjunction between F and U. Then, for each pair of objects $A \in \mathcal{A}$ and $B \in \mathcal{B}$, one gets an ordinary adjunction

$$[FA,B] \underset{T_{A,B}}{\overset{S_{A,B}}{\rightleftarrows}} [A,UB] .$$

Note that $[FA,B] = \mathcal{B}(FA,B)$, etc., so that this gives ordinary adjunctions between the hom-categories (but not the hom-functors.) These adjunctions can be described by universal mapping properties as follows: given $(h\colon FA \longrightarrow B) \in [FA,B]$, then $\psi_h\colon h \longrightarrow TS(h)$ is universal not only in the usual sense in the category $[FA,B]$ but in the broader sense that given any morphism in $[F.\mathcal{B}]$

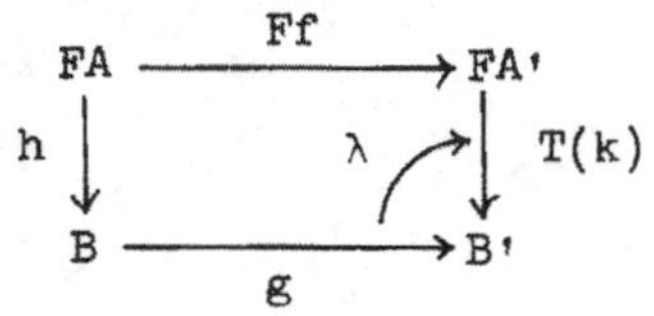

then there is a unique

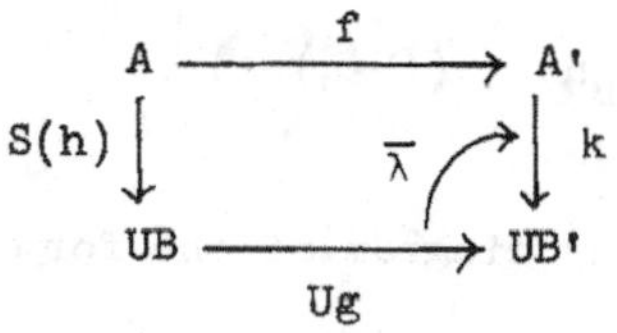

such that $T\bar{\lambda} \boxminus \psi_h = \lambda$. A similar universal property holds for 2-cells. This characterizes S in the usual sense that, given T and the values of S on objects satisfying this universal mapping property, then there is a unique way to extend S to a 2-functor such that $(S,T;\varphi,\psi)$ is a transendental quasi-adjunction.

2) Suppose now that S and T are homomorphic in the sense of I.7.5, so that they are given by composition with quasi-natural transformations $\eta\colon \mathcal{A} \longrightarrow UF$ and $\varepsilon\colon FU \longrightarrow \mathcal{B}$ respectively. The simple, familiar situation above translates into quite interesting properties of ε and η. For simplicity we assume that (ε,η) is a strict quasi-adjunction, and describe the universal property satisfied by η. Let $(h\colon A \longrightarrow UB) \in [\mathcal{A},U]$. Applying ST gives

$$(A \xrightarrow{\eta_A} UFA \xrightarrow{UFh} UFUB \xrightarrow{U\varepsilon_B} UB) \in [\mathcal{A}, U]$$

and the diagram on the left below, which we read as on the right

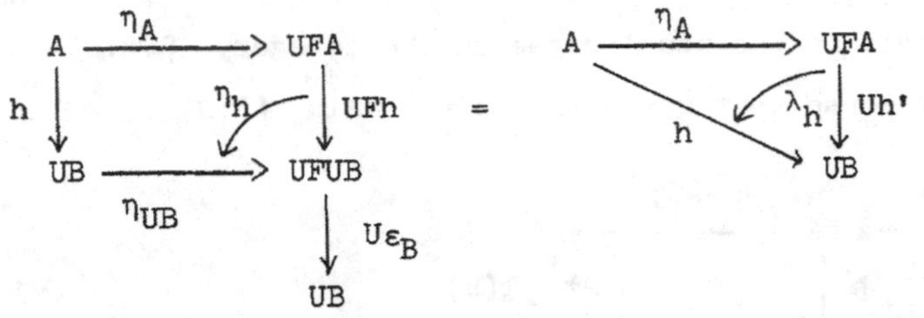

(7.11)

where $h' = \varepsilon_B(Fh)$ and $\lambda_h = (U\varepsilon_B)\eta_h$. Note that if this was only a weak quasi-adjunction, then one would take

$$\lambda_h = (t_B h) \cdot ((U\varepsilon_B)\eta_h) .$$

I.7.8.1 <u>Proposition</u>. Given any diagram of the form

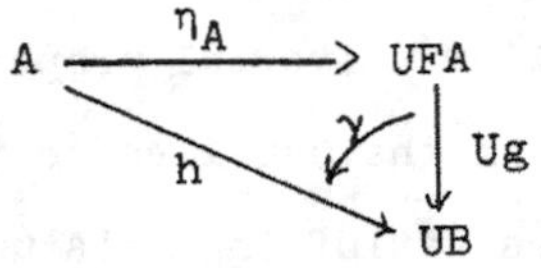

(7.12)

then there is a unique 2-cell $\tau: g \longrightarrow h'$ such that $\gamma = \lambda_h \cdot ((U\tau)\eta_A)$.

<u>Proof</u>: Consider the diagram

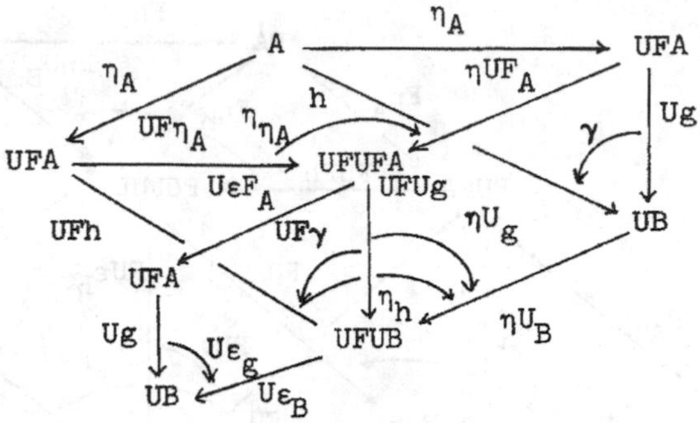

(7.13)

This commutes since η is quasi natural. The adjunction equations give

$$U\varepsilon_g \boxminus \eta U_g = 1$$

while strictness gives

$$(U\varepsilon F_A)\eta_{\eta_A} = 1 .$$

Hence, defining

$$\tau = \varepsilon_g \boxminus F\gamma\colon g \longrightarrow \varepsilon_B(UFh) = h'$$

gives $\gamma = \lambda_h \cdot ((U\tau)\eta_A)$.

To show that τ is unique, suppose $\nu\colon g \longrightarrow h'$ also satisfies this equation. Consider the diagram

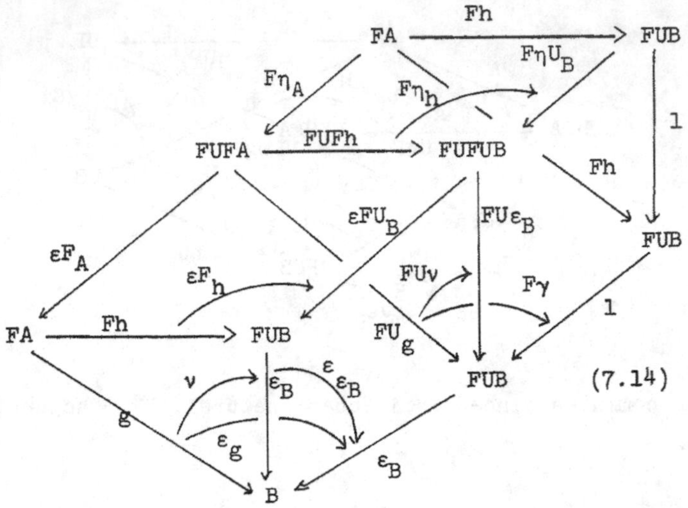

(7.14)

The prism in back commutes by hypothesis and the one in front by quasi-naturality of ε. Hence the other adjunction equation and the other strictness equation give $\nu = \varepsilon_g \boxminus F\gamma$. Note that the same result holds in the weak case; here one takes $\tau = [\varepsilon_g \boxminus F\gamma] \cdot (gs_A)$. Note also that the property satisfied by ε is dual in the sense of reversing both 1-cells and 2-cells.

3) Under appropriate hypotheses, such a universal mapping property gives rise to a quasi-adjunction, except that, in general, the thing constructed is a pseudo-functor rather than a functor. Suppose $U: \mathcal{B} \longrightarrow \mathcal{A}$ is a 2-functor and suppose that for each $A \in \mathcal{A}$ there is an object, denoted by FA, in $\mathcal{B}$ and a morphism $\eta_A: A \longrightarrow UFA$ such that given any $h: A \longrightarrow UB$ there is an $h': FA \longrightarrow B$ and a 2-cell $\lambda_h: (Uh')\eta_A \longrightarrow h$, as illustrated,

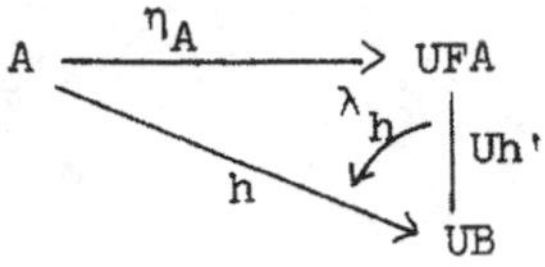

satisfying the universal property that given any other $g: FA \longrightarrow B$
and $\gamma: (Ug)\eta_A \longrightarrow h$, there is a unique 2-cell $\tau: g \longrightarrow h'$ with
$\gamma = \lambda_h \cdot ((U\tau)\eta_A)$. Define $\varepsilon_B: FUB \longrightarrow B$ as the h' corresponding
to $h = 1_{UB}: UB \longrightarrow UB$.

I,7.8.2 <u>Proposition</u>: If for all A and all $h: A \longrightarrow UB$, $h' = \varepsilon_B(Fh)$
(see below) then F extends to a pseudo-functor (I.3.2), η and ε
extend to quasi-natural transformations (I.3.3) and there exist
modifications s and t so that $(\varepsilon, \eta; s, t)$ is a strict weak quasi-
adjunction between F and U.

<u>Proof</u>: The hypothesis means that the correspondence between h and
h' is given by composition with ε in so far as is possible (see
after (7.15)). If $m: A \longrightarrow A'$, define F_m and η_m by the diagram

$$
\begin{array}{ccc}
A & \xrightarrow{\quad \eta_A \quad} & UFA \\
m \downarrow & \overset{\eta_m}{\nearrow} & \downarrow UFm \quad ; \\
A' & \xrightarrow{\quad \eta_{A'} \quad} & UFA'
\end{array}
$$

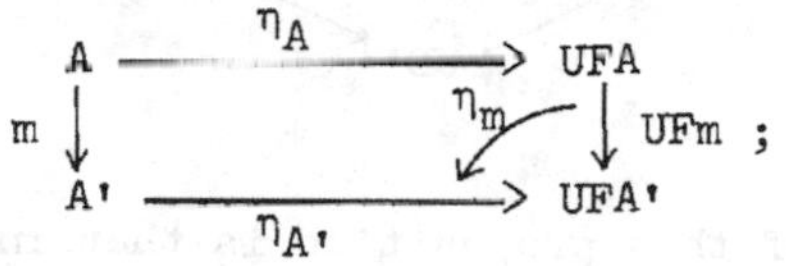

i.e., $F_m = (\eta_{A'}m)'$ and $\eta_m = \lambda_{\eta_A m}$. If $\mu: m \longrightarrow m'$ is a 2-cell,
then by the universal mapping property, m and $(\eta_{A'}\mu) \cdot \eta_m$
determine a unique 2-cell $F\mu: Fm \longrightarrow Fm'$. By uniqueness this gives
a functor $A(A,A') \longrightarrow B(FA,FA')$. Now suppose $n: A' \longrightarrow A''$. Then
$(Fn)(Fm)$ and $\eta_n \boxminus \eta_m$ determine a unique 2-cell

$$\varphi_{m.n}: F(n)F(m) \longrightarrow F(nm) .$$

Similarly, the identity map 1_{FA} and the identity 2-cell η_A
determine a unique 2-cell

$$\varphi_A: 1_{FA} \longrightarrow F(1_A) .$$

One verifies that with this structure, $F: A \longrightarrow B$ is a pseudo
functor and $\eta: A \longrightarrow UF$ is a quasi-natural transformation.

Define ε on objects and t by the diagram

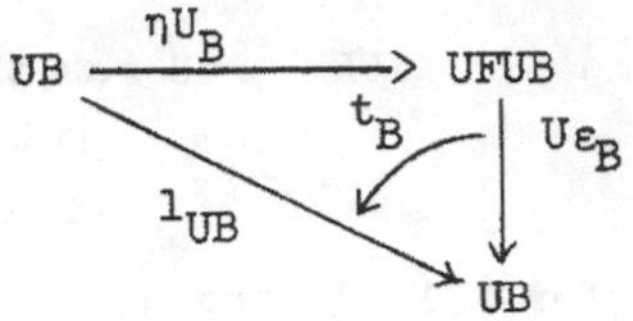

To define ε on morphisms, consider the diagram for $k: B \longrightarrow B'$

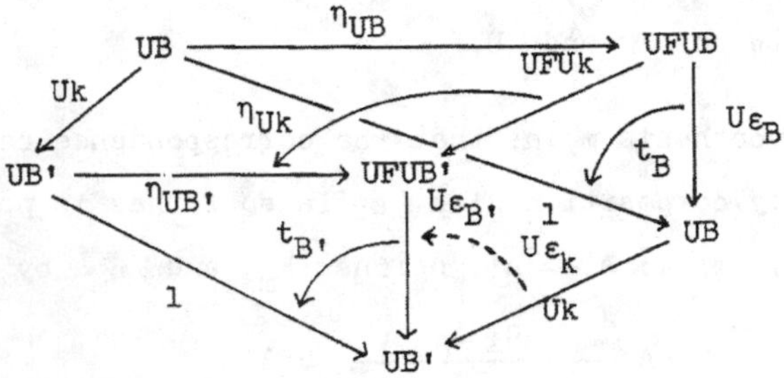

(7.15)

The precise hypothesis of this proposition is that in such a diagram not only is $\varepsilon_{B'}(FUk) = (Uk)'$ but also $\lambda_{Uk} = t_{B'} \boxminus \eta_{Uk}$. Then $k\varepsilon_B$ and kt_B determine a unique 2-cell $\varepsilon_k: k\varepsilon_B \longrightarrow \varepsilon_{B'}(FUk)$, which makes ε a quasi-natural transformation.

Finally, consider the diagram

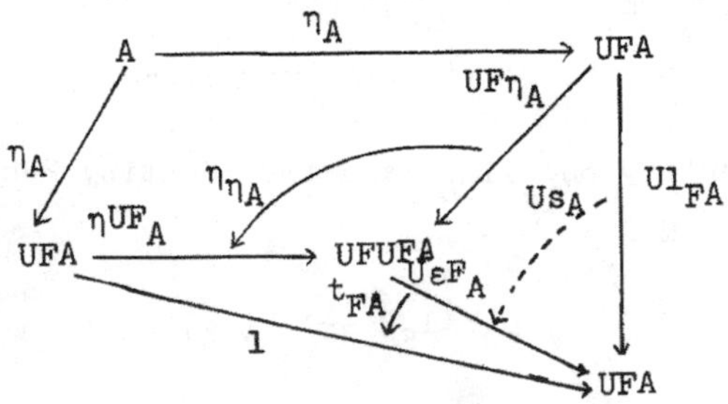

(7.16)

Again, $(\varepsilon F_A)(F\eta_A) = (\eta_A)'$ so 1_{FA} and the identity 2-cell η_A determine a unique 2-cell $s_A\colon 1_{FA} \longrightarrow (\varepsilon F_A)(F\eta_A)$ which satisfies half of strictness automatically and can be shown to be a modification. The other half of strictness follows by putting together the diagrams defining $\varepsilon_{\varepsilon_B}$ and sU_B and taking account of the definition of Ft_B. Note that in this case, (7.1') is to be modified so that the composed 2-cell goes from $\varepsilon_B(1_{FUB})$ to $\varepsilon_B F(1_{UB})$ and is to equal $\varepsilon_B\, \varphi_{UB}$.

Examples.

I,7.9. Some general principles. There are many special kinds of quasi-adjoints and many special situations arise. Some of this bewildering variety is accounted for by the existence of the 2-categories of the form $\mathrm{Fun}(\mathcal{B},\mathcal{B}'_0;\mathcal{A},\mathcal{A}')$ described in I.2.4. There are essentially three possibilities which must be considered.

i) One or both of the adjunction quasi-natural transformations $\varepsilon\colon FU \longrightarrow \mathcal{B}$ and $\eta\colon \mathcal{A} \longrightarrow UF$ may belong to a 2-category of this form.

ii) In situations where $\mathcal{A}$ or $\mathcal{B}$ or both should be of the form $\mathrm{Fun}(\mathfrak{K},\mathcal{M})$ (e.g., Kan extensions or limits), the appropriate 2-categories may actually be of the form $\mathrm{Fun}(\mathfrak{K}.\mathfrak{K}'_0;\mathcal{M}.\mathcal{M}')$.

iii) Cases i) and ii) can be combined so that there are adjunctions of type i) between categories of type ii).

In particular, we shall adopt the following terminology for limits.

I,7.9.1 Definition.

i) Quasi-adjoints to the constant imbedding $\Delta: \mathcal{A} \longrightarrow \mathrm{Fun}(\mathfrak{K},\mathcal{A})$, where $\mathcal{A}$ is a 2-category and $\mathfrak{K}$ is a small 2-category are called quasi-limits and quasi-colimits; written

$$q\text{-}\underrightarrow{\lim}_{\mathfrak{K}} \;\overline{\text{quasi}}\!\!\dashv \Delta \;\overline{\text{quasi}}\!\!\dashv\; q\text{-}\underleftarrow{\lim}_{\mathfrak{K}}$$

Adjectives from I,7.1 will be added as appropriate.

ii) Cat-adjoints (type i), via I,2.4i)) to a constant imbedding as above are called Cartesian quasi-limits and Cartesian quasi-colimits, written

$$\text{Cart } q\text{-}\underrightarrow{\lim}_{\mathfrak{K}} \;\underset{\text{Cat}}{\longrightarrow\!\!\!\dashv}\; \Delta \;\underset{\text{Cat}}{\longrightarrow\!\!\!\dashv}\; \text{Cart } q\text{-}\underleftarrow{\lim}_{\mathfrak{K}} .$$

If there is only an ordinary adjunction at the level of the underlying categories, we write Cart $q_o\text{-}\underrightarrow{\lim}_{\mathfrak{K}}$ and Cart $q_o\text{-}\underleftarrow{\lim}_{\mathfrak{K}}$. (The reason for the word Cartesian is that for Cat-adjoints the functors S and T in I,7.5 are 2-sided homomorphisms, i.e., cleavage and op-cleavage preserving.)

iii) Still more specially, Cat-adjoints to constant imbeddings

$$\mathcal{A} \longrightarrow \mathrm{Fun}(\mathfrak{K}.\mathfrak{K}_o^{!};\, \mathcal{A},\mathcal{A}_o)$$
$$\mathcal{A} \longrightarrow \mathrm{Fun}(\mathfrak{K},\mathfrak{K}_o^{!};\mathcal{A},\mathrm{iso}\,\mathcal{A})$$

(see I,2.4) are written Cart $q\text{-}\underrightarrow{\lim}_{\mathfrak{K}\text{-id }\mathfrak{K}_o^{!}}$; and Cart $q\text{-}\underrightarrow{\lim}_{\mathfrak{K}\text{-iso }\mathfrak{K}_o^{!}}$, respectively; and similarly for inverse limits. Note that the ordinary Cat-enriched colimit is the same as Cart $q\text{-}\underrightarrow{\lim}_{\mathfrak{K}\text{-id}\mathfrak{K}_o}$, while Cart $q\text{-}\underrightarrow{\lim}_{\mathfrak{K}\text{-iso }\mathfrak{K}_o}$ is the Cat-adjoint to

$$\mathcal{A} \longrightarrow \text{Iso - Fun}(\mathbb{R},\mathcal{A})$$

(See I,4.24.)

As with ordinary limits and colimits, cartesian quasi-ones can be defined in a more global fashion as Cat-adjoints. We treat the colimit case. (Cf. I,1.13). Let X be a 2-category and let

$$N: X \longrightarrow {}_s[2\text{-Cat}_{\otimes}, X]_{\otimes} \quad \text{(See I,4.25)}$$

(the s means small 2-categories over X) be the "name" 2-functor taking $X \in X$ to $\ulcorner X \urcorner : \underline{1} \longrightarrow X$, $f: X \longrightarrow Y$ to the 1-cell

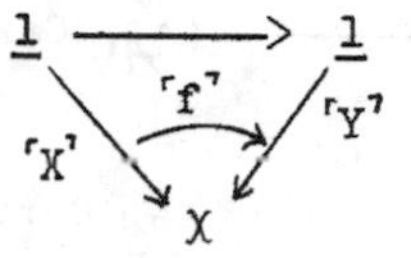

and a 2-cell $\varphi: f \longrightarrow g$ to the 2-cell

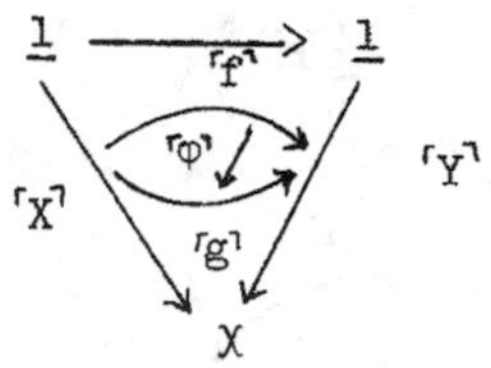

in ${}_s[2\text{-Cat}_{\otimes}, X]_{\otimes}$.

I.7.9.2 <u>Theorem</u>. Let X have small cartesian quasi-colimits. Then

$$\text{Cart q - } \underrightarrow{\lim} : {}_s[2\text{-Cat}_{\otimes}, X]_{\otimes} \longrightarrow X$$

is an enriched functor which is the left 2-$\text{Cat}_{\otimes}$-adjoint to N.

<u>Proof</u>: The main thing is to show that Cart q - $\underrightarrow{\lim}$ is defined here. Its value on an object $F: \mathcal{A} \longrightarrow X$ is, of course, $Q(F) = \text{Cart q} - \underrightarrow{\lim}_{\mathcal{A}} F$. If

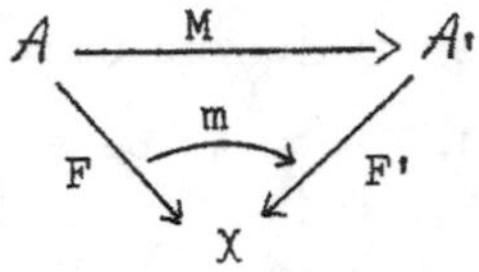

is a morphism (where m is quasi-natural), then the diagram

$$\begin{array}{ccc} F & \xrightarrow{\ \eta_A\ } & \Delta_{\mathcal{A}}\, \text{Cart q} - \underrightarrow{\lim}_{\mathcal{A}} F \\ {\scriptstyle m}\downarrow & & \vdots\downarrow \\ F'M & \xrightarrow{\ \eta_{F'}M\ } & (\Delta_{\mathcal{A}'}\, \text{Cart q} - \underrightarrow{\lim}_{\mathcal{A}'} F') \circ M \\ & & \| \\ & & \Delta_{\mathcal{A}}\, \text{Cart q} - \underrightarrow{\lim}_{\mathcal{A}'} F' \end{array}$$

shows that there is a (unique) induced map

$$Q(M,m): \text{Cart q} - \underrightarrow{\lim}_{\mathcal{A}} F \longrightarrow \text{Cart q} - \underrightarrow{\lim}_{\mathcal{A}'} F' .$$

Similarly, if

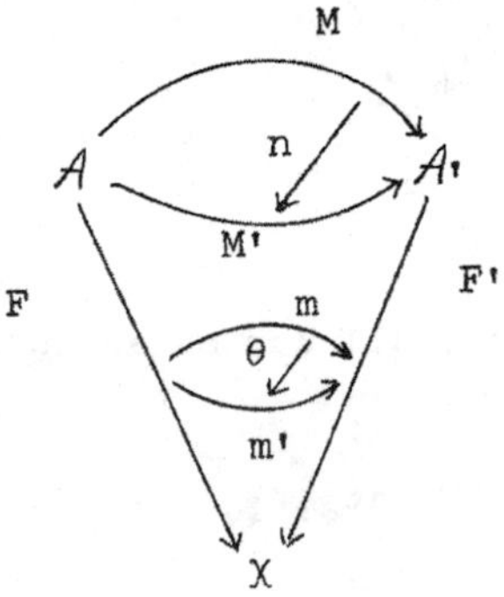

is a 2-cell, then the modification

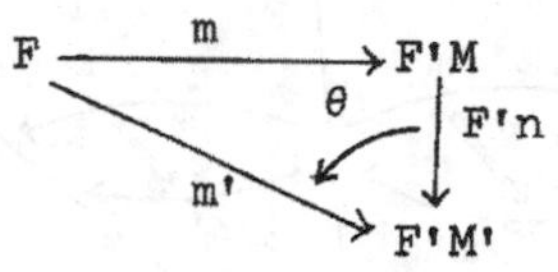

induces a 2-cell

$$Q(n,\theta):Q(M,m) \longrightarrow Q(M',m')$$

2-Cat$_\otimes$-adjointness is immediate, since the universal mapping property satisfied by Cart q - $\underrightarrow{\lim}$ F is easily translated into a functorial isomorphism between the illustrated categories

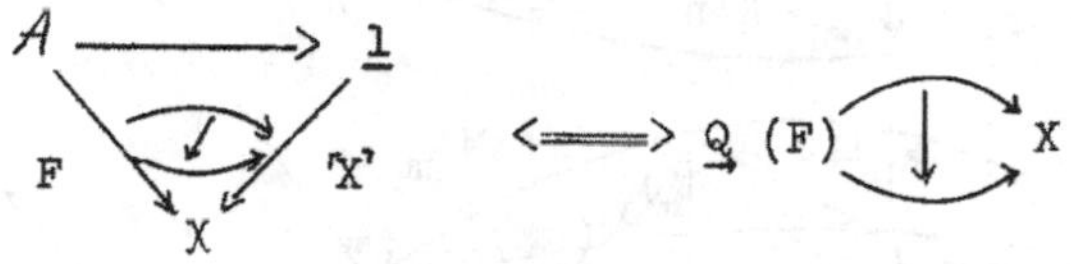

The property which is not immediate is that the above construction yields an enriched functor. The proof is a very complicated calculation and the reader may want to skip ahead to the examples. First, we must describe the composition in ${}_s[2\text{-Cat}_\otimes, X]_\otimes$. The composition of a pair of 2-cells

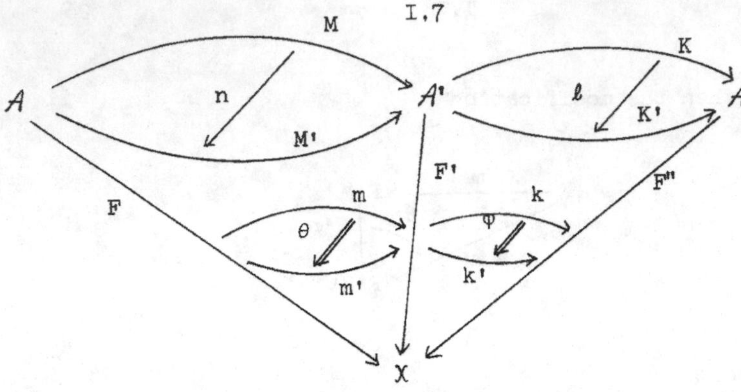

(7.17)

(because we are dealing with a 2-Cat$_\otimes$-category) is a diagram

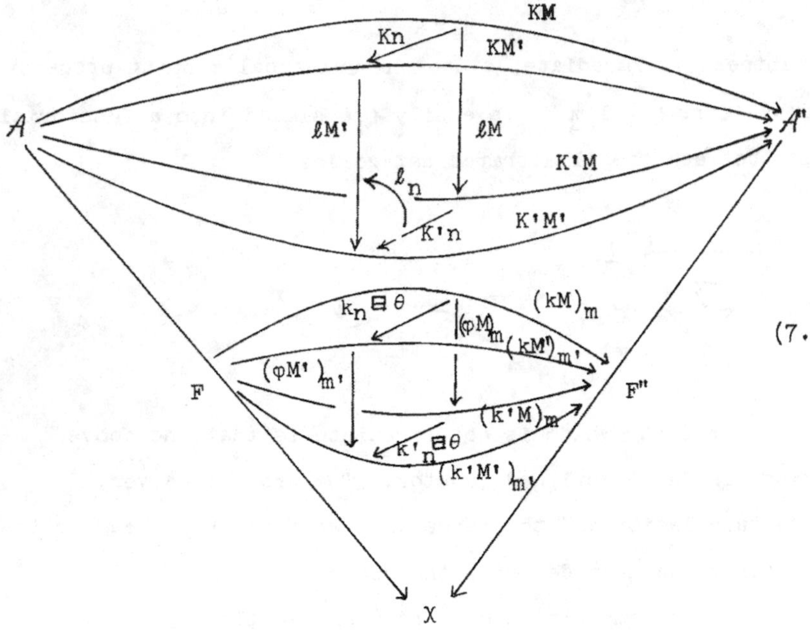

(7.18)

The verification that this is a 3-cell (i.e., that equation 2.19 is satisfied) follows from the diagram

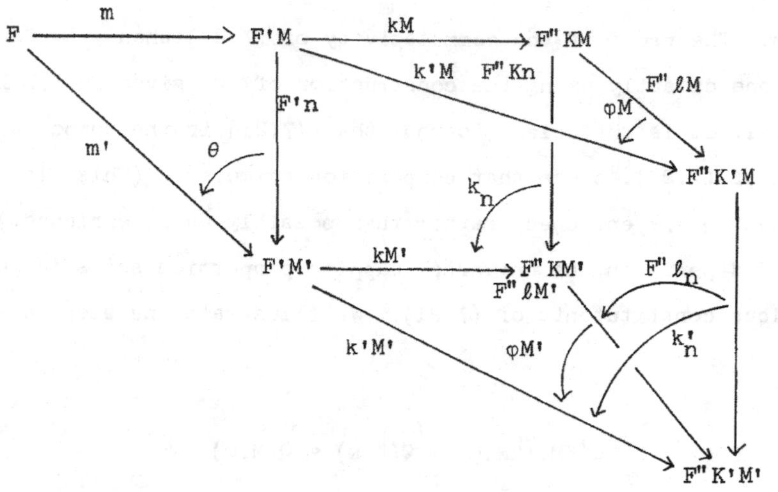

(7.19)

which commutes since φ is a modification. Now an enriched functor from a 2-$Cat_{\otimes}$-category to a 2-category regarded as a "locally, locally discrete" 2-Cat-category and then as a 2-$Cat_{\otimes}$-category must turn (7.18) into a commutative square (since there are no 3-cells). Thus, applying $Q(-) = \text{Cart q-}\underrightarrow{\lim}$ to (7.17) gives

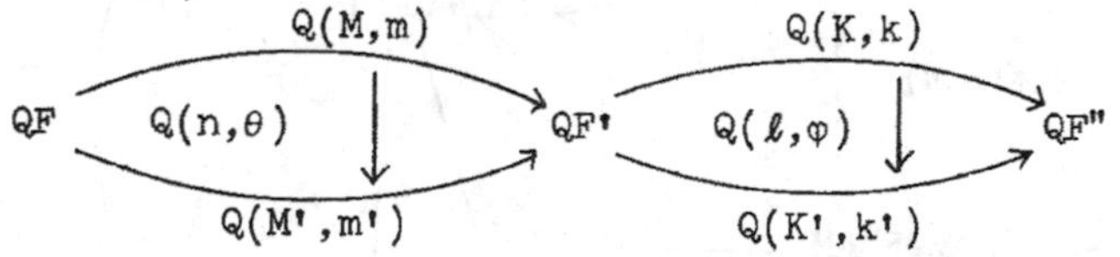

(7.20)

and one must show that the diagram

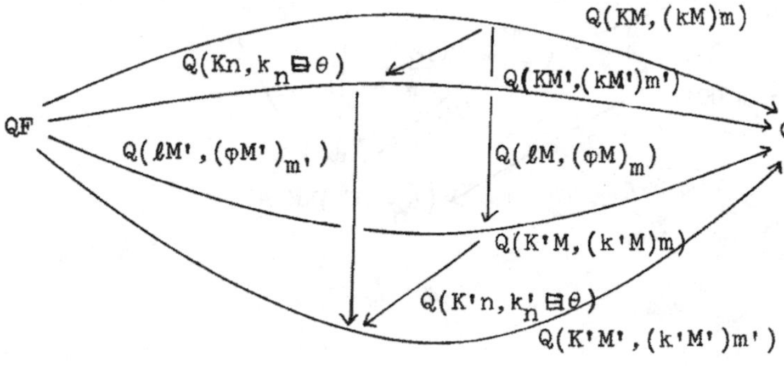

(7.21)

commutes. The proof of the commutativity of (7.21) when χ = Cat can be done directly using the construction of Q given in I,7.11. In general, it is sufficient to show that (7.21) is the composition (in χ) of (7.20), since that composition commutes. (This also shows that Q is enriched, rather than possibly quasi-enriched.) The proof depends on the universal mapping properties satisfied by the various constitutents of (7.21). We illustrate the step showing that

$$Q(KM,(kM)m) = Q(K,k) \circ Q(M,m)$$

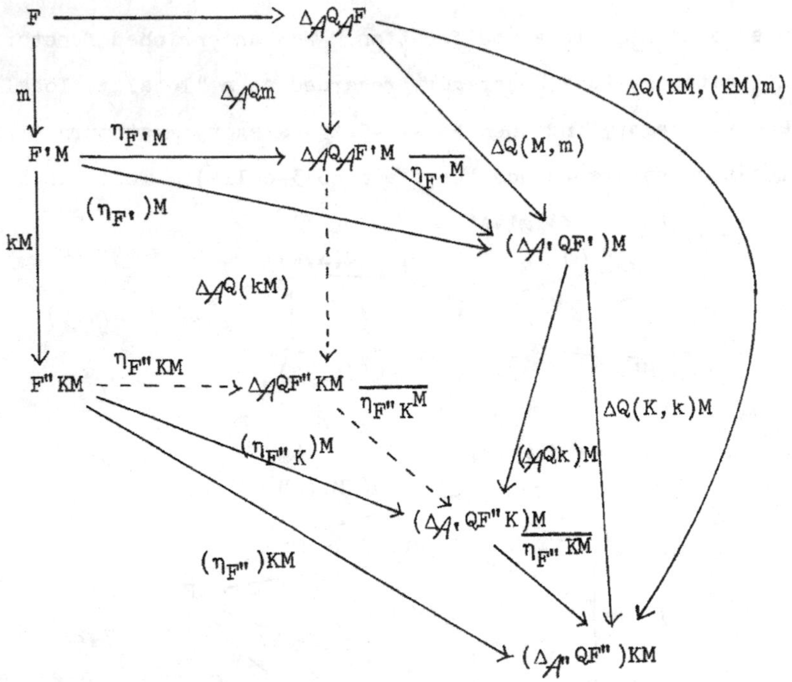

(7.22)

Here a $(\overline{})$ denotes an induced map. Since

$$(\Delta_{\mathcal{A}'}QF')M = \Delta_{\mathcal{A}}QF' \quad , \quad \text{etc.,}$$

this gives the desired result. The diagram to show that

$$[\Delta_{\mathcal{A}'}Q(K.k)M'] \circ [\Delta_{\mathcal{A}}Q(n,\theta)] = \Delta_{\mathcal{A}}Q(Kn.k_n \boxminus \theta)$$

consists of building two interconnected copies of (7.22) on the diagram

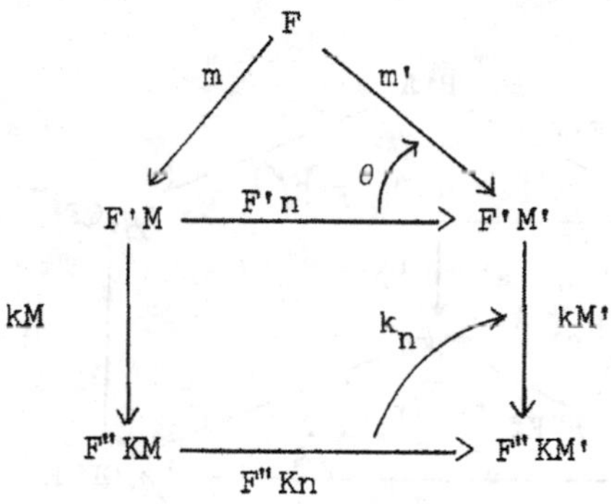

It is sufficient to show that the desired equation holds after composition with η_F. It follows immediately from this big diagram (which the reader must imagine or draw for himself) that

$$\text{i)}\quad [\Delta_{\mathcal{A}'} Q(K,k)M'] \circ [\Delta_{\mathcal{A}} Q(n,\theta)]\eta_F = [\Delta_{\mathcal{A}'} Q(K.k)M'][(\eta_{F'})_n \boxminus \theta]$$
$$= [\overline{\eta_{F''}KM'} \circ (\Delta_{\mathcal{A}} Qk)M][(\eta_{F'})_n \boxminus \theta]$$

while

ii) $\Delta_{\mathcal{A}} Q(Kn, k_n \boxminus \theta)\eta_F = (\eta_{F''})_{Kn} \boxminus k_n \boxminus \theta$.

But now for any $f: A \longrightarrow B$ (in particular, for any $n_A: MA \longrightarrow MA'$), the diagram

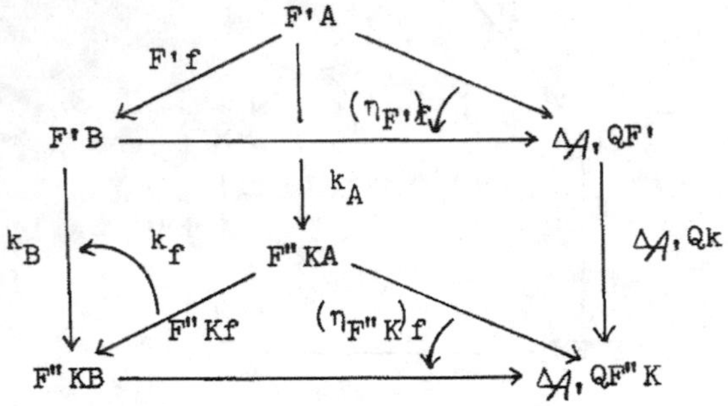

commutes, and hence

$$[(\Delta_{\mathcal{A}'} Qk)M'](\eta_{F'})_n = (\eta_{F''K})_n \boxminus k_n \quad .$$

Similarly,

$$(\overline{\eta_{F''}KM'})\ ((\eta_{F''K})_n) = (\eta_{F''})_{Kn}$$

These two equations show that i) equals ii).

An analogous calculation based on the diagram

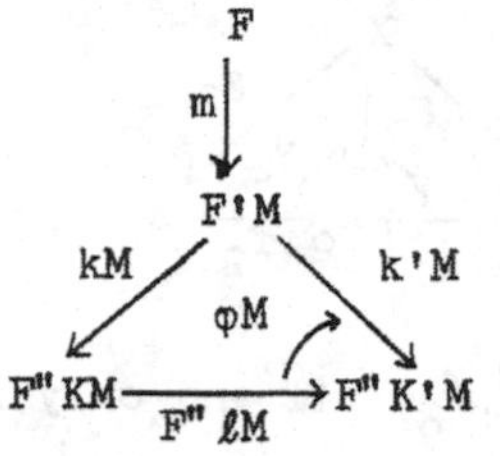

shows that

$$\Delta_{\mathcal{A}} Q(\ell M, (\varphi M)_m) = [\Delta_{\mathcal{A}'} Q(\ell,\varphi)M] \circ [\Delta_{\mathcal{A}} Q(M,m)]$$

which completes the proof.

I,7.10 <u>Some finite quasi-limits</u>.

1) Comma categories are characterized by a diagram

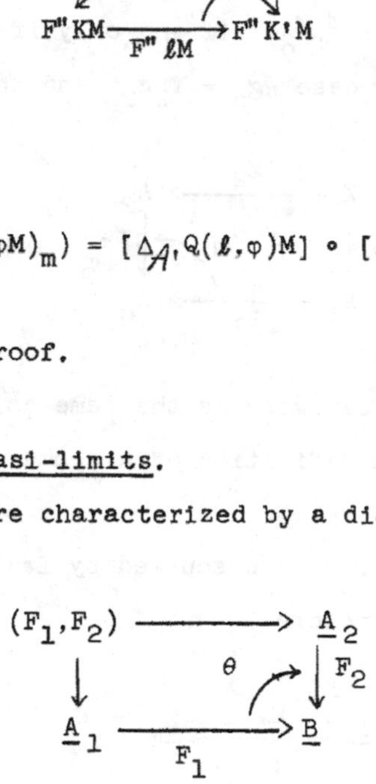

(7.23)

where θ is a natural transformation, satisfying the same universal property as that of I.5.2. This is a Cartesian quasi-limit as follows: let $\underline{P}$ be the category illustrated by

$$1 \xrightarrow{j_1} 0 \xleftarrow{j_2} 2$$

and $\underline{P}'_0$ the subcategory $0 \longleftarrow 2$. A quasi-natural transformation from a constant functor to an arbitrary 2-functor from $\underline{P}$ to $\mathcal{A}$

looks like

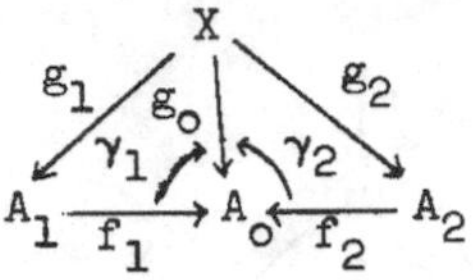

This belongs to $\text{Fun}(\underline{P}, \underline{P}'_o; \mathcal{A}, \mathcal{A}_o)$ if and only if γ_2 is an identity 2-cell, in which case $g_o = f_2 g_2$ and the diagram reduces to

$$
\begin{array}{ccc}
X & \xrightarrow{\;g_2\;} & A_2 \\
{\scriptstyle g_1}\big\downarrow & \gamma_1 \nearrow & \big\downarrow {\scriptstyle f_2} \\
A_1 & \xrightarrow{\;f_1\;} & A_o
\end{array}
$$

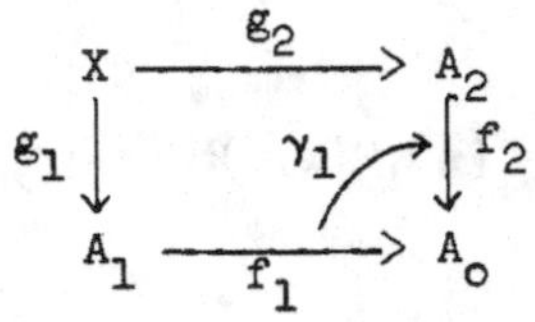

We conclude that a comma category is the same as a Cart $q\text{-}\lim_{\underrightarrow{P}\text{-}id\underline{P}_o}$ in Cat. This serves as a definition of a comma object in an arbitrary 2-category.

      2) Subequalizers are discussed by Lambek in [27]. Here, one is given a pair of functors

$$
\underline{A} \underset{G}{\overset{F}{\rightrightarrows}} \underline{B}
$$

and one looks for a best possible $M: \underline{M} \longrightarrow \underline{A}$ together with a natural transformation $\theta: GM \longrightarrow FM$. If $\underline{E}$ denotes the category

$$
0 \underset{t}{\overset{s}{\rightrightarrows}} 1
$$

and $\underline{E}'_o$ the subcategory $0 \xrightarrow{\;s\;} 1$, then one concludes as above that a subequalizer is the same as a Cart $q\text{-}\lim_{\underrightarrow{\underline{E}}\text{-}id\,\underline{E}_o}$ in Cat. (Cf., I,7.12.3).

3) Products and coproducts, since they are taken over discrete categories, do not involve 2-cells, so Cartesian quasi-products and coproducts coincide with ordinary ones; i.e., if $\mathcal{R}$ is discrete, then $\mathrm{Fun}(\mathcal{R},\mathcal{A}) = \mathcal{A}^{\mathcal{R}}$. On the other hand, strict (weak) quasi-products and coproducts correspond to interesting universal mapping properties. (Cf. I,7.8 (3).) For instance, in the case of coproducts, given $A_1, A_2 \in \mathcal{A}$, then $A_1 \underset{q}{+} A_2$ should satisfy the property that there are maps $i_j\colon A_j \longrightarrow A_1 \underset{q}{+} A_2$, $j = 1,2$, such that given maps $h_j\colon A_j \longrightarrow X$, then there is a map $h\colon A_1 \underset{q}{+} A_2 \longrightarrow X$ and 2-cells λ_j as illustrated

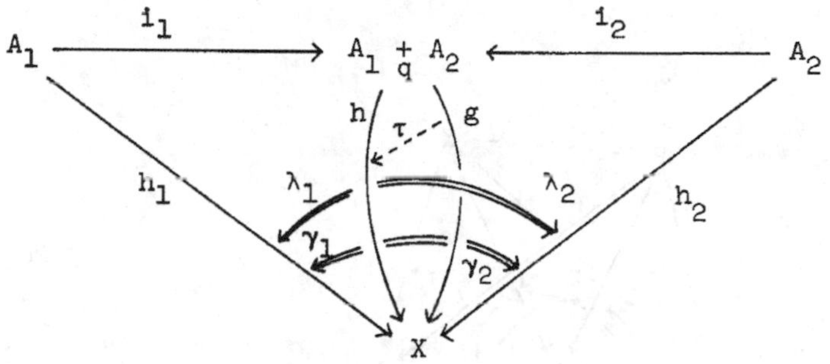

(7.24)

with the property that given any other $g\colon A_1 \underset{q}{+} A_2 \longrightarrow X$ and 2-cells γ_j as indicated, then there is a unique 2-cell $\tau\colon g \longrightarrow h$ with

$$\lambda_j \cdot (\tau i_j) = \gamma_j .$$

Quasi-products satisfy the dual situation in which the 1-cells are reversed. (One can also reverse 2-cells.)

For instance, in the 2-category whose objects are sets, whose 1-cells are relations, and whose 2-cells are inclusions of relations, the ordinary product of sets with its projections becomes such a quasi-product. More generally, in Puppe, Korrespondenzen in abelschen Kategorien, Math. Ann. 148 (1962), p. 1-30, a product is, by definition, such a quasi-product.

4) If $\underline{P}$ is the pullback diagram of part 1) then one can consider quasi-adjoints with domain either $\mathcal{A}^{\underline{P}}$ or Fun($\underline{P}$.$\mathcal{A}$) giving diagrams like

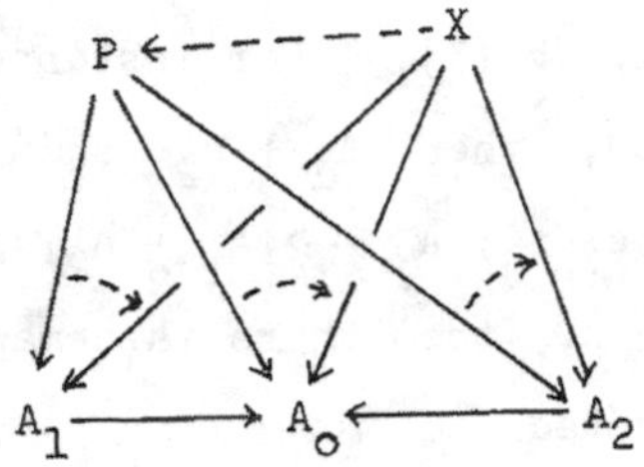

(7.25)

or

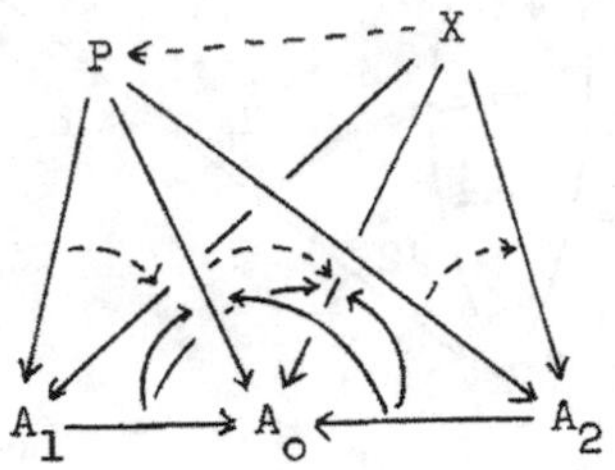

(7.26)

One finds examples in the categories of sets and relations or sets and partial functions.

5) A certain pecularity of the choice of "up" quasi-natural transformations is illustrated by the observation that a cartesian quasi-initial object $0_q \varepsilon \mathcal{A}$ is one such that $\mathcal{A}(0_q.A)$ has a terminal object for all A.

I,7.11 Quasi-colimits in Cat.

Let $F: \mathcal{A} \longrightarrow \text{Cat}$ be a 2-functor from a 2-category $\mathcal{A}$ into the 2-category of small categories. Let $1: \underline{1} \longrightarrow \text{Cat}$ be the functor whose value is the object $\underline{1} \in \text{Cat}$.

I,7.11.1 Proposition. Cart $\text{q-}\varinjlim_{\mathcal{A}} F = L\pi_0[1.F]$. (See I.7.9.1, I,2.3, and I,2.5.)

Remark. If $\mathcal{A}$ is locally discrete (i.e., an ordinary category) then $L\pi_0[1,F] = [1,F]$. It was shown in [CCS] that in this case $[1,F] = \underline{E}_F$ is the opfibred category over $\mathcal{A}$ corresponding to F. In §8 of [CCS] it was shown that then $[1.F] = \text{q} - \varinjlim_{\mathcal{A}} F$. The proof there can be extended to show that the relevant isomorphism between 2-comma categories preserves cocartesian morphisms, and hence $[1,F] = \text{Cart q} - \varinjlim_{\mathcal{A}} F$. However, it was asserted that, in general for a non-locally discrete $\mathcal{A}$, $\text{q} - \varinjlim_{\mathcal{A}} F = |[1.F]|$. This may or may not be so; the difficulty will become apparent in the proof of the proposition here.

Proof: Recall that [1.F] is the 2-category whose objects are pairs (A,a) where $a \in FA$. $A \in \mathcal{A}$, whose morphsism are pairs $(f,\varphi): (A.a) \longrightarrow (B,b)$ where $f: A \longrightarrow B$ in $\mathcal{A}$ and $\varphi: F(f)\, a \longrightarrow b$ in FB, and whose 2-cells are of the form $\overline{\tau}: (f,\varphi) \longrightarrow (f',\varphi')$ where $\tau: f \longrightarrow f'$ is a 2-cell in $\mathcal{A}$ such that

$$\varphi'((F\tau)_a) = \varphi.$$

Composition of 1-cells is given by the formula

$$(g.\psi)(f,\varphi) = (gf.\psi((Fg)\varphi)). \tag{7.27}$$

This 2-category is the 2-opfibration (I,2.9) with locally-discrete fibres associated to F. $L\pi_o[1,F]$ is the category constructed from $[1,F]$ by making all the 2-cells $\overline{\tau}$ identities. Let $Q_F\colon [1,F] \longrightarrow L\pi_o[1,F]$ denote the canonical projection. Since $[1,F]$ is the opfibration associated to F, there are inclusion 2-functors $J_A\colon FA \longrightarrow [1,F]$ of the fibres given by $J_A(a) = (A,a)$ and $J_A(\varphi) = (1_A,\varphi)$. Similarly there are natural transformations $\theta_f\colon J_A \longrightarrow J_B(Ff)$ corresponding to $f\colon A \longrightarrow B$, whose components are the cocartesian morphisms

$$(\theta_f)_a = (f,\ 1_{F(f)a}).$$

Define $\tilde{\eta}$ to be the transformation from the 2-functor F to the constant 2-functor $[1,F]$ whose components are the J_A's and θ_f's. This is not quasi-natural since, given a 2-cell $\overline{\tau}$, the diagram

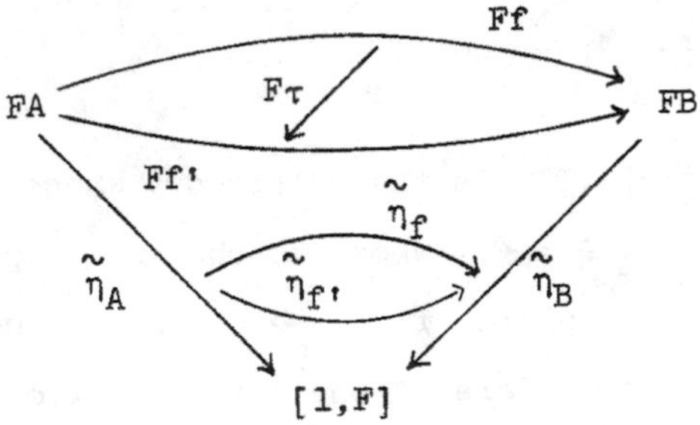

(7.28)

does not commute. One has

$$[(\tilde{\eta}_B\ F\tau)\cdot\tilde{\eta}_f]_a = (f,(F\tau)_a)$$

while

$$(\tilde{\eta}_{f'})_a = (f', 1_{F(f')a}) .$$

However $\bar{\tau}$ is a 2-cell from the first to the second since $\tau\colon f \to f'$ and

$$1_{F(f')a}(F\tau)_a = (F\tau)_a .$$

(Undoubtedly, this is a 3-dimensional quasi-natural transformation, but we are not concerned with such things here.) If $\mathcal{A}$ is locally discrete, then this is quasi-natural; otherwise, define

$$\eta_F = Q_F\tilde{\eta}\colon F \longrightarrow \Delta(L\pi_o[1,F]) .$$

Then η_F is quasi-natural. It is easily checked that η_F is Cat-natural in F, so that one has a Cat-natural transformation

$$\eta\colon \mathrm{Fun}(\mathcal{A},\mathrm{Cat}) \longrightarrow \Delta(L\pi_o[1,-]).$$

It is sufficient to show that this satisfies the universal mapping property making $L\pi_o[1,-]$ the left Cat-adjoint to Δ.

Let $\underline{X} \in$ Cat and let $\varphi\colon F \longrightarrow \Delta\underline{X}$ be a quasi-natural transformation. If $f\colon A \longrightarrow B$ in $\mathcal{A}$, then one has a diagram

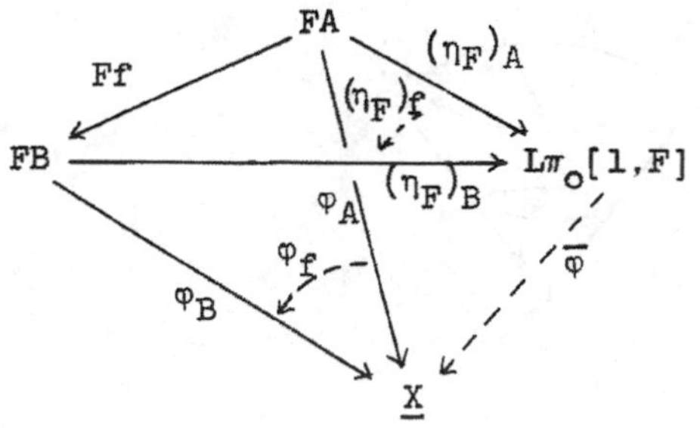

(7.29)

and we must show that there is a unique functor $\bar{\varphi}: L\pi_0[1,F] \longrightarrow \underline{X}$ making all such diagrams commute. Define $\tilde{\varphi}: [1,F] \longrightarrow \underline{X}$ by setting $\tilde{\varphi} \circ J_A = \varphi_A$ (i.e., on the fibre FA, $\tilde{\varphi} = \varphi_A$) and on cocartesian morphisms,

$$\tilde{\varphi}(f, 1_{F(f)a}) = (\varphi_f)_a \qquad (7.30)$$

This determines $\tilde{\varphi}$ uniquely since any morphism in $[1,F]$ has a canonical decomposition

$$(f,\varphi) = (1_B,\varphi)(f,1_{F(f)a}). \qquad (7.31)$$

Since φ is quasi-natural, the diagram corresponding to (7.28) for φ does commute and hence $\tilde{\varphi}$ determines a functor $\bar{\varphi}$ which is obviously unique. (Note that since $\tilde{\eta}$ above is not quasi-natural, $|[1,F]|$ can at best be a transendental quasi-limit. This has not been investigated.)

I,7.11.2 Corollary. The canonical projection $P: [\text{Cat}, \ulcorner\underline{B}\urcorner] \longrightarrow \text{Cat}$ creates cartesian quasi-colimits.

Proof: Here $\ulcorner\underline{B}\urcorner : \underline{1} \longrightarrow \text{Cat}$ is the name of an object $\underline{B} \in \text{Cat}$, $[\text{Cat}, \ulcorner\underline{B}\urcorner]$ has 2-cells that look like

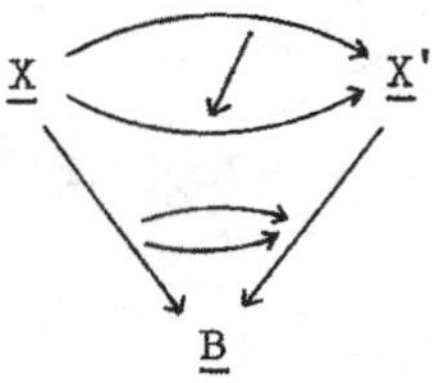

and P reads the top line. An object $\mathfrak{F} \in \mathrm{Fun}(\mathcal{A},[\mathrm{Cat}, \ulcorner\underline{B}\urcorner])$ is the same as a quasi-natural transformation $\varphi\colon F \longrightarrow \Delta\,\underline{B}$, where $F\colon \mathcal{A} \longrightarrow \mathrm{Cat}$. Hence there is an induced $\overline{\varphi}\colon L\pi_o[1,F] \longrightarrow \underline{B}$, and (7.29) (with $\underline{X}$ replaced by $\underline{B}$) can be read as a morphism in $\mathrm{Fun}(\mathcal{A},[\mathrm{Cat}, \ulcorner\underline{B}\urcorner])$ from $\mathfrak{F}$ to $\Delta\,\overline{\varphi}$. It can be easily seen that, in general, if $f\colon \underline{X} \longrightarrow \underline{B}$, then a morphism from $\mathfrak{F}$ to Δf is described by diagrams

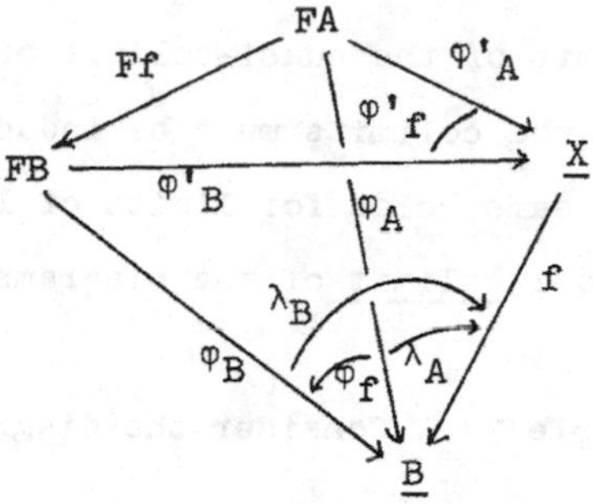

(7.32)

where $\varphi'\colon F \longrightarrow \Delta\underline{X}$ is a quasi-natural transformation and $\lambda\colon (\Delta f)\varphi' \longrightarrow \varphi$ is a modification. φ' induces a unique functor $\overline{\varphi}'\colon L\pi_o[1,F] \longrightarrow \underline{X}$ and λ a unique natural transformation $\overline{\lambda}$ so that

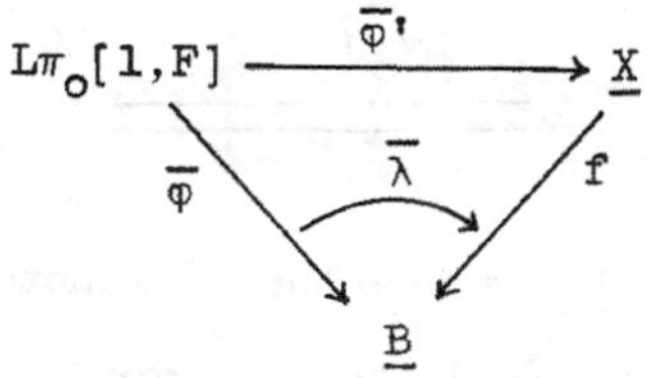

is the desired unique morphism in $[\mathrm{Cat}, \ulcorner\underline{B}\urcorner]$. Hence $\overline{\varphi} = \mathrm{Cart}\ \mathrm{q}\text{-}\varinjlim_{\mathcal{A}} \mathfrak{F}$.

I,7.11.3 <u>Corollary</u>. Let $\mathcal{F}: \mathcal{A} \longrightarrow [\text{Cat}, \ulcorner\underline{B}\urcorner]$ as above, where $\underline{B}$ is cocomplete (resp., complete). Then

$$\varinjlim_{L\pi_0\mathcal{A}} (\varinjlim_{FA} \varphi_A) = \varinjlim_{L\pi_0[1,F]} \overline{\varphi}$$

(resp., $\varprojlim_{L\pi_0\mathcal{A}} (\varprojlim_{FA} \varphi_A) = \varprojlim_{L\pi_0[1,F]} \overline{\varphi}$)

<u>Remark</u>: This says that a colimit of colimits of diagrams can be computed as the single colimit of the quasi-colimit of the diagrams. (Note that the maps between the colimits must be induced by maps between the diagrams.) The same holds for limits of limits, except that one still forms the quasi-<u>colimit</u> of the diagrams.

<u>Proof</u>. (See [CCS], §8, Example 7). Consider the diagram

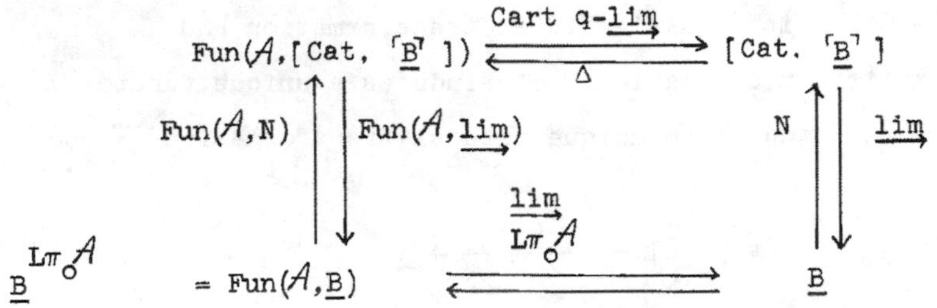

$$\begin{array}{ccc} \text{Fun}(\mathcal{A},[\text{Cat}, \ulcorner\underline{B}\urcorner]) & \underset{\Delta}{\overset{\text{Cart q-}\varinjlim}{\rightleftarrows}} & [\text{Cat}, \ulcorner\underline{B}\urcorner] \\ \text{Fun}(\mathcal{A},N)\uparrow\;\downarrow\text{Fun}(\mathcal{A},\varinjlim) & & N\uparrow\;\downarrow\varinjlim \\ \underline{B}^{L\pi_0\mathcal{A}} = \text{Fun}(\mathcal{A},\underline{B}) & \overset{\varinjlim_{L\pi_0\mathcal{A}}}{\rightleftarrows} & \underline{B} \end{array} \qquad (7.33)$$

Here $N: \underline{B} \longrightarrow [\text{Cat}, \ulcorner\underline{B}\urcorner]$ is the "name" functor taking an object $b \in \underline{B}$ to its name $\ulcorner b\urcorner: \underline{1} \longrightarrow \underline{B}$ and a morphism $\varphi: b \longrightarrow c$ in B to the morphism

$$\begin{array}{ccc} \underline{1} & \xrightarrow{\quad 1 \quad} & \underline{1} \\ & {}_{\ulcorner b\urcorner}\searrow \;\overset{\ulcorner\varphi\urcorner}{\Rightarrow}\; \swarrow_{\ulcorner c\urcorner} & \\ & \underline{B} & \end{array}$$

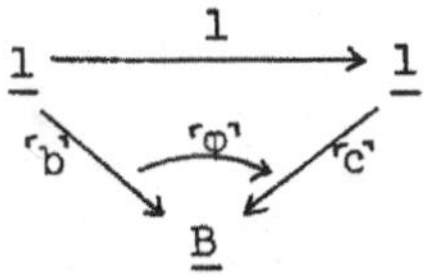

in [Cat, $\ulcorner\underline{B}\urcorner$]. As is well known, $\underline{B}$ is cocomplete if and only if N has a left adjoint $\underrightarrow{\lim}$ which assigns to f: $\underline{X} \longrightarrow \underline{B}$ its colimit in $\underline{B}$, $\underrightarrow{\lim}_{\underline{X}} f$. By I.7.4, Fun($\mathcal{A}$, $\underrightarrow{\lim}$) is then left adjoint to Fun($\mathcal{A}$,N). It is immediate that

$$\Delta N = \mathrm{Fun}(\mathcal{A}, N)\Delta$$

so the diagram of left adjoints commutes, up to an isomorphism. The second formula follows from the first by replacing $\underline{B}$ by $\underline{B}^{op}$.

<u>Remark</u>: There is a problem about size which we have ignored in the above discussion. To take care of it, assume Cat contains a sub-2-category Cat_s of small categories (e.g., take a two-stage universe) such that if $\mathcal{A}$ is a small 2-category and F: $\mathcal{A} \longrightarrow \mathrm{Cat}_s$, then $L\pi_o[1,F] \in \mathrm{Cat}_s$. Then the corollary above should have $[\mathrm{Cat}_s, \ulcorner\underline{B}\urcorner]$ instead of [Cat, $\ulcorner\underline{B}\urcorner$], as well as the hypothesis that $\mathcal{A}$ is small.

From the construction in I.7.11., two other kinds of Cartesian quasi-colimits can be determined. Let F: $\mathcal{A} \longrightarrow$ Cat and let $\mathcal{A}_o'$ be a subcategory of $\mathcal{A}_o$. Let

$$\tilde{\Sigma}_{F,\mathcal{A}_o'} = \{(f, 1_{F(f)a}) \in [1,F] \mid f \in \mathcal{A}_o'\} ;$$

i.e., $\tilde{\Sigma}_{F,\mathcal{A}_o'}$ is the class of all cocartesian morphisms of [1,F] over morphisms in $\mathcal{A}_o'$. Let $\Sigma_{F,\mathcal{A}_o'}$ be the image of $\tilde{\Sigma}_{F,\mathcal{A}_o'}$ in $L\pi_o[1,F]$. If $\mathcal{A}_o' = \mathcal{A}_o$, we omit it from the notation.

I,7.11.4 <u>Corollary</u>.

i) Cart $q - \underrightarrow{\lim}_{\mathcal{A}\text{-iso}\mathcal{A}_o'} F = L\pi_o[1,F][\Sigma^{-1}_{F,\mathcal{A}_o'}]$

ii) $\text{Cart } q - \varinjlim_{\mathcal{A}\text{-iso } \mathcal{A}_o} F = L\pi_o[1,F][\Sigma_F^{-1}]$

iii) $\text{Cart } q - \varinjlim_{\mathcal{A}\text{-id } \mathcal{A}_o'} F = L\pi_o[1,F][[\Sigma_{F,\mathcal{A}_o'}^{-1}]]$

iv) $\text{Cart } q - \varinjlim_{\mathcal{A}\text{-id } \mathcal{A}_o} F = L\pi_o[1,F][[\Sigma_F^{-1}]]$

In particular, if $\mathcal{A}$ is locally discrete then $\varinjlim F = [1,F][[\Sigma_F^{-1}]]$.

Remark: Here $[\Sigma^{-1}]$ denotes the usual category of fractions and $[[\Sigma^{-1}]]$ denotes the category in which the maps in Σ are made identities (i.e., coequalized with their domains.) Case ii) is the one considered by Giraud [19].

Proof. These results are evident from the construction of $\overline{\varphi}$ in the proof of I,7.11.1, by equation (7.30).

I,7.11.5 i) As a simple example, consider the cocomma category construction as described in [FCC]. Given functors $F_i\colon \underline{A}_o \longrightarrow \underline{A}_i$, $i = 1,2$, then the cocomma category $\langle F_1,F_2\rangle$ is the colimit (in Cat) of the diagram

$$\underline{A}_1 \xleftarrow{F_1} \underline{A}_o \xrightarrow{\underline{A}_o \times \partial_o} \underline{A}_o \times 2 \xleftarrow{\underline{A}_o \times \partial_1} \underline{A}_o \xrightarrow{F_2} \underline{A}_2 . \qquad (7.34)$$

It is dual to the comma category as in I,7.10, (7.23), satisfying a universal mapping property of the form

$$\begin{array}{ccc} \underline{A}_o & \xrightarrow{F_2} & \underline{A}_2 \\ {\scriptstyle F_1}\downarrow & \overset{\theta}{\nearrow} & \downarrow \\ \underline{A}_1 & \longrightarrow & \langle F_1,F_2\rangle \end{array} \qquad (7.35)$$

Let $\underline{Q}$ be the category illustrated by

$$1 \longleftarrow 0 \longrightarrow 2$$

and $\underline{Q}'$ the subcategory $1 \longleftarrow 0$. Then $\langle F_1 . F_2 \rangle$ = Cart q - $\underrightarrow{\lim}_{\underline{Q}\text{-id } \underline{Q}'} F$, where $F(i) = \underline{A}_i$, etc. It is an amusing exercise to verify that the description given by I.7.11.4(iii) coincides with that given by (7.34).

ii) Another illustrative example is given by taking the functor $F: \underline{Q} \longrightarrow$ Cat such that $F(0) = \underline{1}$. $F(i) = \underline{Q}'$, i = 1,2. where $\underline{Q}'$ is the object of Cat which looks like $\underline{Q}$, and

$$F(0 \longrightarrow 1) = (\ulcorner 2 \urcorner : \underline{1} \longrightarrow \underline{Q}')$$
$$F(0 \longrightarrow 2) = (\ulcorner 1 \urcorner : \underline{1} \longrightarrow \underline{Q}').$$

Then $[1, F]$ looks like

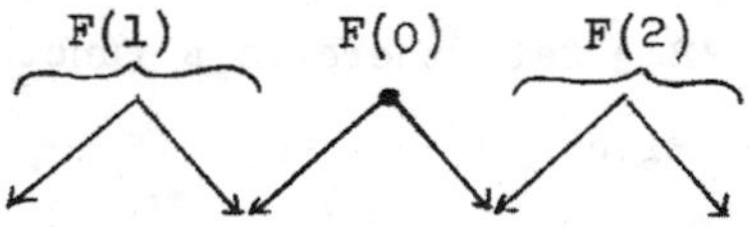

,

$[1,F][\Sigma_F^{-1}]$ looks like

and $\underrightarrow{\lim} F = [1,F][[\Sigma_F^{-1}]]$ looks like

Adapting I,7.11.3 to this case, where $\mathrm{Fun}(\underline{Q},[\mathrm{Cat},\ulcorner\underline{B}\urcorner])$ becomes $[\mathrm{Cat},\ \ulcorner\underline{B}\urcorner\ ]^{\underline{Q}}$, one sees that if $H: \underline{Q} \longrightarrow [\mathrm{Cat},\ \ulcorner\underline{B}\urcorner\ ]$ is a functor such that $PH = F$, then the colimit of a diagram of type $\varinjlim F$ in $\underline{B}$ is given by computing successive pushouts as indicated

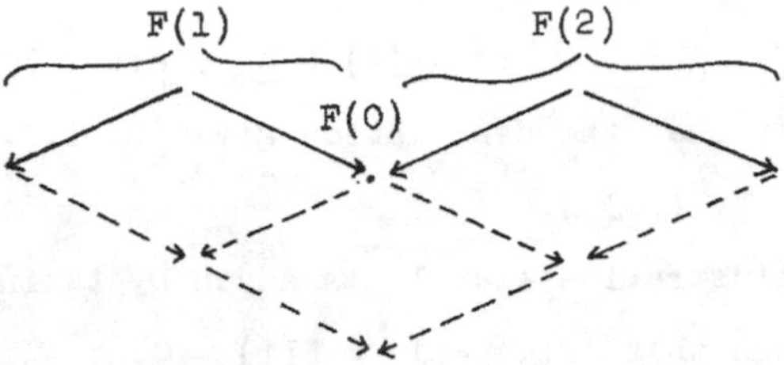

I,7.11.6 __Proposition.__ The closure of Sets $\subset$ Cat under cartesian quasi-colimits is Cat.

__Proof:__ We give two proofs of this important fact. The first shows that finite cartesian quasi-colimits are sufficient if those of type ii) in I,7.9 are allowed, while the second shows that $\{\underline{1}\}$ is "cartesian quasi-codense" in Cat.

1) We shall construct a finite category $\underline{D}$ and a subcategory $\underline{D}_o$ such that, given any $\underline{B}$ ε Cat, there is a functor $F_{\underline{B}}: \underline{D} \longrightarrow$ Sets $\subset$ Cat with

$$\mathrm{Cart}\ q - \varinjlim_{\underline{D}} \text{-id}\ \underline{D}_o\ F_{\underline{B}} = \underline{B}$$

$\underline{D}$ is just an initial part of the category $\underline{\Delta}^{op}$ of ordinals (See I,1.4 and I,1.6). It is generated by objects and maps as illustrated

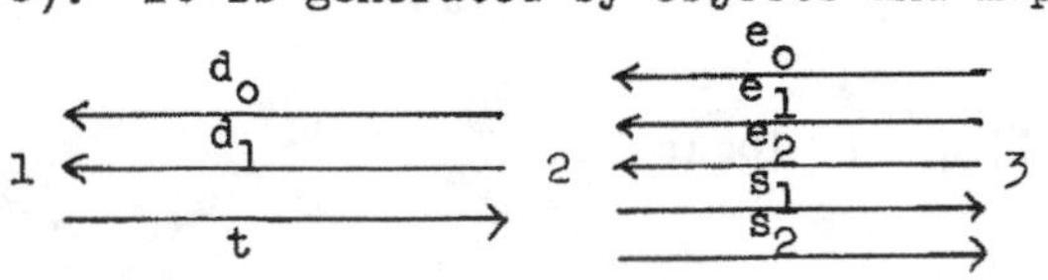

satisfying the following equations (which are chosen to fit categories rather than simplicial objects):

$$
\begin{aligned}
d_o t &= d_1 t = 1 \\
d_o e_o &= d_o e_1 = f_o \\
d_1 e_o &= d_o e_2 = f_1 \\
d_1 e_1 &= d_1 e_2 = f_2 \\
\\
e_1 s_1 &= e_2 s_1 = e_o s_2 = e_1 s_2 = 1 \\
e_o s_1 &= t d_o \qquad e_2 s_2 = t d_1 \\
s_1 t &= s_2 t = u
\end{aligned}
\tag{7.36}
$$

$\underline{D}_o$ is the subcategory consisting of

$$\{d_o, e_o, e_1, f_o, t, s_1, s_2, u\} .$$

Given $\underline{B}$ ε Cat, define $F_{\underline{B}} : \underline{D} \longrightarrow$ Sets by

$$
\begin{aligned}
&F_{\underline{B}}(n) = |\underline{B}^{\underline{n}}| \quad n = 1,2,3, \\
&F_{\underline{B}}(d_i) = |\underline{B}^{\partial_i}| , \ F_{\underline{B}}(t) = |\underline{B}^{\tau}| \\
&F_{\underline{B}}(e_o) = |\underline{B}^{\alpha}|, \ F_{\underline{B}}(e_1) = |\underline{B}^{\gamma}|, \ F_{\underline{B}}(e_2) = |\underline{B}^{\beta}| \\
&F_{\underline{B}}(s_1) = |\underline{B}^{\{\delta_o, \underline{2}\}}|, \ F_{\underline{B}}(s_2) = |\underline{B}^{\{\underline{2}, \delta_1\}}|
\end{aligned}
\tag{7.37}
$$

The notation is as in I.1.6 and I.1.7. The opfibration $[1,F]$ has (three) discrete fibres consisting of the sets of objects, morphisms, and commutative triangles of $\underline{B}$, respectively. There are, for instance, cocartesian morphisms connecting each morphism of $\underline{B}$ with its domain and codomain and each triangle with its three faces. If

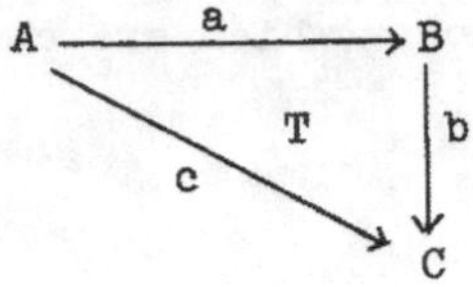

is a commutative triangle in $\underline{B}$, then in $[1,F][[\Sigma^{-1}_{F,\underline{D}_o}]]$ there is a diagram

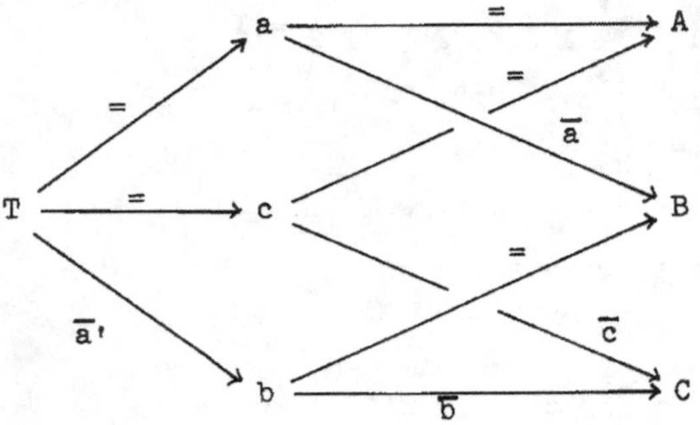

(7.38)

Hence the objects T,a,c, and A are identified, as are b and B. This forces $\overline{a}' = \overline{a}$ and hence $\overline{b}\,\overline{a} = \overline{c}$. Therefore

$$[1,F][[\Sigma^{-1}_{F,\underline{D}_o}]] \simeq \underline{B}.$$

(Note that associativity takes care of itself.) The quasi-natural transformation $\eta_{\underline{B}}$ from $F_{\underline{B}}$ to $\underline{B}$ which is the adjunction morphism is given by

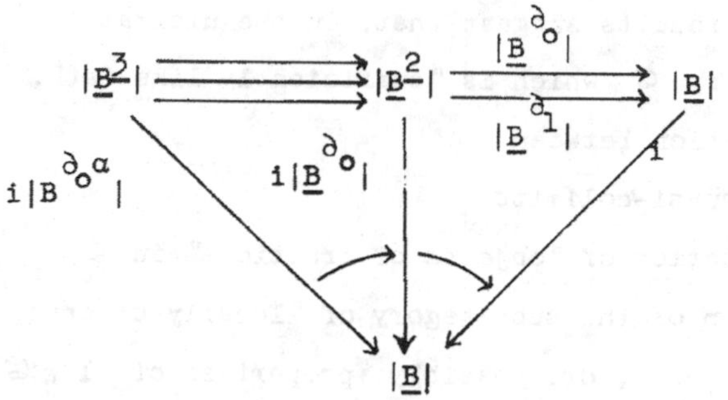

(7.39)

Here all components of $\eta_{\underline{B}}$ on morphisms are identities except $(\eta_{\underline{B}})_{d_1}$, $(\eta_{\underline{B}})_{c_2}$ and their composition. i denotes the inclusion of $|\underline{B}|$ in $\underline{B}$. If $f \in |\underline{B}^2|$, then $[(\eta_{\underline{B}})_{d_1}]_f = f$, while, if $T \in |\underline{B}^3|$ as above, then $[(\eta_{\underline{B}})_{e_2}]_T = a$ and one has

$$[(\eta_{\underline{B}})_{d_1 e_2}]_T = c$$
$$\|$$
$$[(\eta_{\underline{B}})_{d_1} |\underline{B}^{\beta}|]_T \cdot [(\eta_{\underline{B}})_{e_2}]_T = ba \ .$$

2) To see that $\{\underline{1}\} \subset$ Sets $\subset$ Cat is "quasi-codense" in Cat, recall that if $1_{\mathcal{A}}\colon \mathcal{A} \longrightarrow \underline{1} \xrightarrow{1}$ Cat, then $[1, 1_{\mathcal{A}}] \simeq A$. If $\underline{B} \in$ Cat, then the canonical projection $P\colon [1, \ulcorner\underline{B}\urcorner] \longrightarrow \underline{1}$ is of course constant so $(1P\colon [1, \ulcorner\underline{B}\urcorner] \longrightarrow \text{Cat}) = 1_{[1, \ulcorner\underline{B}\urcorner]}$. But, as was observed in [FCC], §6, $[1, \ulcorner\underline{B}\urcorner] \simeq \underline{B}$ and hence

$$\text{Cart } q - \varinjlim_{[1, \ulcorner\underline{B}\urcorner]} 1P = [1, \ulcorner\underline{B}\urcorner] \simeq \underline{B} \ .$$

We retain the quotation marks and forgo a formal definition since there are several other possible meanings for "quasi-codense".

I,7.11.7. Remark. These results suggest that, in the ultimate description of a 2-category $\mathfrak{C}$ which is "sufficiently like Cat", there should be some relation between

i) cartesian quasi-colimits

ii) the construction of "objects of fractions" in $\mathfrak{C}$

iii) the position of the subcategory of "locally discrete objects" in $\mathfrak{C}$, or, possibly, properties of $1 \in \mathfrak{C}$.

I,7.11.8 Corollary. Let $G: \underline{B} \longrightarrow \underline{C}$ be a functor from a small category to a cocomplete category. Then there is a coequalizer diagram

$$\coprod_{|\underline{B}^2|} G(\partial_o f) \rightrightarrows \coprod_{|\underline{B}|} G(B) \longrightarrow \varinjlim G$$

(and dually.)

Proof: Let $F_{\underline{B}}: \underline{D} \longrightarrow$ Sets be the functor constructed in the first proof of I,7.11.6. G determines a diagram

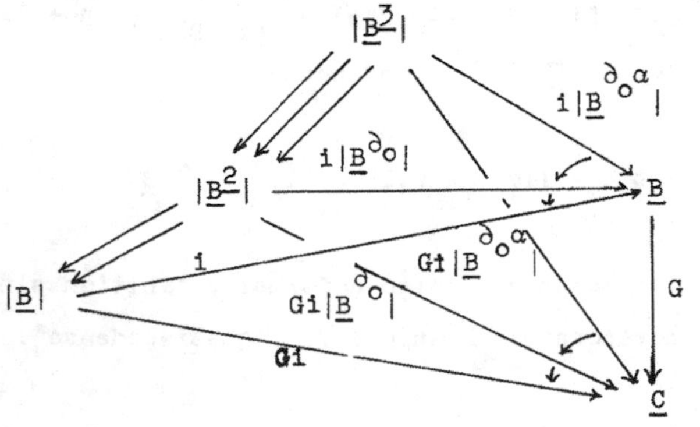

(7.40)

Apply I,7.11.3 to this situation, where $\bar{\varphi} = G$, etc. The categories $F(i)$ are discrete, so the terms $\varinjlim_{F_B(i)} \varphi_i$ are coproducts and hence $\varinjlim_{\underline{B}} G$ is the colimit over $\underline{D}$ of the diagram

$$\coprod_{|\underline{B}^3|} G(\partial_0 \alpha T) \; \substack{\longrightarrow \\ \longrightarrow \\ \longrightarrow \\ \longleftarrow \\ \longleftarrow} \; \coprod_{|\underline{B}^2|} G(\partial_0 f) \; \substack{\longrightarrow \\ \longrightarrow \\ \longleftarrow} \; \coprod_{|\underline{B}|} G(B) \qquad (7.41)$$

It is easily shown that the induced maps are the usual ones and that the subcategory $2 \rightrightarrows 1$ is cofinal in $\underline{D}$, which gives the result.

I,7.11.9 Kleisli categories. The previous examples have all involved functors whose domains are locally discrete. In the following example, which is an adaptation of a result of R. Street, [39], the 2-cells play a crucial role. As at the end of I,4.23, Street considers copseudo-functors from $\underline{1}$ to $\mathcal{A}$ as cotriples on objects in $\mathcal{A}$. These are the same as 2-functors from $\underline{\Delta}^{op}$ (regarded as a 2-category with a single object) to $\mathcal{A}$. We shall show that for $\mathcal{A}$ = Cat, the coKleisli category is the cartesian quasi-colimit of such a functor.

For our purposes the standard presentation of a cotriple is more useful. Let $\mathcal{G}$ be the 2-category with a single object $*$, a 1-cell g and all its powers g^n, and generating 2-cells $\delta: g \longrightarrow g^2$ and $\varepsilon: g \longrightarrow *$ satisfying the usual cotriple equations

$$\delta g \cdot \delta = g\delta \cdot \delta \; . \; g\,\varepsilon \cdot \delta = \varepsilon g \cdot \delta = 1.$$

A 2-functor $\underline{G}: \mathcal{G} \longrightarrow$ Cat is determined by a category $\underline{A} \in$ Cat, a functor $G = \underline{G}(g): \underline{A} \longrightarrow \underline{A}$ and natural transformations $\delta: G \longrightarrow G^2$, $\varepsilon: G \longrightarrow \underline{A}$ as usual. The 2-category $[1,\underline{G}]$ has as objects the objects of $\underline{A}$, as 1-cells from A to B, morphisms

$$G^n A \longrightarrow B$$

in $\underline{A}$. and 2-cells generated by commutative triangles

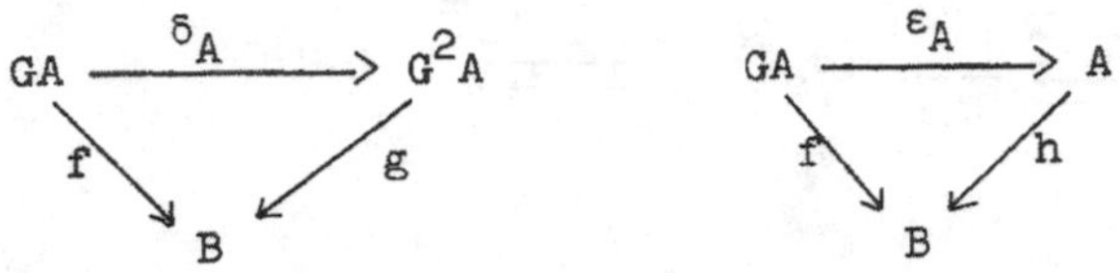

Hence in $L\pi_o[1,\underline{G}]$, every 1-cell from A to B can be expressed as a morphism in $\underline{A}$ from GA to B, ordinary morphisms being identified with $h\,\varepsilon_A$. It is easily check that composition becomes the ordinary Kleisli composition, so

$$\text{Cart q - } \underrightarrow{\lim}\ \underline{G} = L\pi_o[1,\underline{G}]$$

is the coKleisli category of the cotriple $\underline{G}$.

If $\underline{G}(*) = \underline{A}^{op}$, then one obtains a triple in $\underline{A}$ and this construction then yields the Kleisli category for the triple in $\underline{A}$.

I,7.11.10. Associated cofibrations.

Let Φ: $\underline{2} \longrightarrow$ Cat be a functor whose value is a functor F: $\underline{A} \longrightarrow \underline{B}$ in Cat. Since $\underline{2}$ is locally discrete,

$$\text{Cart q - } \underrightarrow{\lim}\ \Phi = [1,F].$$

This category clearly looks like $\underline{A} \amalg \underline{B}$ together with extra hom sets

$$[1,F](A,B) = \underline{B}(FA,B),\ [1,F](B,A) = \phi\ .$$

It can also be described as the pushout in Cat,

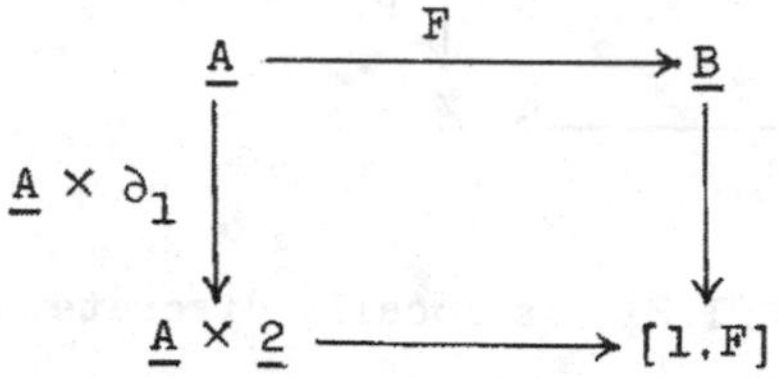

Hence [1,F] = <A,F> is the universal cofibration associated with F, in the sense of [FCC], §5. Thus this universal cofibration is not only the right adjoint to the inclusion functor of split cofibrations into categories under A, but it also satisfies the left adjointness property of a cartesian quasi-colimit, i.e.,

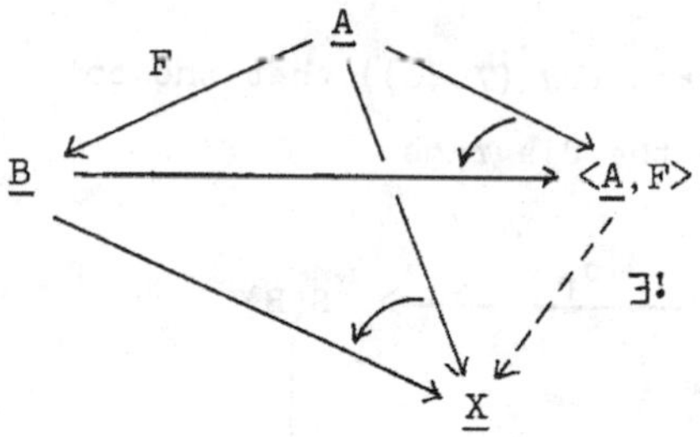

To get the dual construction, one checks that

$$[1, F^{op}]^{op} \simeq \langle F, \underline{A} \rangle .$$

I,7.12. Quasi-limits in Cat.

Let $F: \mathcal{A} \longrightarrow Cat$ be a 2-functor, where $\mathcal{A}$ is a small 2-category. Let $P: [1,F] \longrightarrow \mathcal{A}$ be the canonical projection and define the category of sections of [1,F] to be the pullback (in 2-Cat)

$$\begin{array}{ccc} \Gamma[1,F] & \longrightarrow & [1,F]^{\mathcal{A}} \\ \downarrow & & \downarrow P^{\mathcal{A}} \\ \underline{1} & \xrightarrow[1_{\mathcal{A}}]{} & \mathcal{A}^{\mathcal{A}} \end{array} \tag{7.42}$$

It is easily checked that $\Gamma[1,F]$ is locally discrete. Its objects are 2-functors $S: \mathcal{A} \longrightarrow [1,F]$ such that $PS = \text{id}$ and its morphisms are natural transformations $\mu: S \longrightarrow S'$ such that $P\mu = \text{id}$. Note that if $A \in \mathcal{A}$, then $S(A) \in F(A)$ and if $f: A \longrightarrow B$, then

$$S(f) = (f, \sigma_f): (A, S(A)) \longrightarrow (B, S(B))$$

where

$$\sigma_f: (Ff)(S'(A)) \longrightarrow S(B).$$

Naturality for $\mu: S \longrightarrow S'$ means (by (7.21)) that the components μ_A are maps in $F(A)$ and make the diagrams

$$\begin{array}{ccc} (Ff)(S(A)) & \xrightarrow{\sigma_f} & S(B) \\ {\scriptstyle (Ff)\mu_A}\downarrow & & \downarrow{\scriptstyle \mu_B} \\ (Ff)(S'(A)) & \xrightarrow{\sigma_f'} & S'(B) \end{array}$$

commute.

If $\mathcal{A}_o'$ is a subcategory of $\mathcal{A}$, then we write

$$\text{Cart}_{\mathcal{A}_o'}\, \Gamma[1,F] \quad (\text{resp.,}\ \ \text{Cl}_{\mathcal{A}_o'}\, \Gamma[1,F])$$

for the full subcategory of $\Gamma[1,F]$ determined by those sections S

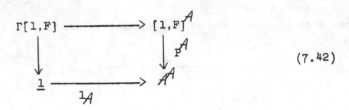

such that σ_f is an isomorphism (resp., the identity) for all $f \in \mathcal{A}_o'$. Note that the morphisms (f,σ) with σ an isomorphism are precisely the cartesian morphisms in $[1,F]$ while those with $\sigma = id$ are the chosen cartesian morphisms in the given cleavage of $[1,F]$. Hence the first subcategory above consists of cartesian sections while the second consists of cleavage preserving ones.

I,7.12.1 Proposition.

i) $\text{Cart } q - \varprojlim_{\mathcal{A}} F = \Gamma[1,F]$

ii) $\text{Cart } q - \varprojlim_{\mathcal{A}-\text{iso}\mathcal{A}_o'} F = \text{Cart}_{\mathcal{A}_o'} \Gamma[1,F]$

iii) $\text{Cart } q - \varprojlim_{\mathcal{A}-\text{id}\mathcal{A}_o'} F = \text{Cart}_{\mathcal{A}_o'} \Gamma[1,F]$.

In particular, if $\mathcal{A}$ is locally discrete, then

$$\varprojlim F = \text{Cl}_{\mathcal{A}_o} \Gamma[1,F] .$$

Proof: Define $\varepsilon_F\colon \Delta\,\Gamma[1,F] \longrightarrow F$ to be the transformation whose components are

$$(\varepsilon_F)_A = ev_A\colon \Gamma[1,F] \longrightarrow F(A)$$

$$[(\varepsilon_F)_f]_S = \sigma_f .$$

Since S is a 2-functor, ε_F is quasi-natural. Suppose $\varphi\colon \Delta X \longrightarrow F$ is a quasi-natural transformation. Consider the diagram

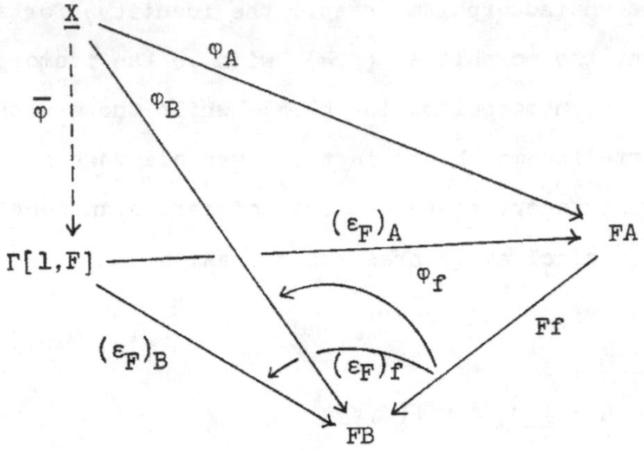

(7.43)

If $X \in \underline{X}$, define

$$(\overline{\varphi}(X))(A) = (A, \varphi_A(X))$$
$$(\overline{\varphi}(X))(f) = (f.(\varphi_f)_X) \ . \qquad (7.44)$$

Then $\overline{\varphi}(X)$ is a section. If $t\colon X \longrightarrow Y$ in $\underline{X}$, define $\overline{\varphi}(t)_A = \varphi_A(t)$. Then $\overline{\varphi}(t)$ is a natural transformation. Clearly $\varepsilon_F\overline{\varphi} = \varphi$ and $\overline{\varphi}$ is the unique functor with this property. Hence

$$\text{Cart } q - \underleftarrow{\lim}_{A}\, F = \Gamma[1,F] \ .$$

The other cases follow immediately from the formulas given above.

I,7.12.2 Remark. There is a certain analogy with the case of ordinary limits and colimits which is worth observing and which probably generalizes.

i) Let $F\colon \underline{A} \longrightarrow \text{Sets}$. Then $[1,F]$ is an opfibration over $\underline{A}$ with discrete fibres. $\underrightarrow{\lim}\, F$ is constructed by making all morphisms (1-cells) in [1,F] identities, while $\varprojlim F$ consists of all sections of [1,F].

ii) Let $F: \mathcal{A} \longrightarrow Cat$. Then [1,F] is an opfibration over $\mathcal{A}$ with locally discrete fibres. Cart q - $\varinjlim_{\mathcal{A}} F$ is constructed by making all 2-cells in [1,F] identities, while Cart q - $\varprojlim_{\mathcal{A}} F$ consists of all sections of [1,F].

Corollary I,7.11.2 has an obvious analogue but, as far as we know, this does not lead to anything useful along the lines of I,7.11.3. One of the examples in I,7.11.5 dualizes, as we shall see, but I,7.11.6 does not. In fact, if $F: \mathcal{A} \longrightarrow Sets$, then $\Gamma[1,F]$ is obviously discrete, so Sets $\subset$ Cat is closed under Cartesian quasi-limits. Finally, I,7.11.9 dualizes nicely.

I,7.12.3 We can now calculate directly the Cartesian quasi-limits giving comma categories in I.7.10. Using the notation there, if $F: \underline{P} \longrightarrow Cat$ with $F(i) = \underline{A}_i$ and $F(j_i) = F_i$, then $Cl_{\underline{P}_o}\Gamma[1,F]$ consists of sections $S: \underline{P} \longrightarrow [1,F]$ such that $S(i) = A_i \in \underline{A}_i$, $S(j_1) = (j_1, f)$, $S(j_2) = (j_2, id)$. Here $f: F_1(A_1) \longrightarrow A_o$ and $F_2(A_2) = A_o$, so objects are exactly maps $f: F_1(A_1) \longrightarrow F_2(A_2)$ in $\underline{A}_o$. It is easily checked that morphisms of sections are the usual commutative diagrams.

A similar calculation shows that a subequalizer of F and G as in I.7.10, has as objects an object $A \in \underline{A}$ together with a morphism $f: GA \longrightarrow FA$ with the obvious notion of morphism.

I,7.12.4 <u>Eilenberg-Moore categories</u>. Let $\mathcal{G}$ be as in I.7.11.9, and let $\mathcal{T} = {}^{op}\mathcal{G}$, so that 2-functors from $\mathcal{T}$ to Cat are triples. We denote the 1-cells of $\mathcal{T}$ by t^n and the generating 2-cells by $\eta: 1 \longrightarrow t$ and $\mu: t^2 \longrightarrow t$. Let $\underline{T}: \mathcal{T} \longrightarrow Cat$ be a

2-functor. We shall show that the cartesian quasi-limit of $\underline{T}$ is the category of Eilenberg-Moore algebras of the triple $\underline{T}$. For, consider the category $\Gamma[1,\underline{T}]$. A section $S: \mathcal{T} \longrightarrow [1,\underline{T}]$ is given by $S(*) = A \varepsilon \underline{A}$, where $\underline{T}(*) = \underline{A}$, and morphisms $S(t^n) = (id,\xi_n)$ where $\xi_n: T^nA \longrightarrow A$. The formula for composition in $[1,\underline{G}]$ (7.27) shows that if $\xi = \Gamma(t): TA \longrightarrow A$, then $\xi_2 = \xi\ T(\xi)$. etc., so that all ξ_n's are determined by ξ. Since S has to be a 2-functor, the diagrams

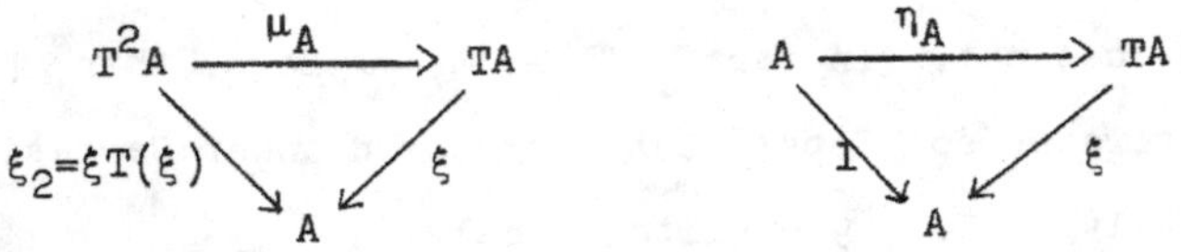

must commute. Hence $(\xi: TA \longrightarrow A)$ is a $\underline{T}$-algebra. Morphisms of sections are clearly the same as morphisms of algebras.

Replacing $\underline{A}$ by $\underline{A}^{op}$ yields the Eilenberg-Moore category of coalgebras of the corresponding cotriple. One easily verifies that the cartesian quasi-colimit of $\underline{T}$ is uninteresting (it is $\underline{A}$), as is the cartesian quasi-limit of a functor $\underline{G}: \mathcal{D} \longrightarrow Cat$.

I,7.12.5. Associated fibrations

As in I,7.11.10, let $\Phi: \underline{2} \longrightarrow Cat$ have as its value $F: \underline{A} \longrightarrow \underline{B}$. It is easily checked that

$$\text{Cart } q - \varprojlim \Phi = \underline{\Gamma}[1,F] = (F.\underline{B})$$

since a section of $[1,F] \longrightarrow \underline{2}$ is described by a morphism $FA \longrightarrow B$, while a natural transformation of sections is the same as a morphism in the indicated comma category. Thus $(F,\underline{B})$ has three different universal properties; first, it is the pullback in Cat

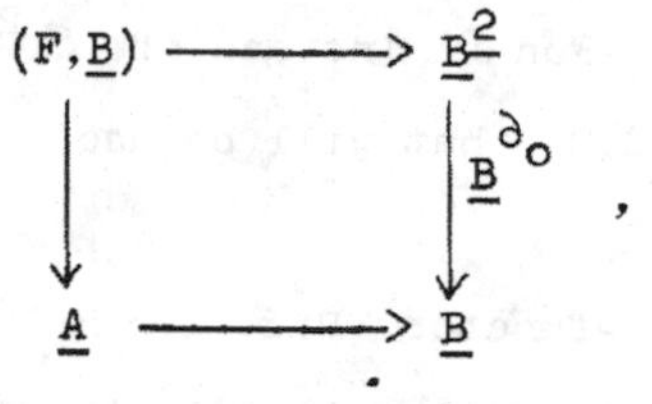

,

second, $(F,\underline{B}) \longrightarrow \underline{B}$ is the universal opfibration associated to F and hence left adjoint to an inclusion functor, and third it satisfies the right adjointness property of a cartesian quasi-limit

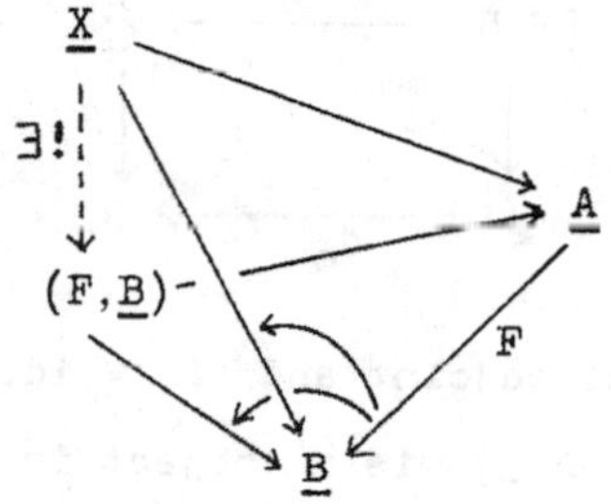

The dual construction is given by

$$\underline{\Pi}[1.F^{op}]^{op} = (\underline{B},F) .$$

I,7.13 <u>Quasi-fibrations</u>. Besides the notion of 2-fibrations, as in I,2.9, there are various notions of quasi-fibrations; the essential difference is the use of Fun $\mathcal{E}$ instead of $\mathcal{E}^{\underline{2}}$. We treat here the case (in dual form to I,2.9) that will be used in the next section.

I,7.13.1 <u>Definition</u>. A 2-functor $P: \mathcal{E} \longrightarrow \mathcal{B}$ between 2-categories is called a <u>Cartesian quasi-opfibration</u> if there exists a 2-functor L, as illustrated,

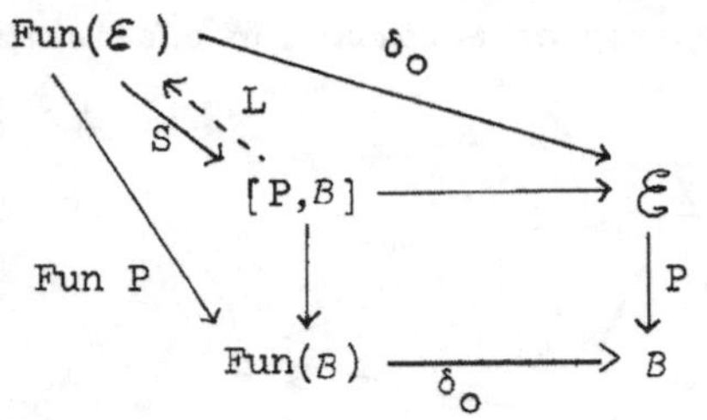

having S as a right Cat-adjoint and SL = id.

If $(E, f: PE \longrightarrow B)$ is an object in $[P,\mathcal{B}]$ then $L(E,f)$ is a morphism in $\mathcal{E}$ satisfying the following universal mapping property:

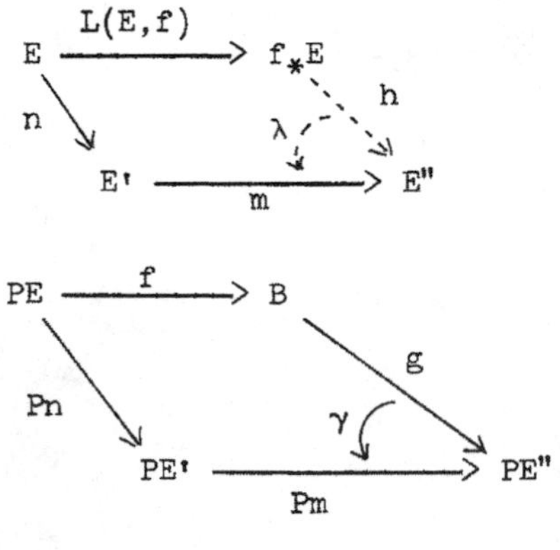

(7.45)

for any $m: E' \longrightarrow E''$ in $\mathcal{E}$, given $n: E \longrightarrow E'$ in $\mathcal{E}$ and a map (Pn,γ,g) in Fun $\mathcal{B}$ as illustrated, there is a unique map (n,λ,h) in Fun $\mathcal{E}$ with Fun $P(n,\lambda,h) = (Pn,\gamma,g)$. Similarly, given a 2-cell $\sigma: n \longrightarrow n'$ in $\mathcal{E}$ and a 2-cell $(P\sigma,\tau): (Pn,\gamma,g) \longrightarrow (Pn',\gamma',g')$, there is a unique 2-cell $(\sigma,\tilde{\tau}): (n,\lambda,h) \longrightarrow (n',\lambda',h')$ in Fun $\mathcal{E}$ with Fun $P(\sigma,\tilde{\tau}) = (P\sigma,\tau)$. In particular, taking m and g identities, given $n: E \longrightarrow E'$ in $\mathcal{E}$, there is a unique diagram

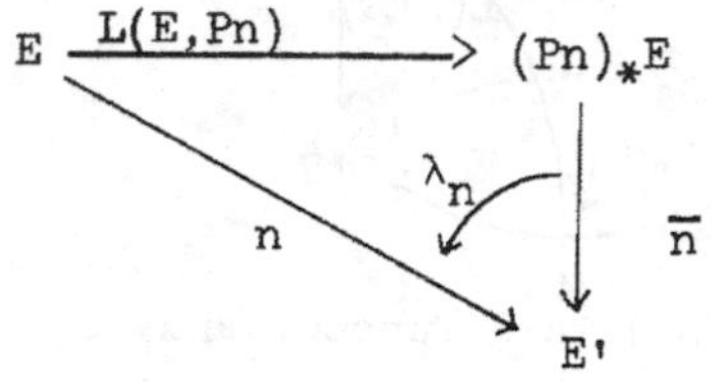

(7.46)

with $P(\bar{n}) = \text{id}$, $P(\lambda_n) = \text{id}$. Similarly, given a 2-cell $\sigma: n \longrightarrow n'$, then taking $m = 1_{(Pn')_*E}$ and $g = 1$, and using the fact above about 2-cells, there is a unique diagram

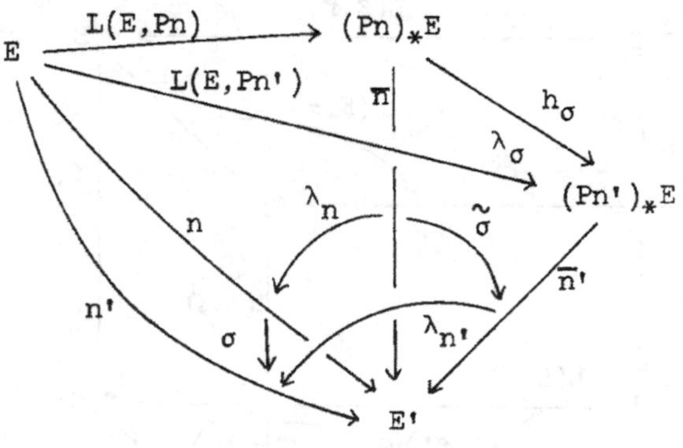

(7.47)

where $P(h_\sigma) = 1_{PE'}$, $P(\lambda_\sigma) = P(\sigma)$, and $P(\tilde{\sigma}) = 1_{1_{PE'}}$.

On the other hand, taking $m = L(E',f')$ for $f\colon PE' \longrightarrow B'$ shows that there are uniquely defined 1-cells and 2-cells in the diagram

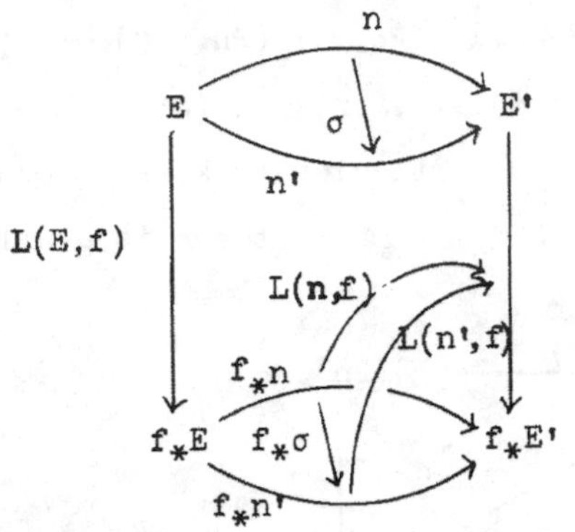

(7.48)

making $f_*\colon P^{-1}(B) \longrightarrow P^{-1}(B')$ a 2-functor and $L(-,f)\colon J_B \longrightarrow J_{B'}f_*$ a quasi-natural transformation. Here $P^{-1}(B)$ is the fibre over B; i.e., 0-cells, 1-cells and 2-cells mapping to B, and $J_B\colon P^{-1}(B) \longrightarrow \mathcal{E}$ is the inclusion 2-functor. Similarly, if $\tau\colon f \longrightarrow f'$ is a 2-cell in $\mathcal{B}$, then there are uniquely defined 1-cells and 2-cells in the diagram

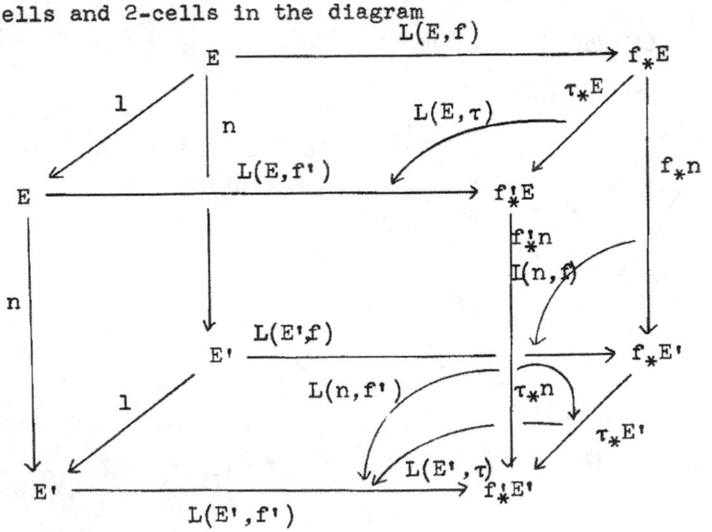

(7.49)

making $\tau_*: f_* \longrightarrow f'_*$ a quasi-natural transformation and $L(-,\tau)$ a modification in the diagram

$$\begin{array}{ccc} J_B & \xrightarrow{L(-,f)} & J_{B'}f_* \\ & \searrow^{L(-,f')} \quad \overset{L(-,\tau)}{\swarrow} & \downarrow J_B\tau_* \\ & & J_B f'_* \end{array} \tag{7.50}$$

In particular, this gives a functor

$$(-)_*: \mathcal{B}(B,B') \longrightarrow \mathrm{Fun}(P^{-1}(B),P^{-1}(B'))$$

for each pair of objects in $\mathcal{B}$. Globalized statements about various kinds of functors from $\mathcal{B}$ to 2-$\mathrm{Cat}_{\otimes}$ are too complicated to go into here.

A choice of L is called a <u>cleavage</u> for P. If $P: \mathcal{E} \to \mathcal{B}$ and $P': \mathcal{E}' \longrightarrow \mathcal{B}$ are cartesian quasi-opfibrations, a 2-functor $M: \mathcal{E} \longrightarrow \mathcal{E}'$ which satisfies

i) $P'M = P$

ii) $L'(M,1,1) = \mathrm{Fun}(M)L$ (i.e., M "commutes with L)

is called <u>cleavage preserving.</u> If L is chosen (when possible) so that

i) $L(id) = id$

ii) $L(f_*E,g) \circ L(E,f) = L(E,gf)$

then P together with L is called a <u>split-normal</u> <u>cartesian</u> <u>quasi-opfibration</u>. We denote the 2-category of such with cleavage preserving morphisms and Cat-natural transformations over $\mathcal{B}$ by

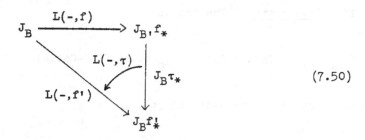

$$\text{Cart q-Split } (\mathcal{B})_o .$$

I,7.13.2 <u>Proposition</u>. The inclusion

$$\text{Cart q-Split } (\mathcal{B})_o \xhookrightarrow{\ I\ } [{}^{op}2\text{-Cat},\mathcal{B}]$$

has a strict left quasi-adjoint Φ.

<u>Proof</u>. We verify the conditions of I,7.8.2, showing that the adjoint is a 2-functor and that t and s are identities.

i) Let $F: \mathcal{A} \longrightarrow \mathcal{B}$ be a 2-functor. Then the projection $P_F: [F,\mathcal{B}] \longrightarrow \mathcal{B}$ is a split-normal cartesian quasi-opfibration; for, given $(FA \xrightarrow{h} B) \in [F,\mathcal{B}]$ and $B \xrightarrow{f} B'$ in $\mathcal{B}$, define

$$L(h,f): \quad \begin{array}{ccc} FA & \xrightarrow{\ 1\ } & FA \\ & & \downarrow h \\ h \downarrow & & B \\ & & \downarrow f \\ B & \xrightarrow[\ f\ ]{} & B' \end{array}$$

with the obvious extension to 1-cells and 2-cells in $[P_F,\mathcal{B}]$. The required conditions are easily verified. Define $\Phi(H) = P_F$.

ii) Define $\eta_F: F \longrightarrow P_F$ to be the map

$$\begin{array}{ccc} \mathcal{A} & \xrightarrow{\ \tilde{j}_F\ } & [F,\mathcal{B}] \\ & F \searrow \quad \swarrow P_F & \\ & \mathcal{B} & \end{array}$$

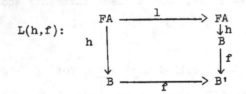

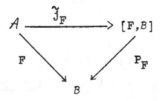

where $\tilde{j}_F$ is as in (7.6'). Then η_F satisfies the required universal mapping property. For, let $(P: \mathcal{E} \longrightarrow \mathcal{B},L)$ be a split-normal cartesian quasi-opfibration and consider a map $(M,m)\colon F \longrightarrow P$ in $[{}^{op}2\text{-Cat}, \mathcal{B}]$

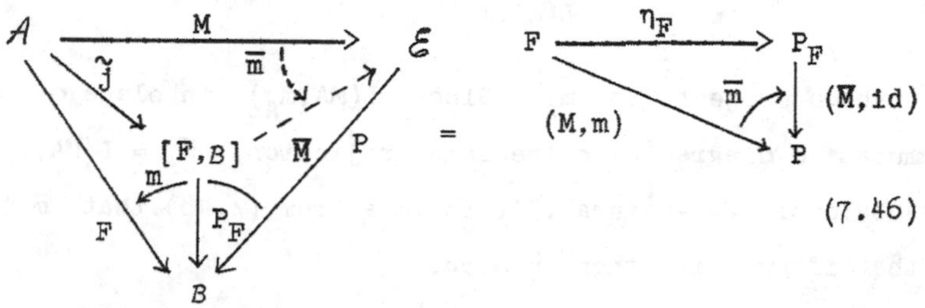

(7.46)

We must first construct $\overline{M}$ and $\overline{m}$ as illustrated with $P\overline{M} = P_F$ and $P\overline{m} = m$. These are given by the diagram

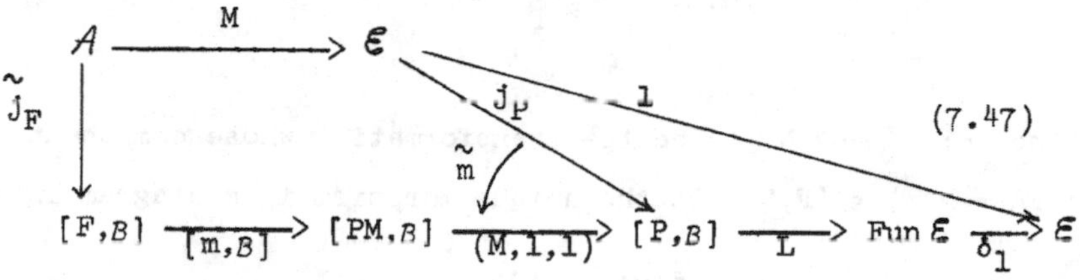

(7.47)

where $\tilde{m}$ is given by

$$\tilde{m}_A = \begin{pmatrix} PMA \\ \downarrow 1 \\ PMA \end{pmatrix} \xrightarrow[\;m_A\;]{1} \begin{pmatrix} PMA \\ \downarrow m_A \\ FA \end{pmatrix}$$

$\overline{M}$ is the composition of the bottom row of (7.47), and

$$\overline{m} = \delta_1\, L(M,1,1)\tilde{m}\ . \tag{7.48}$$

Thus $\overline{m}_A$ is the unique (by (7.45)) morphism in a diagram

$$MA \xrightarrow{\quad 1 \quad} MA$$

$$1 \downarrow \qquad\qquad 1 \downarrow \quad \downarrow \bar{m}_A$$

$$MA \xrightarrow[L(MA,m_A)]{} \bar{M}(1_{HA})$$

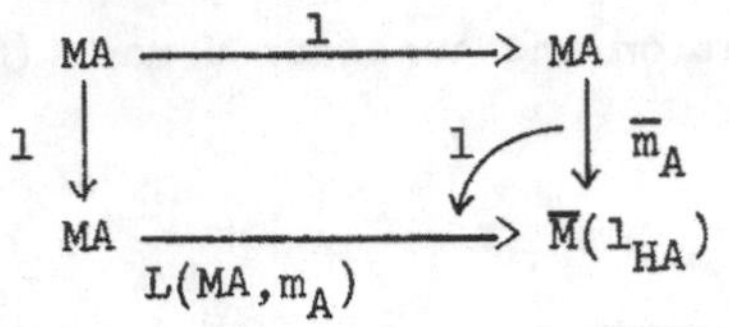

in $\mathcal{E}$ which projects to $\tilde{m}_A$. Since $L(MA,m_A)$ in place of $\bar{m}_A$ gives a commutative diagram with the same projection, $\bar{m}_A = L(MA,m_A)$. Since $\tilde{m}$ is obviously Cat-natural, it follows from (7.48) that $\bar{m}$ is also. Note that if $m = id$ then $\bar{m} = id$.

Now suppose that there is given another

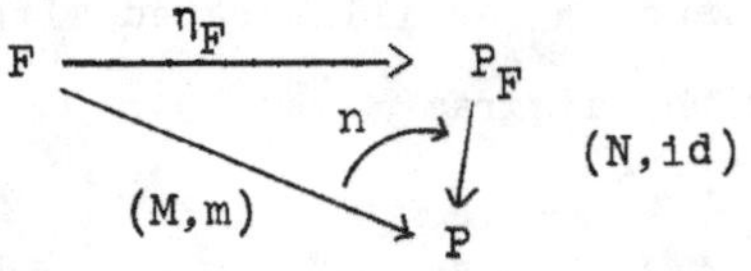

$$F \xrightarrow{\quad \eta_F \quad} P_F$$

$$n \searrow \quad \downarrow \qquad (N,id)$$

$$(M,m) \searrow P$$

Define τ: $\bar{M} \longrightarrow N$ to be the transformation whose component τ_h at $(h: FA \longrightarrow B) \in [F,B]$ is the unique morphism in a diagram in $\mathcal{E}$

$$MA \xrightarrow{L(MA,hm_A)} \bar{M}(h)$$

$$n_A \searrow \qquad \qquad \tau_h$$

$$N(1_{FA}) \qquad$$

$$N(1,1,h) \quad N(h)$$

where $(1,1,h): \begin{pmatrix} FA \\ \downarrow 1 \\ FA \end{pmatrix} \xrightarrow[h]{1} \begin{pmatrix} FA \\ \downarrow h \\ B \end{pmatrix}$.

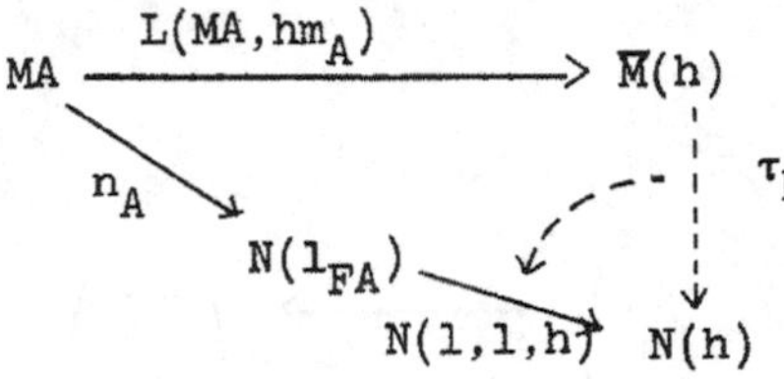

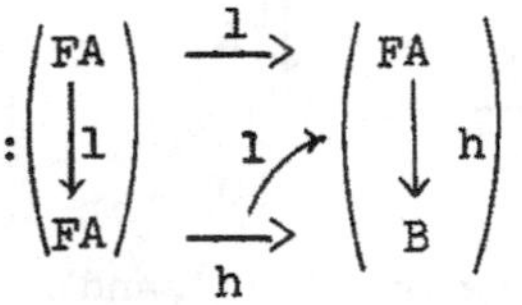

The diagram

$$\begin{pmatrix} MA \\ 1 \downarrow \\ MA \end{pmatrix} \xrightarrow{\ 1\ } \begin{pmatrix} MA \\ \bar{m}_A \downarrow \\ \bar{M}(1_{FA}) \end{pmatrix} \xrightarrow{\ 1\ } \begin{pmatrix} MA \\ n_A \downarrow \\ N(1_{FA}) \end{pmatrix}$$

$$\text{(with } L(MA,\bar{m}_A)\text{, } \tau_{1_{FA}}\text{, } n_A\text{)}$$

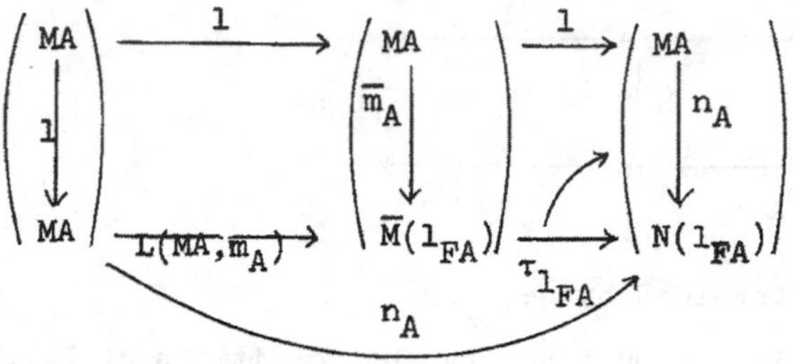

in $\mathcal{E}$ over the diagram

$$\begin{pmatrix} PMA \\ 1 \downarrow \\ PMA \end{pmatrix} \xrightarrow{\ 1\ } \begin{pmatrix} PMA \\ m_A \downarrow \\ FA \end{pmatrix} \xrightarrow{\ 1\ } \begin{pmatrix} PMA \\ m_A \downarrow \\ FA \end{pmatrix}$$

$$\text{(with } m_A\text{, } 1\text{)}$$

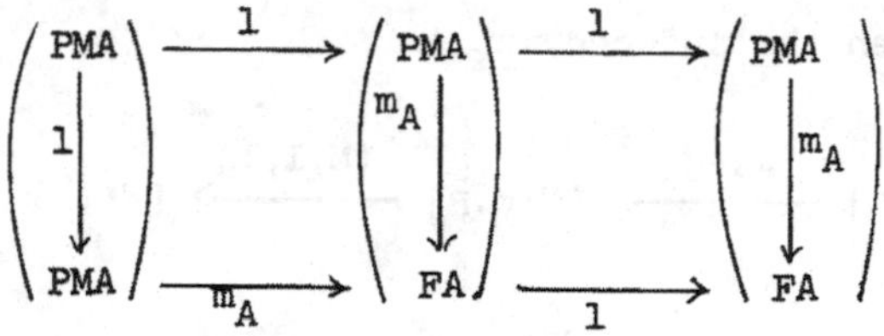

in $\mathcal{B}$ shows that

$$\tau_{1_{HA}} \bar{m}_A = \tau_{1_{FA}} L(MA, m_A) = n_A$$

as required. Furthermore, the components $\tau_{1_{FA}}$ are uniquely determined by this. Finally the diagram

$$\begin{array}{ccccc} MA & \xrightarrow{\ L\ } & \bar{M}(1_{HA}) & \xrightarrow{\ \tau_{1_{FA}}\ } & N(1_{HA}) \\ {\scriptstyle 1}\downarrow & & {\scriptstyle \bar{M}(1,1,h)}\downarrow & & {\scriptstyle N(1,1,h)}\downarrow \\ MA & \xrightarrow{\ L\ } & \bar{M}(h) & \xrightarrow{\ \tau_h\ } & N(h) \end{array}$$

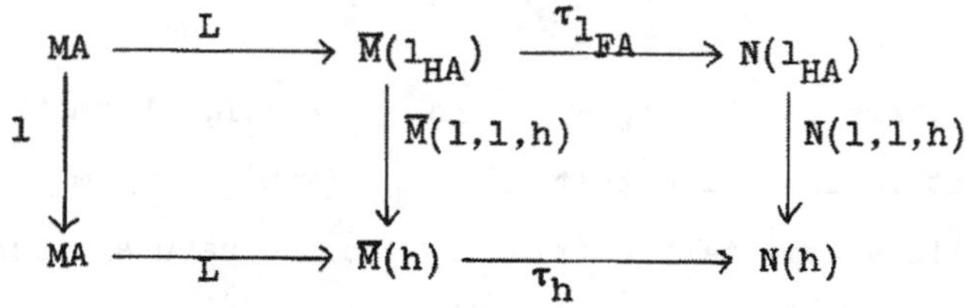

over

$$\begin{array}{ccccc} PMA & \xrightarrow{m_A} & FA & \xrightarrow{1} & FA \\ {\scriptstyle 1}\downarrow & & \downarrow{\scriptstyle h} & & \downarrow{\scriptstyle h} \\ PMA & \xrightarrow[hm_A]{} & B & \xrightarrow[1]{} & B \end{array}$$

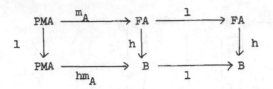

shows that τ_h is determined by $\tau_{1_{FA}}$.

iii) Finally, we must verify the conditions of I,7.8.2 and show that t and s are identities. It is clear, either directly, or by verifying the universal mapping property, that if $(M,m)\colon F \longrightarrow F'$, then the induced map is

$$\Phi(M,m)\colon [F,\mathcal{B}] \xrightarrow{[m,\mathcal{B}]} [F'M,\mathcal{B}] \xrightarrow{(M,1,1)} [F',\mathcal{B}] \qquad (7.49)$$

which defines a 2-functor from $[{}^{op}2\text{-Cat},\mathcal{B}]$ to Cart q-Split $(\mathcal{B})_o$.

By definition, if $P\colon \mathcal{E} \longrightarrow \mathcal{B}$ is such an opfibration then ε_P and t_P are the $\overline{M}$ and $\overline{m}$ in the diagram

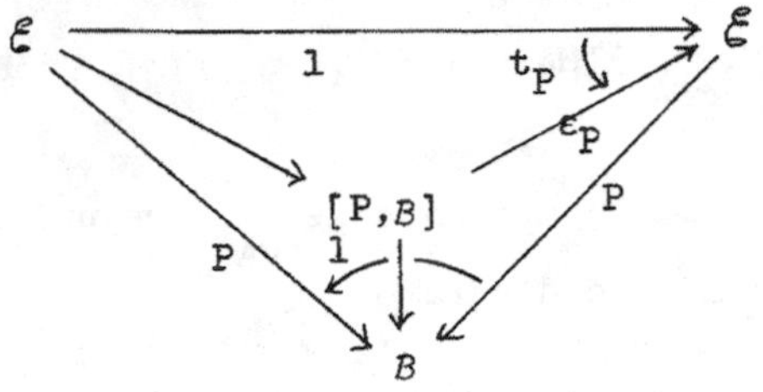

(7.50)

Since the m here is 1, $t_P = 1$ and $\varepsilon_P = \delta_1 L$. From this, (7.49) and (7.47), it is immediate that $\overline{M} = \varepsilon_P\, \Phi(M,m)$ and $\overline{m} = t_P \circledcirc \eta_{(M,m)}$. The map η_F is a commutative triangle, so the natural transformation η_{η_F} is the identity, which by (7.16) implies that $s = 1$. Finally, we observe that, since maps between opfibrations commute with the L's, ε is in fact natural. From this, using I,7.7, we get the following result.

I,7.13.3 **Corollary**: In the induced ordinary adjunction (I,7.5),

$$[F,\mathrm{Cart\ q\text{-}Split}\ \mathcal{B}_o] \underset{T}{\overset{S}{\rightleftarrows}} [[{}^{op}2\text{-}\mathrm{Cat},\mathcal{B}],I]$$

we have

i) T is a bihomomorphism

ii) TS = id.

In the ordinary case, the associated opfibration has another important property (I,1.11) which also has an analogue in this situation.

I,7.13.4. **Proposition**. There is a diagram

$$\begin{array}{ccc} \mathcal{A} & \underset{P}{\overset{Q_F}{\rightleftarrows}} & [F,\mathcal{B}] \\ & {\scriptstyle F}\searrow \quad \swarrow{\scriptstyle P_F} & \\ & \mathcal{B} & \end{array}$$

in which

i) $P_F Q_F = F$

ii) P is left inverse, strict quasi-right adjoint to Q_F.

Proof: Here $Q_F = \mathfrak{J}_F$ and P is the projection of $[F,\mathcal{B}]$ on $\mathcal{A}$. It is obvious that $P_F Q_F = F$ and $PQ_F = 1$. Define $\eta = \mathrm{id}\colon \mathcal{A} \to PQ_F$ and $\varepsilon\colon Q_F P \longrightarrow [F,\mathcal{B}]$ by the formulas

$$\varepsilon_h = \left(\begin{array}{c} FA \\ \downarrow \\ FA \end{array}\right) \begin{array}{c} \xrightarrow{\ 1\ } \\ \\ \xrightarrow[\ h\ ]{} \end{array} \left(\begin{array}{c} FA \\ \downarrow{\scriptstyle h} \\ B \end{array}\right) \quad \text{for } (h\colon FA \longrightarrow B) \in [F,\mathcal{B}]$$

$$\varepsilon_{(f,\theta,g)} = (1,\theta) \quad \text{for} \quad (f,\theta,g)\colon h \longrightarrow h' \ \text{ in } [F,\mathcal{B}].$$

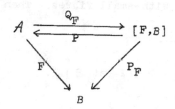

One verifies immediately that ε is quasi-natural and that (ε,η) is a strict quasi-adjunction. One can also check the universal mapping properties (I,7.8.2) for ε and η, both of which are non-trivial.

Adjoint functor theorems based on this will be discussed elsewhere.

Finally, in the discussion of Kan extensions as outlined in I,1.13, there is one other important aspect of fibrations which also has an analogue in this case.

I,7.13.5 Proposition. Let $(P\colon \mathcal{E} \longrightarrow \mathcal{B}, L)$ be a split-normal Cartesian quasi-opfibration with small fibres. Then there is a 2-$\mathrm{Cat}_{\otimes}$-enriched imbedding

$$J\colon \mathcal{B} \longrightarrow {}_s[2\text{-}\mathrm{Cat}_{\otimes}, \mathcal{E}]_{\otimes} .$$

Proof: If $B \in \mathcal{B}$, let $P^{-1}(B)$ be the fibre over B (i.e., objects, 1-cells and 2-cells which project to the identity of B) and let $J_B\colon P^{-1}(B) \longrightarrow \mathcal{E}$ be the inclusion 2-functor, regarded as an object of ${}_s[2\text{-}\mathrm{Cat}_{\otimes}, \mathcal{E}]_{\otimes}$. If $f\colon B \longrightarrow C$ in $\mathcal{B}$ and $n\colon E \longrightarrow E'$ in $P^{-1}(B)$ then there is a unique diagram

$$\begin{array}{ccc} E & \xrightarrow{\;L(E,f)\;} & f_*E \\ {\scriptstyle n}\downarrow & \swarrow_{L(n,f)} & \downarrow{\scriptstyle f_*n} \\ E' & \xrightarrow[\;L(E',f)\;]{} & f_*E' \end{array} \qquad (7.51)$$

such that $P(f_*n) = B$ and $P(n,f) = f$. The uniqueness (or the fact that L is a 2-functor in I,7.13.1) shows that this gives rise to a morphism

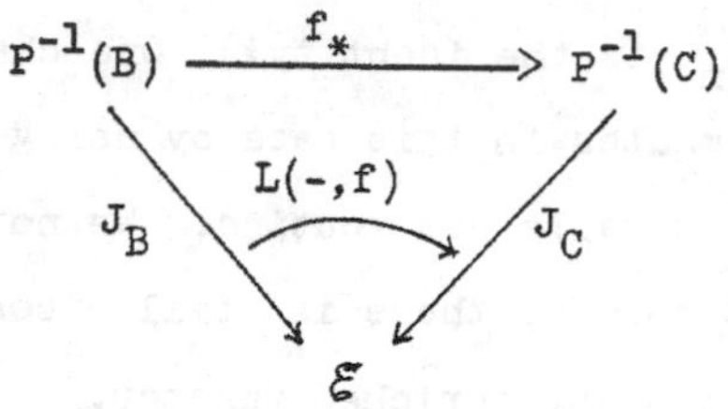

(7.52)

in ${}_s[2\text{-Cat}_{\otimes},\mathcal{E}]_{\otimes}$, where f_* is a 2-functor and $L(-,f)$ is a quasi-natural transformation. Finally, if

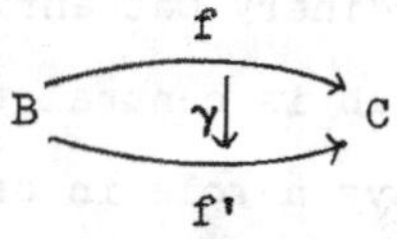

is a 2-cell in $\mathcal{B}$, then there is a unique diagram

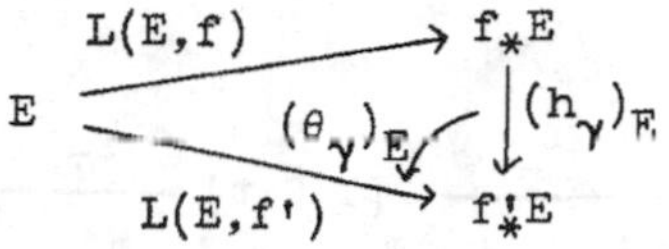

(7.53)

such that $P((h_\gamma)_E) = C$ and $P((\theta_\gamma)_E) = \gamma$. One checks easily that this gives a 2-cell

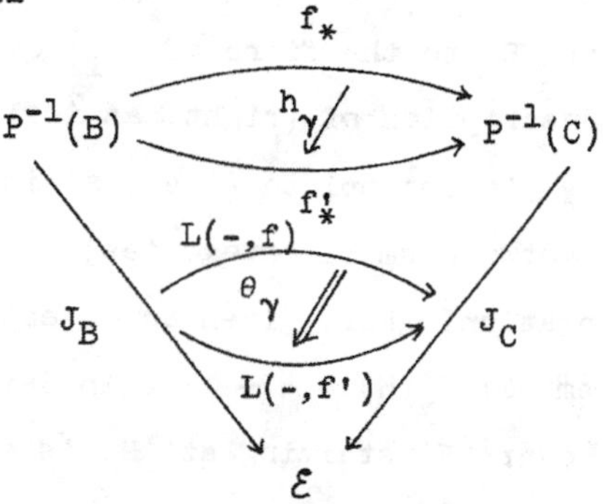

(7.54)

in ${}_s[2\text{-Cat}_{\otimes},\mathcal{E}]_{\otimes}$ in which h_γ is Cat-natural and θ_γ is a modification. A composition of 2-cells in $\mathcal{B}$ gives rise to diagrams like (7.17) and (7.18), except that the arrow labeled ℓ_n in (7.18) is the identity since the h_γ's are identities (i.e., in (7.19), the arrow labeled $F''\ell_n$ is the identity.) One checks that the diagram like (7.19) commutes in this case by using the universal mapping property. The details are routine. We note that if the fibration is not split-normal, there is still a construction as above, but what it yields is a quasi-enriched functor.

I,7.13.6 Remark: Given $F\colon \mathcal{A} \longrightarrow \mathcal{B}$, it is easily checked that $P_F\colon [F,\mathcal{B}] \longrightarrow \mathcal{B}$ is also an ordinary Cat-enriched opfibration, as in [FCC]. This construction, which is central to the discussion of quasi-Kan extensions, also plays a role in ordinary Kan extensions. For, let $I\colon \mathrm{Cat} \longrightarrow \widetilde{\mathrm{Cat}}$ be an inclusion of the category of small categories into some category of big categories (e.g., for a higher universe), and consider

$$\mathrm{Cat} \xleftarrow{\quad P \quad} [I,\widetilde{\mathrm{Cat}}] \xrightarrow{\quad P_I \quad} \widetilde{\mathrm{Cat}}.$$

Then, as above, P_I is a Cat-enriched opfibration and P is a Cat-enriched fibration. Furthermore, for fixed $\chi \in \mathrm{Cat}$, the restriction $P_\chi\colon [I, \ulcorner\chi\urcorner] \longrightarrow \mathrm{Cat}$ of P to the fibre of P_I over χ is a fibration. The usual description of (right, as in I,6.12) Kan-extensions shows that χ is cocomplete (i.e., admits Kan-extensions for arbitrary functors between small categories) if and only if P_χ is a Cat-enriched opfibration. For, given an object $H\colon \mathcal{A} \longrightarrow \chi$ in $[I, \ulcorner\chi\urcorner]$ and a morphism $F\colon P_\chi(H) = \mathcal{A} \longrightarrow \mathcal{B}$ in Cat, a cocartesian morphism in $[I, \ulcorner\chi\urcorner]$ over F starting at H is a diagram

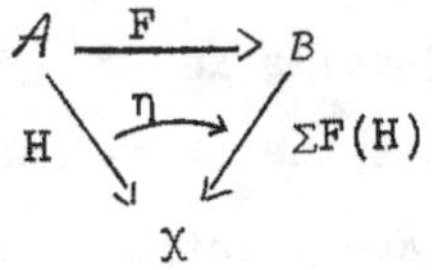

such that given any other map of the form $(F,\gamma): H \longrightarrow K$ (i.e., over F), there is a unique map of the form $(1,\gamma'): \Sigma F(H) \longrightarrow K$ such that $(1,\gamma')(F,\eta) = (F,\gamma)$; i.e., $\Sigma F(H)$ is the Kan extension of H along F.

I,7.14 Quasi-Kan extensions.

Let $F: \mathcal{A} \longrightarrow \mathcal{B}$ be a 2-functor between small 2-categories and let χ be a 2-category which has cartesian quasi-colimits. If

$$F^* = \mathrm{Fun}(F,\chi): \mathrm{Fun}(\mathcal{B},\chi) \longrightarrow \mathrm{Fun}(\mathcal{A},\chi)$$

then we would like a left quasi-adjoint

$$\Sigma_q F \;\underset{?}{\longmapsto}\; F^*$$

is as good a sense as possible. This was discussed briefly without proof in [CCS]. The general idea is to try to follow the scheme of I,1.13, using I,2.9, I,7.9.2 and I,7.13 to replace everything by the appropriate quasi-constructions. It turns out that quasi-opfibrations are very well behaved, while general functors are rather poorly behaved.

I,7.14.1. Theorem. If $P: \mathcal{E} \longrightarrow \mathcal{B}$ is a split-normal Cartesian quasi-opfibration with small fibres ($\mathcal{E}$ and $\mathcal{B}$ need not be small), then P^* has a left Cat-adjoint $\Sigma_q P$, given by "integration along the fibres."

<u>Proof</u>: The left adjoint is denoted by $\Sigma_q P$ even though it is a Cat-adjoint since it goes between 2-categories of the form $\mathrm{Fun}(\mathcal{A}, \mathcal{C})$ rather than of the form $\mathcal{C}^{\mathcal{A}}$. Define $\Sigma_q P$ as follows. If $H\colon \mathcal{E} \longrightarrow \chi$ is a 2-functor, then $\Sigma_q P(H)$ is the composition

$$\begin{array}{ccccc} B & \xrightarrow{\ J\ } & {}_s[2\text{-Cat}_{\otimes}, \mathcal{E}]_{\otimes} & \xrightarrow{\ H_*\ } & {}_s[2\text{-Cat}_{\otimes}, \chi]_{\otimes} \\ & \searrow^{\Sigma_q P(H)} & & & \Big\downarrow {\scriptstyle \text{Cart } q\text{-}\varinjlim_{\chi} = Q} \\ & & & & \chi \end{array} \qquad (7.55)$$

Here H_* denotes composition with H; i.e., abbreviating Cart q - $\varinjlim_{\chi}$ by Q as in I,7.9.2, and using the notation of I,7.13.5,

$$\begin{aligned} [\Sigma_q P(H)](B) &= Q(HJ_B) \\ [\Sigma_q P(H)](f) &= Q(f_*, H\,L(-,f)) \\ [\Sigma_q P(H)](\gamma) &= Q(h_\gamma, H\theta_\gamma) \end{aligned} \qquad (7.56)$$

In the standard case, it is evident how this is functoral in H. In our case, we must specifically indicate how a quasi-natural transformation $\psi\colon H \longrightarrow H'$ gives rise (remarkably) to a quasi-natural transformation

$$\Sigma_q P(\psi)\colon \Sigma_q P(H) \longrightarrow \Sigma_q P(H');$$

namely,

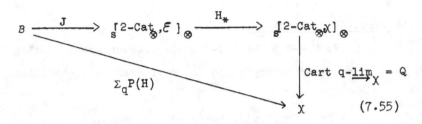

$$\Sigma_q P(\psi)_B = Q(1,\psi J_B)$$

$$\Sigma_q P(\psi)_f = Q(1,\psi_{L(-,f)}) \qquad (7.57)$$

(The reader should draw the appropriate figure like (7.54)). Finally, if $s\colon \psi \longrightarrow \psi'$ is a modification, then $\Sigma_q P(s)$ is the modification with components

$$\Sigma_q P(s)_B = Q(1,sJ_B) \quad .$$

This gives a 2-functor

$$\Sigma_q P\colon \mathrm{Fun}(\mathcal{E},\chi) \longrightarrow \mathrm{Fun}(\mathcal{B},\chi).$$

To show that $\Sigma_q P$ is the left Cat-adjoint to P^*, let $H\colon \mathcal{E} \longrightarrow \chi$, $K\colon \mathcal{B} \longrightarrow \chi$, and let $\varphi\colon H \longrightarrow KP$ be a quasi-natural transformation. Then, for $B \in \mathcal{B}$,

$$\varphi J_B\colon HJ_B \longrightarrow K\,P\,J_B = \Delta\,K(B)$$

induces a unique morphism

$$\tilde{\varphi}_B\colon\ \underset{\displaystyle [\Sigma_q P(H)](B)}{\underset{\|}{Q(HJ_B)}} \longrightarrow K(B)$$

such that, whenever $n\colon E \longrightarrow E'$ in $P^{-1}(B)$, then the diagram

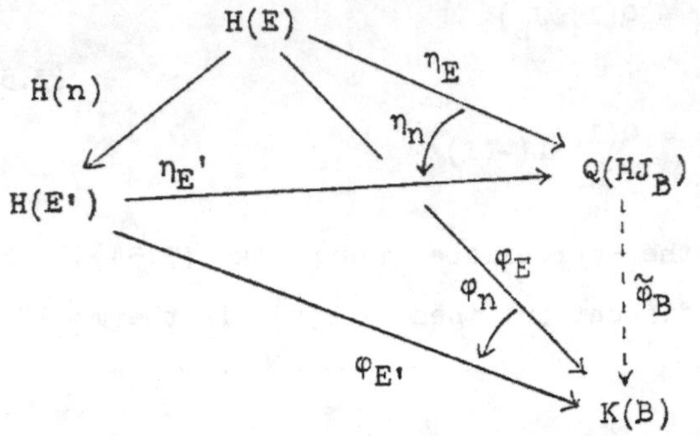

(7.58)

commutes. If $f \colon B \longrightarrow C$ in $\mathcal{B}$, then there is a 2-cell $\tilde{\varphi}_f$ in the diagram

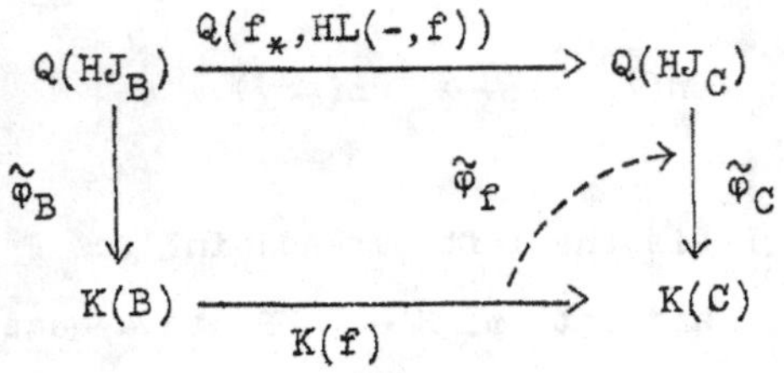

(7.59)

constructed as follows. For each E, we have

$$K(f)\tilde{\varphi}_B\eta_E = K(f)\varphi_E$$

$$\tilde{\varphi}_C Q(f_*, HL(-,f))\eta_E = \tilde{\varphi}_C \eta_{f_*E} HL(E,f)$$

$$= \varphi_{f*E} HL(E,f)$$

and one easily checks that the 2-cells

$$\varphi_{L(E,f)} \colon K(f)\eta_E \longrightarrow \varphi_{f*E}\ HL(E,f)$$

are the components of a modification

$$\varphi_{L(-,f)}\colon K(f)\tilde{\varphi}_B\eta_{(-)} \longrightarrow \tilde{\varphi}_C\, Q(f_*,HL(-,f))\eta_{(-)}$$

and hence induce the uniquely determined 2-cell $\tilde{\varphi}_f$, as indicated, satisfying $\tilde{\varphi}_f\eta_E = \varphi_{L(E,f)}$. A modification $s\colon \varphi \longrightarrow \varphi'$ clearly determines a modification $\tilde{s}\colon \tilde{\varphi} \longrightarrow \tilde{\varphi}'$ and by uniqueness everywhere, the construction is natural with respect to quasi-natural transformations and modifications in both variables H and K.

Conversely, if $\psi\colon \Sigma_q P(H) \longrightarrow K$ is a quasi-natural transformation, let $\tilde{\psi}\colon H \longrightarrow KP$ be the quasi-natural transformation whose components are given as follows. If $E \in \mathcal{E}$, then

$$\tilde{\psi}_E = \psi_{PE}\eta_E, \tag{7.60}$$

while if $g\colon E \longrightarrow E'$ in $\mathcal{E}$, then $\tilde{\psi}_g$ is the composed 2-cell in the diagram

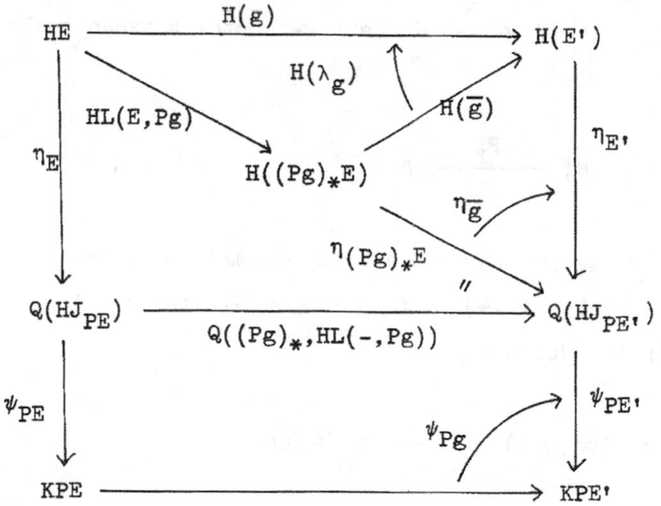

(7.61)

i.e., $\tilde{\psi}_g = (\psi_{PE'}\eta_{E'}H(\lambda_g))(\psi_{PE'}\eta_{\bar{g}}\, HL(E,Pg))(\psi_{Pg}\eta_E)$

It is easily checked that $\tilde{\psi}$ is quasi-natural and that $\tilde{\tilde{\psi}} = \psi$. It is also clear that on objects $\tilde{\tilde{\varphi}}_E = \tilde{\varphi}_{PE}\eta_E = \varphi_E$. To calculate $\tilde{\tilde{\varphi}}_g$, observe that $\tilde{\varphi}_{Pg}\eta_E = \varphi_{L(E,Pg)}$ and $\tilde{\varphi}_{PE'}\eta_{\bar{g}} = \varphi_{\bar{g}}$, while λ_g satisfies $P(\lambda_g) = \mathrm{id}$ (7.46). Hence, by quasi-naturality of φ, the above formula yields $\tilde{\tilde{\varphi}}_g = \varphi_g$.

The transformation

$$\bar{\eta}\colon \mathrm{Id} \longrightarrow P^* (\Sigma_q P)$$

is given by taking ψ the identity in (7.60) and (7.61); i.e., it is the top half of (7.61). One can show directly that it is Cat-natural by using a quasi-natural transformation $\sigma\colon H \longrightarrow H'$ to construct a cube with two sides like the top of (7.61) and with the bottom given by (7.57).

I,7.14.2 <u>Definition.</u> Let $F\colon \mathcal{A} \longrightarrow \mathcal{B}$ be a 2-functor between small 2-categories and let

$$\mathcal{A} \underset{P}{\overset{Q_F}{\rightleftarrows}} [F,\mathcal{B}] \xrightarrow{\;P_F\;} \mathcal{B}$$

be its universal factorization through a split-normal cartesian quasi-opfibration. (See I,7.13.4). If χ is a 2-category with cartesian quasi-colimits, define

$$\Sigma_q F = (\Sigma_q P_F)P^*\colon\ \mathrm{Fun}(\mathcal{A},\chi) \longrightarrow \mathrm{Fun}(\mathcal{B},\chi).$$

I,7.14.3 <u>Theorem.</u> ${}^{op}\Sigma_q F$ is a strict quasi left adjoint to

$$^{op}F^*\colon {}^{op}Fun(B,X) \longrightarrow {}^{op}Fun(A,X) .$$

Proof: By I,7.4.1, $\Sigma_q P_F$ is left Cat-adjoint to P_F^*. Since weak dualization does not affect Cat-adjoints, $^{op}(\Sigma_q P_F)$ is left Cat-adjoint to $^{op}P_F^*$. By I,7.13.4, P is left inverse, strict quasi-right adjoint to Q_F, so, by I,7.4 and its proof, $^{op}P^*$ is right inverse, strict quasi-left adjoint to $^{op}(Q_F)^*$, with adjunction morphisms $(^{op}\varepsilon^*,1)$ where ε is a given in I,7.4. Thus we are in a situation to apply I,7.3, ii), c). In the relevant diagrams, (7.5) and (7.5'), we must calculate $(^{op}Q_F^*\bar{\eta})_{(^{op}\varepsilon^*)(^{op}P_F^*)}$ and $[^{op}(\Sigma_q P_F)^*\ {}^{op}\varepsilon^*]_{\bar{\eta}\,{}^{op}P^*}$, where $\bar{\varepsilon}$ and $\bar{\eta}$ are the Cat-natural transformations giving the adjunction between $\Sigma_q P_F$ and P_F^*. From the description of $^{op}\varepsilon^*$ in I,7.4, the second modification (as well as the first) has as its components values of $\bar{\eta}$ for 1-cells. Since $\bar{\eta}$ is Cat-natural these 2-cells are all identities and hence the conditions for a composition of quasi-adjunctions to be a quasi-adjunction are satisfied.

I,7.14.4. Remarks. 1) Since $\bar{\eta}$ is Cat-natural and since $PQ_F = id$, the adjunction transformation

$$Id \longrightarrow {}^{op}F^*({}^{op}\Sigma_q F)$$

is also Cat-natural. The other adjunction transformation, however, is only quasi-natural. It follows that in the associated transendental quasi-adjunction (I,7.6 and I,7.5),

$$[{}^{op}(\Sigma_q F),\ {}^{op}Fun(A,X)] \underset{T}{\overset{S}{\rightleftarrows}} [{}^{op}Fun(B,X),{}^{op}F^*],$$

we have ST = id; so that this Cat-natural adjunction is a reflection. However TS $\neq$ id contrary to what is asserted in [CCS].

ii) Part of the reason for the failure of $\Sigma_q F$ to be a Cat-adjoint lies in the observation that if ψ: H ⟶ H' is a quasi-natural transformation in Fun($\mathcal{A}$,X), then $\Sigma_q F(\psi)$: $\Sigma_q F(H) \longrightarrow \Sigma_q F(H')$ is a Cat-natural transformation in Fun($\mathcal{B}$,X). This follows because $\Sigma_q F(\psi) = \Sigma_q P_H(\psi P)$. But, by 7.57, if f is a morphism in $\mathcal{B}$, then $\Sigma_q P_H(\psi P)_f = Q(1,\psi P_{L(-,f)})$. However, by step i) in the proof of I,7.13.2, it is immediate that for any object (h: FA ⟶ B) $\in$ [F,$\mathcal{B}$], we have

$$P(L(h,f)) = id_A$$

so $\psi P_{L(h,f)}$ is an identity 2-cell. Intuitively, P is transversal to the fibres in [F,$\mathcal{B}$]. One can show that in the above transendental situation the restriction of TS to the full subcategory determined by Cat-natural transformations is the identity.

I,7.14 <u>The Categorical Comprehension Scheme</u>. This construction and its meaning are discussed thoroughly in [CCS] and a proof of the transcendental quasi-adjunction asserted there is given in [21], the last twenty pages of which consist of a construction of the required Cat-adjoints between 2-comma categories. We shall show here that there is a much shorter construction of a strict quasi-adjunction, which, by I,7.5,implies the transendental quasi-adjunction. We use the notation of [21], p. 466-471, which explicitely describes the 2-functors

$$[{}^{op}Cat,X] \underset{[1,-] = \mathcal{U}}{\overset{\Sigma_q(-)(1_{(-)}) = \mathcal{F}}{\rightleftarrows}} {}^{op}Fun(X,Cat)$$

and will construct quasi-natural transformations

$$\varepsilon: \mathcal{F}\mathcal{U} \longrightarrow 1, \quad \eta: 1 \longrightarrow \mathcal{U}\mathcal{F}$$

which satisfy the adjunction equations as well as the equations

$$\varepsilon_\varepsilon = 1, \quad \eta_\eta = 1. \tag{7.62}$$

i) If $K: \mathcal{X} \longrightarrow Cat$ is a functor, then $\mathcal{U}(K)$ is the associated opfibration $P_K: [1,K] \longrightarrow \mathcal{X}$ and $\mathcal{F}\mathcal{U}(K)$ is the functor $(P_K,-): \mathcal{X} \longrightarrow Cat$. Let $\varepsilon_K: (P_K,-) \longrightarrow K$ be the morphism (i.e., quasi-natural transformation) in ${}^{op}Fun(\mathcal{X},Cat)$ whose components are the functors

$$(\varepsilon_K)_X: (P_K,X) \longrightarrow K(X)$$

given as follows. An object of (P_K,X) consists of a pair $((X',x'), f': X' \longrightarrow X)$ where $x' \in K(X')$, so we set $(\varepsilon_K)_X((X',x'),f') = (Kf')x' \in K(X)$. A morphism in (P_K,X) is a pair

$$(g,m): ((X',x'), f': X' \longrightarrow X) \longrightarrow ((X'',x''), f'': X'' \longrightarrow X)$$

where $g: X' \longrightarrow X''$ satisfies $f''g = f'$ and $m: (Kg)x' \longrightarrow x''$. We set $(\varepsilon_K)_X(g,m) = (Kf'')m: (Kf')x' \longrightarrow (Kf'')x''$ in $K(X)$. It is easily verified that this is a functor and that if $k: X \longrightarrow Y$, then the diagram

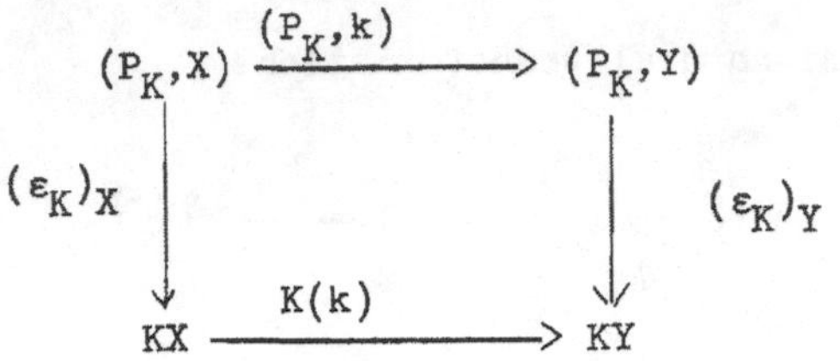

(7.63)

commutes, so ε_K is a natural transformation.

This defines the 1-cells $\varepsilon_K\colon \mathcal{F}\mathcal{U}(K) \longrightarrow K$. We must also describe 2-cells ε_τ corresponding to quasi-natural transformations $\tau\colon K \longrightarrow K'$, which map as indicated.

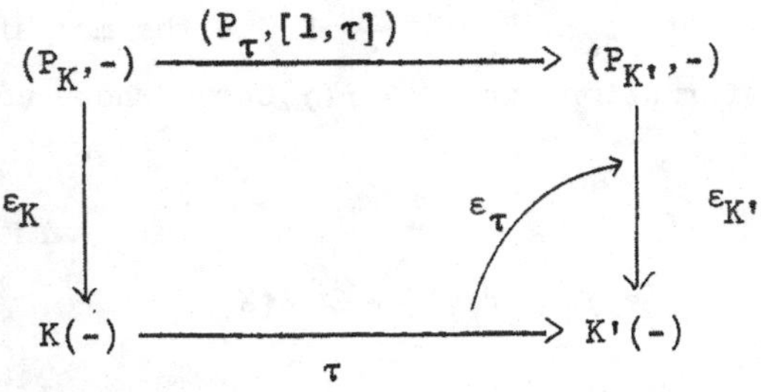

(7.64)

Here $[1,\tau]\colon [1,K] \longrightarrow [1,K']$ takes (X,x) to $(X,\tau_X(x))$ and (f,m) to $(f,\tau_Y(m)\cdot(\tau_f)_x)$ where $f\colon X \longrightarrow Y$ and $m\colon K(f)x \longrightarrow y$. Hence $(P_\tau,[1,\tau])$ takes $((X',x'),\ f'\colon X' \longrightarrow X)$ to $((X',\tau_{X'}(x')),f')$ and (g,m) to $(g,\tau_{X''}(m)\tau_g)$. Hence an object $((X',x'),f') \in (P_K,X)$ is taken clockwise in (7.64) to $(K'f')(\tau_{X'}(x'))$ and counterclockwise to $(\tau_X(Kf))(x')$. Thus we can set

$$((\varepsilon_\tau)_X)_{((X',x'),f')} = (\tau_{f'})_{x'} \ .$$

One verifies that $(\varepsilon_\tau)_X$ is a natural transformation, that ε_τ is a 2-cell in ${}^{op}\mathrm{Fun}(\chi,\mathrm{Cat})$, and that the ε_K's and ε_τ's define a quasi-natural transformation $\varepsilon: \mathcal{F}\mathcal{U} \longrightarrow 1$. It is immediate from the definition of ε_τ that if τ is a natural transformation, then $\varepsilon_\tau = 1$. Since each ε_K is natural, $\varepsilon_{\varepsilon_K} = 1$, which is the first relation in (7.62).

ii) If $F: \mathcal{A} \longrightarrow \chi$ is an object in $[{}^{op}Cat, \chi]$ then $\mathcal{F}(F)$ is the functor $(F,-): \chi \longrightarrow Cat$ and $\mathcal{U}\mathcal{F}(F)$ is the functor $P_F: (F,\chi) \longrightarrow \chi$. Let η_F be the morphism

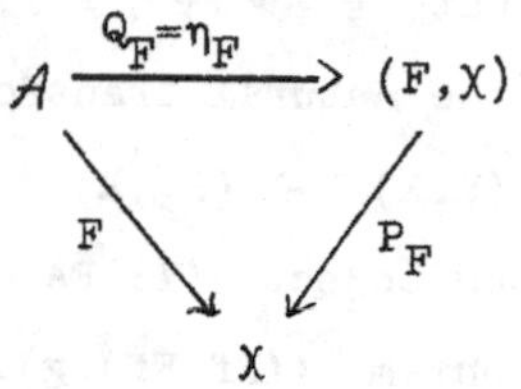

(7.65)

as in I,1.11, where $Q_F(A) = 1_{FA}: FA \longrightarrow FA$ and $Q_F(f) = (f,Ff)$. There is no 2-cell indicated since the triangle commutes. On a morphism $\langle M,m\rangle: F \longrightarrow F'$ in $[{}^{op}Cat,\chi]$, $\eta_{\langle M,m\rangle}$ is the diagram

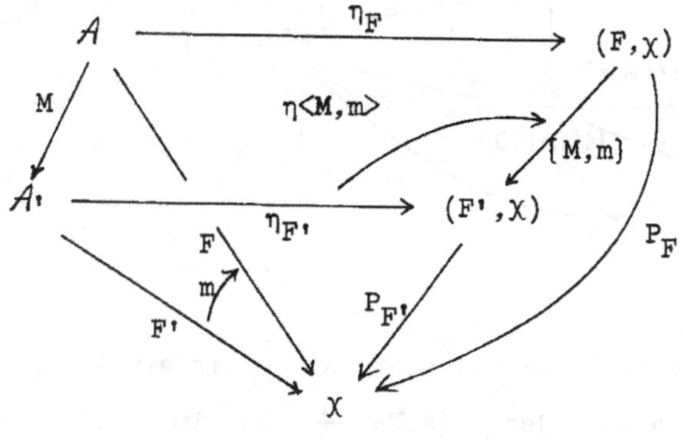

(7.66)

where $\eta_{\langle M,m\rangle}$ is the natural transformation whose component at $A \in \mathcal{A}$ is the morphism

$$\begin{array}{ccc} F'MA & \xrightarrow{\quad 1 \quad} & F'MA \\ {\scriptstyle 1}\downarrow & & \downarrow{\scriptstyle m_A} \\ F'MA & \xrightarrow[\quad m_A \quad]{} & FA \end{array}$$

in (F',χ). In particular, if $m = 1$, then $\eta_{\langle M,m\rangle} = 1$. Hence $\eta_{\eta_F} = 1$, which is the second relation in (7.62).

iii) Next we show that $\varepsilon\mathcal{F} \circ \mathcal{F}\eta = 1$. First observe that $(\mathcal{F}\eta)_F\colon (F,-) \longrightarrow (P_F,-)$ is the natural transformation whose component at X, $((\mathcal{F}\eta)_F)_X\colon (F,X) \longrightarrow (P_F,X)$, is the functor taking an object $(A, FA \longrightarrow X)$ to the object $(1\colon FA \longrightarrow FA,\ FA \longrightarrow X)$ and a morphism (f,g) to the morphism $((Ff,Ff),g)$. Furthermore, if $\langle M,m\rangle\colon F \longrightarrow F'$ is a morphism in $[{}^{op}Cat,\chi]$, then $\eta_{\langle M,m\rangle}$ is given by (7.66) and $(\mathcal{F}\eta)_{\langle M,m\rangle}$ is the diagram

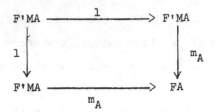

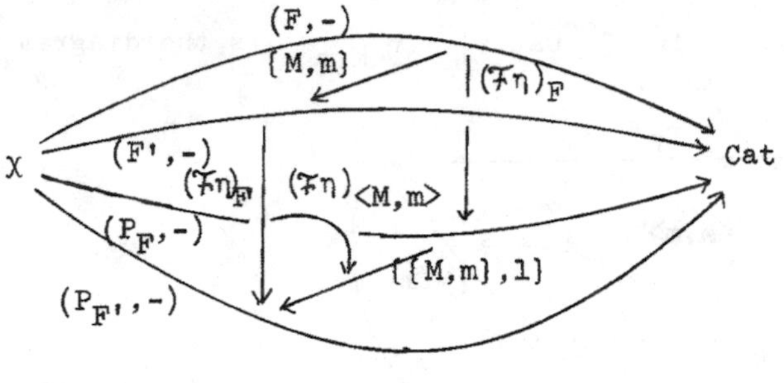

; (7.67)

i.e., it is a modification whose value at $X \in \chi$ is a natural transformation whose value on an object $(A, FA \xrightarrow{h} X)$ in (F,X) is the morphism

$$\begin{array}{ccc|ccccc}
F'MA & \xrightarrow{1} & F'MA & F'MA & \xrightarrow{m_A} & FA & \xrightarrow{h} & X \\
{\scriptstyle 1}\downarrow & & \downarrow{\scriptstyle m_A} & {\scriptstyle m_A}\downarrow & & & & \downarrow{\scriptstyle 1} \\
F'MA & \xrightarrow[m_A]{} & FA & FA & & \xrightarrow[h]{} & & X
\end{array}$$

in $(P_{F'},X)$.

On the other hand,

$$((\varepsilon\mathcal{F})_F)_X\colon (P_F,X) \longrightarrow (F,X)$$

is just composition; i.e., it takes an object

$$(h\colon FA \longrightarrow X',\ f'\colon X' \longrightarrow X) \quad \text{in} \quad (P_F,X) \quad \text{to}$$

the object $(A, f'h\colon FA \longrightarrow X)$ in (F,X) and a morphism $(m,g,g,1)$ to $(m,1)$. Hence

$$\begin{aligned}
&[((\varepsilon\mathcal{F})_F)_X \circ ((\mathcal{F}\eta)_X)](A, FA \longrightarrow X) \\
&\qquad = ((\varepsilon\mathcal{F})_F)_X(1\colon FA \longrightarrow FA,\ FA \longrightarrow X) = (A, FA \longrightarrow X)
\end{aligned}$$

so the first adjunction equation holds on objects.

To calculate it on morphisms, observe that $\mathcal{F}$ takes morphisms in $[{}^{op}Cat,\chi]$ to natural transformations and hence, by i); $\varepsilon\mathcal{F}$ is natural. Thus we need only calculate $(\varepsilon\mathcal{F})_{F'}$ applied to the morphism in (7.68), which gives the identity map

$$\begin{array}{ccc}
F'MA & \xrightarrow{\quad 1 \quad} & F'MA \\
& {\scriptstyle hm_A}\searrow \quad \swarrow{\scriptstyle hm_A} & \\
& X &
\end{array}$$

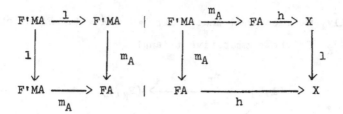

iv) Finally, we must show that $\mathcal{U}\varepsilon \circ \eta\mathcal{U} = 1$. First observe that $(\eta\mathcal{U})_K$ is the commutative triangle

$$\begin{array}{ccc} [1,K] & \xrightarrow{\eta_{P_K}} & (P_K,\chi) \\ & {\scriptstyle P_K}\searrow \quad \swarrow{\scriptstyle P_{(P_K)}} & \\ & \chi & \end{array}$$

and $(\eta\mathcal{U})_\tau$, for a quasi-natural transformation $\tau: K \longrightarrow K\,'$, is the identity, since $\mathcal{U}$ applied to a morphism gives a commutative triangle (cf. (7.66)). On the other hand, $(\mathcal{U}\varepsilon)_K$ is the commutative triangle

$$\begin{array}{ccc} (P_K,\chi) & \xrightarrow{(\mathcal{U}\varepsilon)_K} & [1,K] \\ & {\scriptstyle P_{(P_K)}}\searrow \quad \swarrow{\scriptstyle P_K} & \\ & \chi & \end{array}$$

in which $(\mathcal{U}\varepsilon)_K$ is the functor taking an object $((X',x'),f': X' \to X)$ to the object $(X,(Kf')x')$ and a morphism $((f,\varphi),f,h)$ to $(h,(Kg')\varphi)$, where one has

$$\begin{array}{ccc} X' & \xrightarrow{f'} & X \\ {\scriptstyle f}\downarrow & & \downarrow{\scriptstyle h} \\ Y' & \xrightarrow[g']{} & Y \end{array} .$$

If $\tau: K \longrightarrow K'$, then, as in (7.64), $(\mathcal{U}\varepsilon)_\tau$ is the natural transformation whose component at the object $((X',x'),f': X' \longrightarrow X)$ in (P_K,χ) is the morphism $(1_X,(\tau_{f'})_{x'})$ in $[1,K']$. Hence

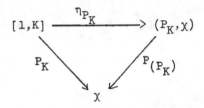

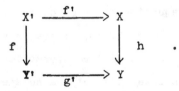

$$[(\mathcal{U}\varepsilon)_K \circ (\eta\mathcal{U})_K](X,x) = (\mathcal{U}\varepsilon)_K((X,x),\ 1\colon X \longrightarrow X) = (X,x)$$

$$[(\mathcal{U}\varepsilon)_K \circ (\eta\mathcal{U})_K](f,\varphi) = (\mathcal{U}\varepsilon)_K((f,\varphi),f,f) = (f,\varphi)$$

and, on morphisms, since $(\eta\mathcal{U})_\tau = 1$, we have

$$[(\mathcal{U}\varepsilon)_\tau \circ \eta_{P_K}](X,x) = ((\mathcal{U}\varepsilon)_\tau)_{((X,x),1:X \to x)} = (1_X,(\tau_1)_x) = 1.$$

Hence the other adjunction equation holds.

One can verify that the transcendental quasi-adjunction of [21] is derived from the strict quasi-adjunction given here by using I,7.5.

I,7.16. <u>The Quasi-Yoneda Lemma</u>.

This example illustrates another aspect of quasi-adjointness. We first describe the situation in general and then prove both local and global analogues of the Yoneda lemma.

The general situation is as follows: $\mathcal{A}$ and $\mathcal{B}$ are 2-categories, $F\colon \mathcal{A} \longrightarrow \mathcal{B}$ and $U\colon \mathcal{B} \longrightarrow \mathcal{A}$ are 2-functors which are bijections on the objects, and $\varepsilon\colon FU \longrightarrow \mathcal{B}$ and $\eta\colon \mathcal{A} \longrightarrow UF$ are quasi-natural transformations whose components at objects are identity morphisms. If we identify the objects of $\mathcal{A}$ and $\mathcal{B}$ via F and U, then all that is left are the components of ε and η at morphisms, which look like

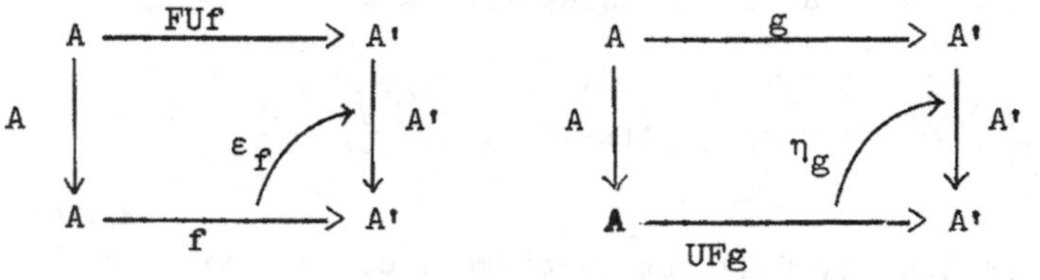

(7.69)

where $f \in \mathcal{B}$ and $g \in \mathcal{A}$.

I,7.16.1. Proposition. ε and η define a quasi-adjunction $F \xrightarrow[\text{quasi}]{} \!\!| \; U$ if and only if for each pair of objects A, A', the functors $F_{A,A'}: \mathcal{A}(A,A') \longrightarrow \mathcal{B}(A,A')$, $U_{A,A'}: \mathcal{B}(A,A') \longrightarrow \mathcal{A}(A,A')$ are adjoint (in the ordinary sense) via adjunction morphisms, ε_f and η_g as above, which are multiplicative in the sense that

$$\varepsilon_{f'f} = \varepsilon_{f'} \boxplus \varepsilon_f \quad , \quad \eta_{g'g} = \eta_{g'} \boxplus \eta_g \tag{7.70}$$

Proof. Clear.

If F and U are pseudo-functors then these nice equations are spoiled by having the appropriate φ's, α's, etc., inserted in them, but the principle is the same. Also, if one has $F: {}^{op}\mathcal{A} \longrightarrow {}^{op}\mathcal{B}$, $U: {}^{op}\mathcal{B} \longrightarrow {}^{op}\mathcal{A}$, then $F_{A,A'} \dashv U_{A,A'}$, so the order of adjointness is preserved.

For instance, if a cartesian closed category is regarded as a bicategory $\mathcal{C}$ with a single object, then $A\times -$ is a copseudo-functor via the diagonal and $(-)^A$ is a strict copseudo-functor. Regarded as pseudo-functors from ${}^{op}\mathcal{C}$ to itself, one has $A\times - \xrightarrow[\text{quasi}]{} \!\!| \; (-)^A$.

The local quasi-Yoneda lemma is concerned with describing the situation of I,5.10 i) in terms of quasi-adjoints as above. In what follows, $\mathcal{A}$ and $\mathcal{B}$ are 2-categories and

$$(2\text{-}_{\ell}\mathrm{Cat}, \mathcal{B} \times \mathcal{A})' \;\; ; \text{ resp.,} \quad '(2\text{-}_{\ell}\mathrm{Cat}, \mathcal{A} \times \mathcal{B})$$

is the full and locally full sub 2-category of $(2\text{-}_{\ell}\mathrm{Cat}, \mathcal{B} \times \mathcal{A})$; resp., $(2\text{-}_{\ell}\mathrm{Cat}, \mathcal{A} \times \mathcal{B})$, determined by objects of the form

$$[\mathcal{B},F] \longrightarrow \mathcal{B} \times \mathcal{A}; \text{ resp., } [F,\mathcal{B}] \longrightarrow \mathcal{A} \times \mathcal{B}$$

where $F: \mathcal{A} \longrightarrow \mathcal{B}$ is a 2-functor. The idea is that the operation taking a 2-functor $F: \mathcal{A} \longrightarrow \mathcal{B}$ into the objects in Spans (2-Cat)

$$[\mathcal{B},F] \longrightarrow \mathcal{B} \times \mathcal{A} ,$$

which can be thought of as the bifibration corresponding to $\mathrm{Hom}_{\mathcal{B}}(-,F(-)): \mathcal{B}^{op} \times \mathcal{A} \longrightarrow \mathrm{Cat}$, should be full and faithful. However, it is not. Rather, it is faithful onto a quasi-reflective subcategory in the sense made precise by the following two theorems.

I,7.16.2 <u>Theorem</u> (The local quasi-Yoneda lemma). There exists a quasi-adjunction

$$\hat{\Psi} \dashv_{\text{quasi}} \Psi: \ {}^{op}\mathrm{Fun}(\mathcal{A},\mathcal{B}) \longrightarrow {}^{op}(2\text{-}_{\ell}\mathrm{Cat}, \mathcal{B}\times\mathcal{A})' ,$$

resp.,

$$\Psi \dashv_{\text{quasi}} \check{\Psi}: \ '(2\text{-}_{\ell}\mathrm{Cat}, \mathcal{A} \times \mathcal{B}) \longrightarrow \mathrm{Fun}(\mathcal{A},\mathcal{B})^{op}$$

such that

i) Ψ is a 2-functor and $\hat{\Psi}$ (resp., $\check{\Psi}$) is a pseudo-functor.

ii) $\hat{\Psi}\Psi = \mathrm{id}$ and $\Psi,\hat{\Psi}$ and $\eta: \mathrm{Id} \longrightarrow \Psi\hat{\Psi}$ are identities on objects. (Resp., $\check{\Psi}\Psi = \mathrm{id}$ and $\Psi,\check{\Psi}$ and $\varepsilon: \Psi\check{\Psi} \longrightarrow \mathrm{Id}$ are identities on objects.) (Cf., I,7.16.1).

iii) Furthermore, if $T: [\mathcal{B},F] \longrightarrow [\mathcal{B},G]$ is over $\mathcal{B} \times \mathcal{A}$ then the following are equivalent

a) $\Psi\hat{\Psi}(T) = T$

b) T is a left $[\mathcal{B},\mathcal{B}]$-homomorphism in the sense of I,5.9,10

c) T applied to a morphism of the form

$$\begin{array}{ccc} B & \xrightarrow{\ h\ } & FA \\ {\scriptstyle h}\big\downarrow & & \big\downarrow{\scriptstyle FA} \\ FA & \xrightarrow[\ FA\ ]{} & FA \end{array}$$

is a commutative square. (Resp., for $T: [F,\mathcal{B}] \longrightarrow [G,\mathcal{B}]$ over $\mathcal{A} \times \mathcal{B}$; a') $\Psi\check{\Psi}(T) = T$, b') T is a right $[\mathcal{B},\mathcal{B}]$-homomorphism; c') T applied to a morphism of the form

$$\begin{array}{ccc} FA & \xrightarrow{\ FA\ } & FA \\ {\scriptstyle FA}\big\downarrow & & \big\downarrow{\scriptstyle h} \\ FA & \xrightarrow[\ h\ ]{} & B \end{array}$$

is a commutative square.)

<u>Proof</u>: i) If $F: \mathcal{A} \longrightarrow \mathcal{B}$ then

$$\Psi(F) = \{P_{\mathcal{B}}, P_{\mathcal{A}}\}: [\mathcal{B},F] \longrightarrow \mathcal{B} \times \mathcal{A} .$$

If $\varphi: F \longrightarrow G$ is a quasi-natural transformation, then $\Psi(\varphi)$ is the morphism (I,5.3,ii) and I,5.4 c).)

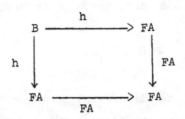

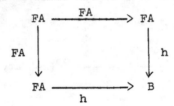

$$\Psi(\varphi):\quad \begin{array}{ccc} [\mathcal{B},F] & \xrightarrow{[\mathcal{B},\varphi]} & [\mathcal{B},G] \\ & \searrow \quad \swarrow & \\ & \mathcal{B}\times\mathcal{A} & \end{array}$$

in the comma category. Finally, if $u\colon \varphi \longrightarrow \varphi'$ is a modification, then $\Psi(u)$ is the 2-cell (I,5.3 ii) and I,5.4,d).)

$$\Psi(u):\quad \begin{array}{ccc} [\mathcal{B},F] & \overset{[\mathcal{B},\varphi]}{\underset{[\mathcal{B},\varphi']}{\underset{\Downarrow\,[\mathcal{B},u]}{\longrightarrow}}} & [\mathcal{B},G] \\ & \searrow \quad \swarrow & \\ & \mathcal{B}\times\mathcal{A} & \end{array}$$

Note that $[\mathcal{B},u]$ is a natural transformation. It is a special case of I,5.3,ii), that Ψ is a 2-functor. We regard it as a 2-functor between the weak duals in order to have the adjointness come out correctly.

Conversely, define

$$\hat{\Psi}([\mathcal{B},F] \longrightarrow \mathcal{B}\times\mathcal{A}) = F$$

and given a morphism

$$\begin{array}{ccc} [\mathcal{B},F] & \xrightarrow{T} & [\mathcal{B},G] \\ & \searrow \quad \swarrow & \\ & \mathcal{B}\times\mathcal{A} & \end{array} \tag{7.71}$$

then, following the proceedure of I,5.9, ii), let $\mathbb{T}$ be the unique 2-functor over $\mathcal{A}\times\mathcal{A}$ making the diagram

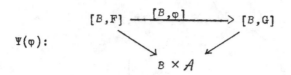

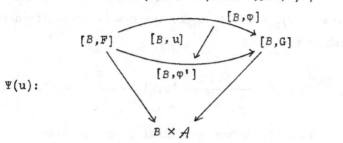

$$\begin{array}{ccc} [F,F] & \xrightarrow{(F,1,1)} & [\mathcal{B},F] \\ \bar{T}\downarrow & & \downarrow T \\ [F,G] & \xrightarrow[(F,1,1)]{} & [\mathcal{B},G] \end{array}$$

commute and define $\hat{\Psi}(T): F \longrightarrow G$ to be the quasi-natural transformation corresponding to

$$\overline{\hat{\Psi}(T)} = \mathcal{A} \xrightarrow{j_F} [F,F] \xrightarrow{\bar{T}} [F,G] \quad (7.72)$$

via I,5.2, iii). Similarly, given a natural transformation $\lambda: T \longrightarrow T'$ over $\mathcal{B} \times \mathcal{A}$, it determines a natural transformation $\bar{\lambda}: \bar{T} \longrightarrow \bar{T}'$ by I,5.6, iii) and $\hat{\Psi}(\lambda): \hat{\Psi}(T) \longrightarrow \hat{\Psi}(T')$ is the modification corresponding to $\overline{\hat{\Psi}}(\lambda) = \bar{\lambda}_{j_F}$ via I,5.2, iii). It follows from the uniqueness part of I,5.6 together with I,5.1 that this defines a functor

$$\hat{\Psi}_{F,G} \text{ for fixed } F \text{ and } G.$$

$\hat{\Psi}$ itself is only a copseudo-functor since, given a composition

$$[\mathcal{B},F] \xrightarrow{T} [\mathcal{B},G] \xrightarrow{T'} [\mathcal{B},H] \qquad (7.73)$$

over $\mathcal{B} \times \mathcal{A}$, there is a modification

$$s_{T',T}: \hat{\Psi}(T'T) \longrightarrow \hat{\Psi}(T')\hat{\Psi}(T) \qquad (7.74)$$

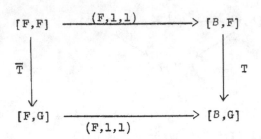

between quasi-natural transformations from F to H whose component at $A \in \mathcal{A}$ is the indicated 2-cell:

$$T'\left(\begin{array}{ccc} FA & \xrightarrow{T(id_{FA})} & GA \\ {\scriptstyle T(id_{FA})}\downarrow & & \downarrow{\scriptstyle id_{GA}} \\ GA & \xrightarrow[GA]{} & GA \end{array}\right) = \left(\begin{array}{ccc} FA & \xrightarrow{T(id_{FA})} & GA \\ {\scriptstyle T'T(id_{FA})}\downarrow & (s_{T',T})_A \nearrow & \downarrow{\scriptstyle T'(id_{GA})} \\ HA & \xrightarrow[HA]{} & HA \end{array}\right)$$

since, by definition $(\hat{\Psi}T)_A = T(id_{FA})$, etc. To see that this is a modification, observe that the modification $\gamma_{T'}$ of I,5.10 induces a modification $\overline{\gamma}_{T'}$ in

$$\begin{array}{ccc} [F,G]\underset{\mathcal{A}}{\times}[G,G] & \xrightarrow{\circ} & [F,G] \\ {\scriptstyle 1\underset{\mathcal{A}}{\times}\overline{T}'}\downarrow & \overline{\gamma}_{T'} \swarrow & \downarrow{\scriptstyle \tilde{T}'} \\ [F,G]\underset{\mathcal{A}}{\times}[G,H] & \xrightarrow[\circ]{} & [F,H]\end{array}$$

where $\tilde{T}'$ is induced by $T'(F,1,1)$, by I,5.6,i), and that $s_{T'T}$ is the composition of this with

$$\{\overline{T}\, j_F, j_G\} : \mathcal{A} \longrightarrow [F,G] \underset{\mathcal{A}}{\times} [G,G] \;.$$

There should be a proof based on I,5.9 that $(\hat{\Psi}, s_{(-,-)}, id)$ is a copseudo-functor, but we have not been able to find it. Instead, observe directly that given

$$T'' : [\mathcal{B},H] \longrightarrow [\mathcal{B},K]$$

over $\mathcal{B} \times \mathcal{A}$, then T'' applied to

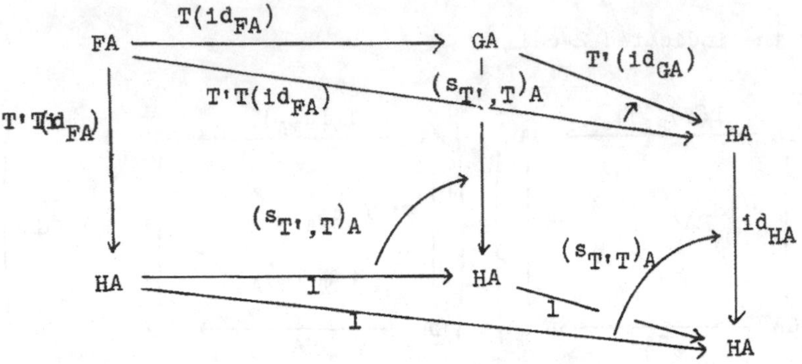

yields

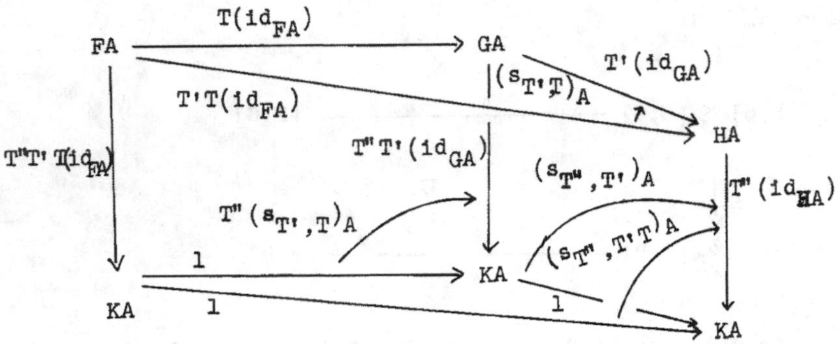

But $T''(s_{T',T})_A = (s_{T''T',T})_A$ and $T(id_{FA}) = \hat{\Psi}(T)_A$, etc., so

$$(s_{T'',T'})_A\ \hat{\Psi}(T)_A \cdot (s_{T''T',T})_A = \hat{\Psi}(T'')_A (s_{T',T})_A \cdot (s_{T'',T'T})_A$$

which shows that $\hat{\Psi}$ is a copseudo-functor. We choose to regard $\hat{\Psi}$ as a pseudo-functor between the weak duals as indicated.

By definition, Ψ and $\hat{\Psi}$ are mutual inverses on objects. Furthermore, it is evident that on 1-cells and 2-cells, as well $\hat{\Psi}\Psi = id$. To show that $\hat{\Psi} \;\underset{\text{quasi}}{\longrightarrow}\; \Psi$ as indicated, we must describe a quasi-natural transformation η: Id $\longrightarrow \Psi\hat{\Psi}$ whose components are

natural transformations as illustrated

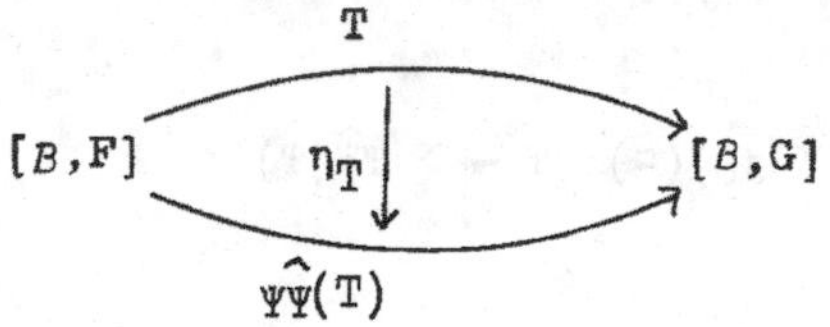

over $\mathcal{B} \times \mathcal{A}$, satisfying $\eta_{\Psi(\varphi)} = \mathrm{id} = \hat{\Psi}(\eta_T)$ and "$\eta_{T'T} = \eta_{T'}\eta_T$". Now by the definition of $\hat{\Psi}(T)$ and by I,5.4, c), $\Psi\hat{\Psi}(T)$ is given by the composition

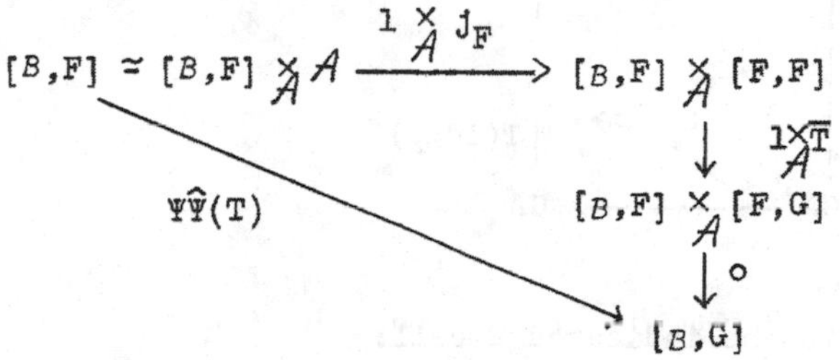

Hence there is a diagram (from I,5.10)

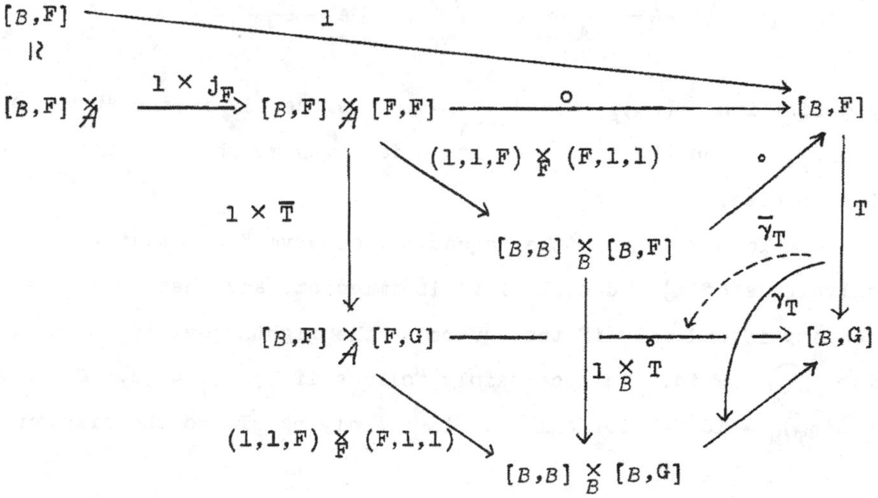

where $\overline{\gamma}_T = \gamma_T[(1,1,F) \underset{F}{\times} (F,1,1)]$, and we set

$$\eta_T = \overline{\gamma}_T(1 \times j_F)(\simeq): \ T \longrightarrow \Psi\widehat{\Psi}(T) \tag{7.75}$$

Since γ_T is a natural transformation, so is η_T. The component of η_T at $(h: B \longrightarrow FA) \ \varepsilon \ [B,F]$ is easily calculated to be the morphism

$$(\eta_T)_h: \quad T(h) \quad \begin{array}{ccc} B & \xrightarrow{\ \ B\ \ } & B \\ & & \downarrow h \\ & \widetilde{(\eta_T)_h} & FA \\ & \nearrow & \downarrow T(\mathrm{id}_{FA}) \\ GA & \xrightarrow{\ \ GA\ \ } & GA \end{array}$$

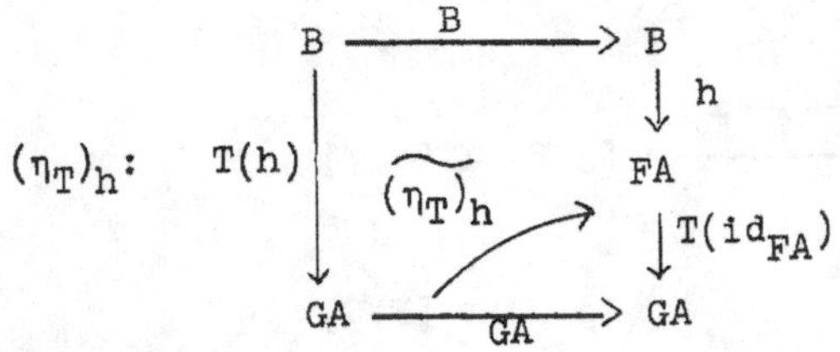

in which $\widetilde{(\eta_T)_h}$ is the indicated 2-cell:

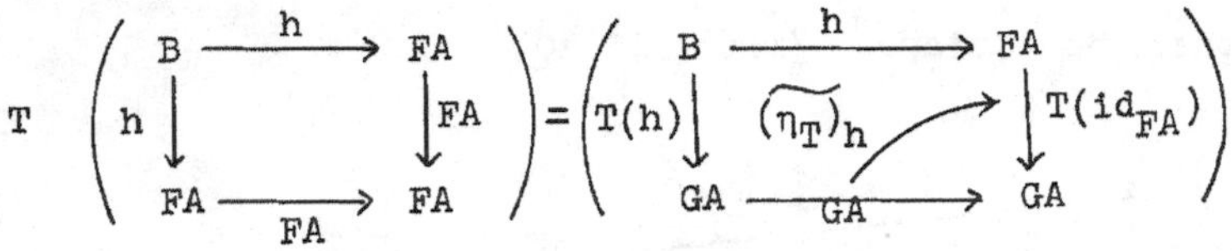

$$T \left(\begin{array}{ccc} B & \xrightarrow{\ h\ } & FA \\ h\downarrow & & \downarrow FA \\ FA & \xrightarrow{\ FA\ } & FA \end{array} \right) = \left(T(h)\downarrow \begin{array}{ccc} B & \xrightarrow{\ h\ } & FA \\ & \widetilde{(\eta_T)_h}\nearrow & \downarrow T(\mathrm{id}_{FA}) \\ GA & \xrightarrow{\ GA\ } & GA \end{array} \right)$$

By definition $\widehat{\Psi}(\eta_T)_A = (\eta_T)_{\mathrm{id}_{FA}} = \mathrm{id}$, since, if in the above square, $h = \mathrm{id}_{FA}$, then the left side is an identity morphism so the right side is also.

To derive the other equation, observe first that the equivalence of a) and c) in iii) is immediate and that η_T is the identity if and only if the square on the right above is commutative, with $\widetilde{(\eta_T)_h} = \mathrm{id}$. This certainly follows if b), $\gamma_T = \mathrm{id}$. Conversely, if $\widetilde{(\eta_T)_h} = \mathrm{id}$ for all h, then applying T to the diagram

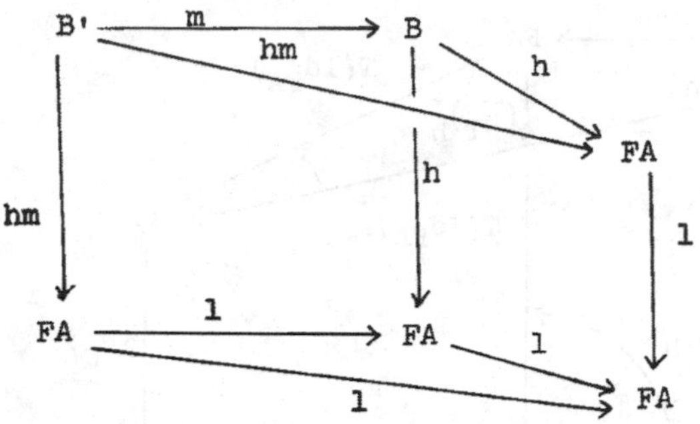

and using $\widetilde{(\eta_T)}_h = \mathrm{id} = \widetilde{(\eta_T)}_{hm}$ shows that $(\gamma_T)_{m,h} = \mathrm{id}$ for all $(m,h) \in [B,B] \underset{B}{\times} [B,F]$. Hence the conditions in iii) are equivalent. Furthermore, $\eta_{\Psi(\varphi)} = \mathrm{id}$ is evident since $\gamma_{\Psi(\varphi)} = \mathrm{id}$ follows from I,5.10.

Finally, we must show that η is multiplicative as in I,7.16.1. This can be done using I,5.9, but the requisite diagram is quite complicated. In terms of components, the following must be shown: given a composition as in (7.73), then, taking account of the pseudo-functoriality of $\hat{\Psi}$ in (7.70),

$$(\eta_{T'})(\eta_T) = \Psi(s_{T',T}) \cdot \eta_{T'T} \;;$$

i.e., given $(h\colon B \longrightarrow FA) \in [B,F]$, then

$$T'(\mathrm{id}_{GA})(\bar{\eta}_T)_h \cdot (\bar{\eta}_{T'})_{T(h)} = (s_{T',T})_A h \cdot (\bar{\eta}_{T'T})_h \,.$$

This follows by applying T' to the diagram

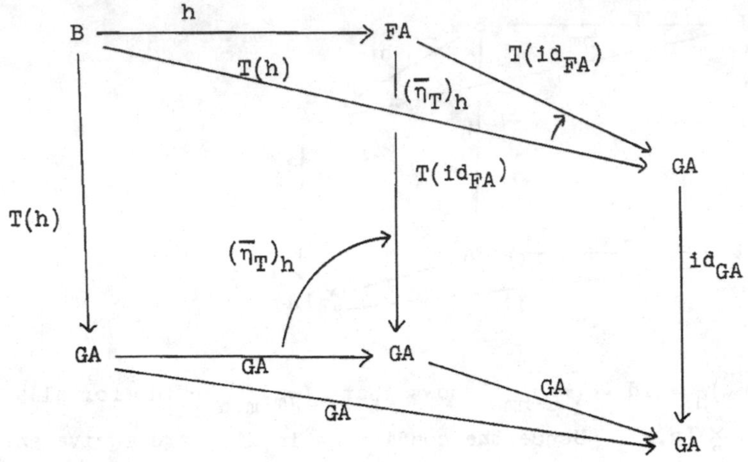

and observing that

$$\Psi\hat{\Psi}(T)(h) = B \xrightarrow{h} FA \xrightarrow{T(id_{FA})} GA$$

etc., so that $\Psi\hat{\Psi}(T)(id_{FA}) = T(id_{FA})$, $\Psi(s_{T',T})_A = (s_{T',T})_A h$ and $T'(\bar\eta_T)_h = (\bar\eta_{T'T})_h$. This finishes the proof of the local quasi-Yoneda lemma.

I,7.16.3 Theorem. (The global quasi-Yoneda lemma). The functors Ψ above for each $\mathcal{A}$ and $\mathcal{B}$ are parts of a quasi-functor

$$\overline{\Psi}\colon 2\text{-Cat}_{\otimes} \longrightarrow \text{Spans (2-Cat)}$$

between $2\text{-Cat}_{\otimes}$-categories.

Proof: Define

$$\overline{\Psi}(\mathcal{A}) = \mathcal{A}$$

$$\overline{\Psi}(F\colon \mathcal{A} \longrightarrow \mathcal{B}) = \mathcal{A} \longleftarrow [\mathcal{B},F] \longrightarrow \mathcal{B}$$

$$\overline{\Psi}(\varphi\colon F \longrightarrow G) = \begin{array}{ccccc} & \swarrow & [\mathcal{B},F] & \searrow & \\ \mathcal{A} & & \big\downarrow{\scriptstyle [\mathcal{B},\varphi]} & & \mathcal{B} \\ & \nwarrow & [\mathcal{B},G] & \nearrow & \end{array}$$

$$\overline{\Psi}(u\colon \varphi \longrightarrow \varphi') = [\mathcal{B},u]$$

We must describe what happens to compositions in $2\text{-Cat}_{\otimes}$. Given

$$\mathcal{A} \xrightarrow{F} \mathcal{B} \xrightarrow{G} \mathcal{C}$$

then there is a 2-functor

$$\varphi_{F,G}\colon [\mathcal{B},F] \underset{\mathcal{B}}{\times} [\mathcal{C},G] \longrightarrow [\mathcal{C},GF]$$

giving a 2-cell in Spans (2-Cat)$(\mathcal{A},\mathcal{C})$ from $\overline{\Psi}(G)\overline{\Psi}(F)$ to $\overline{\Psi}(GF)$. For instance,

$$\varphi_{F,G}(B \longrightarrow FA,\ C \longrightarrow GB) = C \longrightarrow GB \longrightarrow GFA, \quad \text{etc.}$$

This is clearly compatible with associativity. The interesting part of the structure occurs for a pair of quasi-natural transformations $\sigma\colon F \longrightarrow F'$, $\tau\colon G \longrightarrow G'$. This gives rise to a commutative cube (which is compatible with everything else).

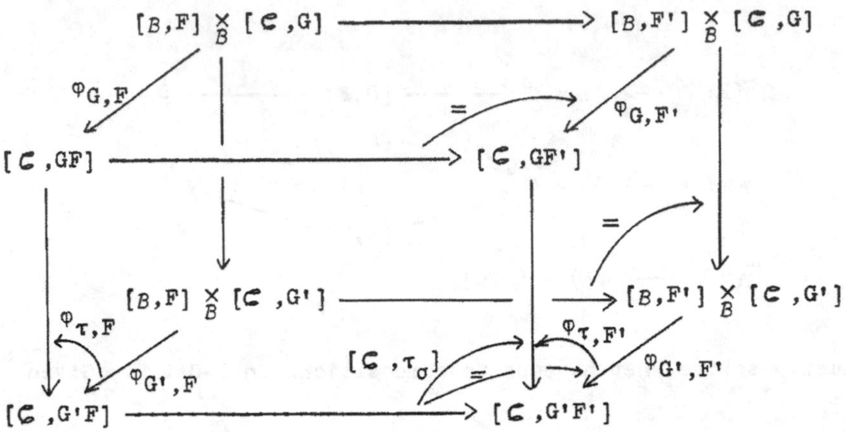

Here $\varphi_{\tau,F}$ is the natural transformation whose component at
(f: B ⟶ FA,   g: C ⟶ GB) is the morphism

$$
\begin{array}{c}
\boxed{ C \xrightarrow{g} GB \xrightarrow{\tau_B} G'B \xrightarrow{G'f} G'FA } \\[4pt]
\downarrow 1 \qquad\qquad\qquad\quad \tau_f \nearrow\quad \downarrow 1 \\[4pt]
\boxed{ C \xrightarrow{g} GB \xrightarrow{Gf} GFA \xrightarrow[\tau_{FA}]{} G'FA }
\end{array}
$$

in [C,G'F], and $\varphi_{F,\sigma} = id$.   This is the structure required by
I,4.25.

The dual situation is given by the quasi-functor

$$\bar{\Phi}\colon {}^{op}2\text{-}Cat^{op}_\otimes \longrightarrow \text{Spans (2-Cat)}$$

where

$$\bar{\Phi}(A) = A$$

$$\bar{\Phi}(F\colon A \longrightarrow B) = B \longleftarrow [F,B] \longrightarrow A$$

$$\bar{\Phi}(\varphi\colon F \longrightarrow G) = B \overset{[F,B]}{\underset{[G,B]}{\lessgtr}} \Big\uparrow [\varphi,B] \quad A$$

$$\Phi(u\colon \varphi \longrightarrow \varphi') = [u,B] .$$

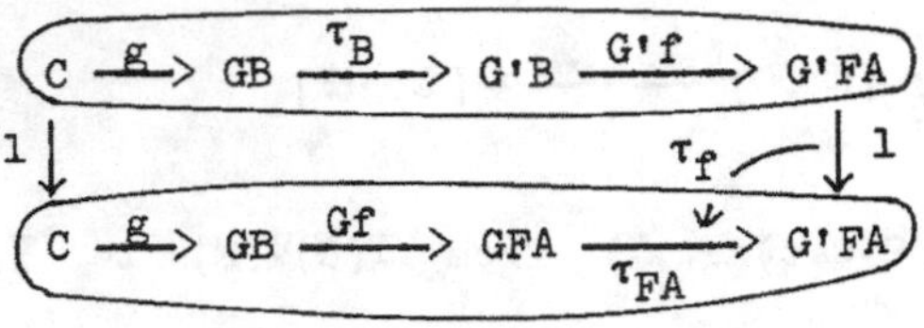

I,7.17. Globalized adjunction morphisms

In this last example, we want to examine the correspondence in I,5.10,ii) from the stand point of 2-$Cat_{\otimes}$-categories and quasi-adjoints. In what follows, $F: \mathcal{A} \longrightarrow \mathcal{B}$ and $U: \mathcal{B} \longrightarrow \mathcal{A}$ are 2-functors, and we consider mainly the correspondence between quasi-natural transformations $\varepsilon: FU \longrightarrow \mathcal{B}$ and 2-functors $T: [\mathcal{A},U] \longrightarrow [F,\mathcal{B}]$ over $\mathcal{A} \times \mathcal{B}$. At first we consider only a very local result, since we cannot describe anything like I,7.16.2 here.

I,7.17.1. Proposition. There exist adjoint functors (between categories)

$$\hat{\Xi} \dashv \Xi: [\mathrm{Fun}(\mathcal{B},\mathcal{B})](FU,\mathcal{B}) \longrightarrow (2\text{-Cat},\mathcal{A} \times \mathcal{B})([\mathcal{A},U],[F,\mathcal{B}]) \;;$$

resp.,

$$\Xi \dashv \check{\Xi}: (2\text{-Cat}, \mathcal{A} \times \mathcal{B})([F,\mathcal{B}],[\mathcal{A},U]) \longrightarrow [\mathrm{Fun}(\mathcal{A},\mathcal{A})](\mathcal{A},UF)$$

such that

i) $\hat{\Xi}\Xi = \mathrm{id}$; resp., $\check{\Xi}\Xi = \mathrm{id}$

ii) If $T: [\mathcal{A},U] \longrightarrow [F,\mathcal{B}]$ is over $\mathcal{A} \times \mathcal{B}$ then the following are equivalent:

a) $\Xi\hat{\Xi}(T) = T$

b) T is a left homomorphism with respect to the change of monoids F_*, in the sense of I,5.10.

c) T applied to a morphism of the form

$$\begin{array}{ccc} A & \xrightarrow{f} & UB \\ {\scriptstyle f}\downarrow & & \downarrow{\scriptstyle UB} \\ UB & \xrightarrow[UB]{} & UB \end{array}$$

is a commutative square. (Resp., for $T'\colon [F,\mathcal{B}] \longrightarrow [\mathcal{A},U]$ over $\mathcal{A} \times \mathcal{B}$; a') $\Xi\check{\Xi}(T') = T'$, b') T' is a right U_*-homomorphism, c') T' applied to a square of the form

$$\begin{array}{ccc} FA & \xrightarrow{FA} & FA \\ {\scriptstyle FA}\downarrow & & \downarrow{\scriptstyle g} \\ FA & \xrightarrow[g]{} & B \end{array}$$

is a commutative square.)

<u>Proof</u>: We treat only the first case. As in I,5.10, ii), Ξ is the functor taking $\varepsilon\colon FU \longrightarrow \mathcal{B}$ into

$$T = \Xi(\varepsilon) = [F,\varepsilon] \circ F_*\colon [\mathcal{A},U] \longrightarrow [F,\mathcal{B}]$$

and taking a modification $u\colon \varepsilon \longrightarrow \varepsilon'$ to $\Xi(u) = [F,u] \circ F_*$. On the other hand, as in I,7.6 and I,5(5.33), $\hat{\Xi}$ is the functor taking $T\colon [\mathcal{A},U] \longrightarrow [F,\mathcal{B}]$ over $\mathcal{A} \times \mathcal{B}$ to

$$\bar{\varepsilon} = \hat{\Xi}(T) = (\mathcal{B} \xrightarrow{j_U} [U,U] \xrightarrow{\bar{T}} [FU,\mathcal{B}].)$$

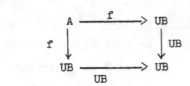

and taking a natural transformation $\varphi\colon T \longrightarrow T'$ over $\mathcal{A} \times \mathcal{B}$ to $\hat{\Xi}(T) = j_U\bar{\varphi}$. Ξ and $\hat{\Xi}$ are clearly functors, and it follows from I,5.2 that $\hat{\Xi}\Xi =$ identity. Thus it sufficies to describe a natural transformation $\eta\colon Id \longrightarrow \Xi\hat{\Xi}$ such that $\eta\Xi = id = \hat{\Xi}\eta$. As in the proof of I,7.16.2, (7.75), one defines $\bar{\gamma}_T = \gamma_T[(1,1,U) \times_U (U,1,1)]$ and $\eta_T = \bar{\gamma}_T(1 \times j_u)(\cong)\colon T \longrightarrow \Xi\hat{\Xi}(T)$ and verifies the equations as is done there.

I,7.17.3 <u>Definition</u>. Let $\mathcal{E}$ consist of the following data :

i) 0-cells are 2-categories $\mathcal{A}, \mathcal{B}$, etc.

ii) a 1-cell from $\mathcal{B}$ to $\mathcal{A}$ is a triple $(F,\alpha,U)\colon \mathcal{B} \longrightarrow \mathcal{A}$ where $F\colon \mathcal{A} \longrightarrow \mathcal{B}$ and $U\colon \mathcal{B} \longrightarrow \mathcal{A}$ are 2-functors and $\alpha\colon FU \longrightarrow \mathcal{B}$ is a quasi-natural transformation.

iii) 2-cells are triples $(\sigma,u,\tau)\colon (F,\alpha,U) \longrightarrow (G,\beta,V)$ where $\sigma\colon G \longrightarrow F$ and $\tau\colon U \longrightarrow V$ are quasi-natural transformations and $u\colon \alpha(\sigma U) \longrightarrow \beta(G\tau)$ is a modification, as indicated

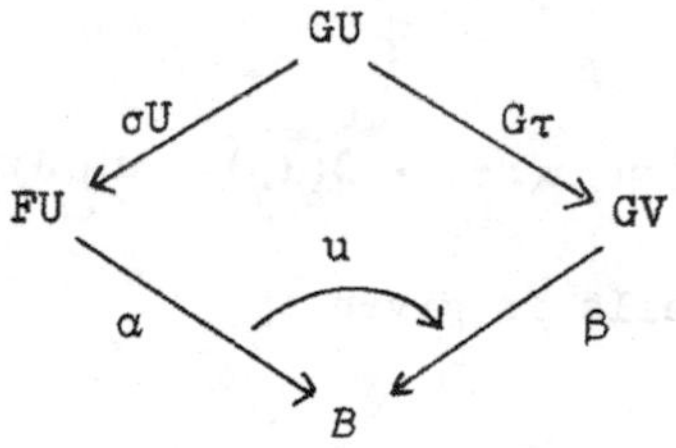

iv) 3-cells are pairs $(s,t)\colon (\sigma,u,\tau) \longrightarrow (\sigma',u',\tau')$ where $s\colon \sigma \longrightarrow \sigma'$ and $t\colon \tau \longrightarrow \tau'$ are modifications such that $u' \cdot (\alpha(sU)) = (\beta(Gt)) \cdot u$.

Compositions are defined as follows:

a) 1-cells $(F,\alpha,U)\colon \mathcal{C} \longrightarrow \mathcal{B}$ and $(H,\gamma,W)\colon \mathcal{B} \longrightarrow \mathcal{A}$ have as composition

$$(FH,\ \gamma(H\alpha W), WU): \mathcal{C} \longrightarrow \mathcal{A}$$

b) The weak composition of 2-cells is indicated by the diagram, where $(\sigma,u,\tau): (F,\alpha,U) \longrightarrow (G,\beta,V)$ and $(\mu,v,\nu): (G,\beta,V) \to (H,\gamma,W)$.

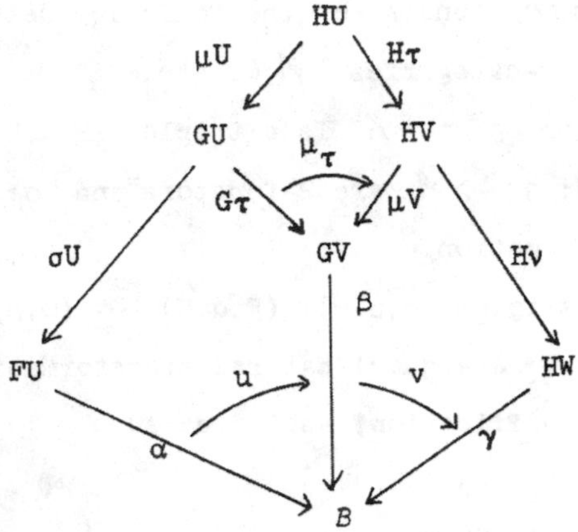

;

i.e., $(\mu,v,\nu) \cdot (\sigma,u,\tau) = (\sigma\mu,\ \nu(H\tau) \cdot \beta(\mu_\tau) \cdot u(\mu U), \nu\tau)$

c) The composition of 3-cells is given by

$$(s',t')(s,t) = (s's,t't)$$

d) The strong composition of 1-cells and 2-cells is given as follows: Consider

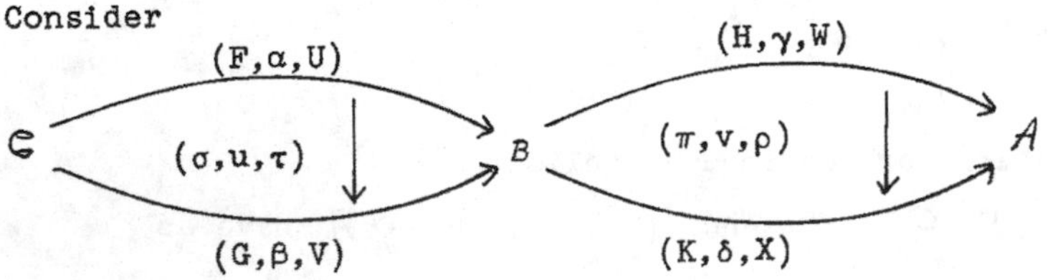

(7.77)

Then

$$(H,\gamma,W)(\sigma,u,\tau) = (\sigma H,\quad u*\gamma,\quad W\tau)$$

$$(\pi,v,\rho)(F,\alpha,U) = (F\pi,\quad \alpha(FvU),\rho U)$$

where

$$u*\gamma = \beta(G\gamma_\tau) \cdot u(G\gamma U) \cdot \alpha(\sigma_{\gamma u})$$

as illustrated.

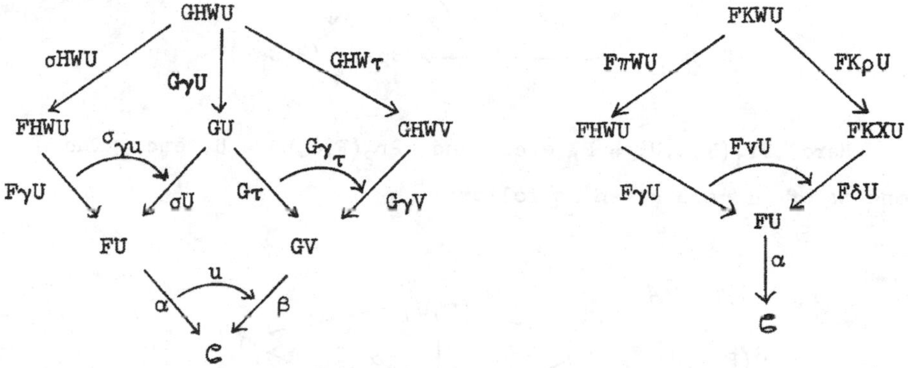

I,7.17.4. __Proposition.__ $\mathcal{E}$ has the structure of a 2-Cat$_\otimes$-category.

__Proof:__ In the situation illustrated in (7.77), there is a diagram

$$(H,\gamma,W)(F,\alpha,U) \xrightarrow{\;\;(H,\gamma,W)(\sigma,u,\tau)\;\;} (H,\gamma,W)(G,\beta,V)$$

$$(FH,\dots,WU) \longrightarrow (GH,\dots,WV)$$

$$(\pi,v,\rho)(F,\alpha,U) \qquad\qquad (\sigma_\pi,\rho_\tau) \qquad\qquad (\pi,v,\rho)(G,\beta,V)$$

$$(FK,\dots,XU) \longrightarrow (GK,\dots,XV)$$

$$(K,\delta,X)(F,\alpha,U) \xrightarrow{\;\;(K,\delta,X)(\sigma,u,\tau)\;\;} (K,\delta,X)(G,\beta,V)$$

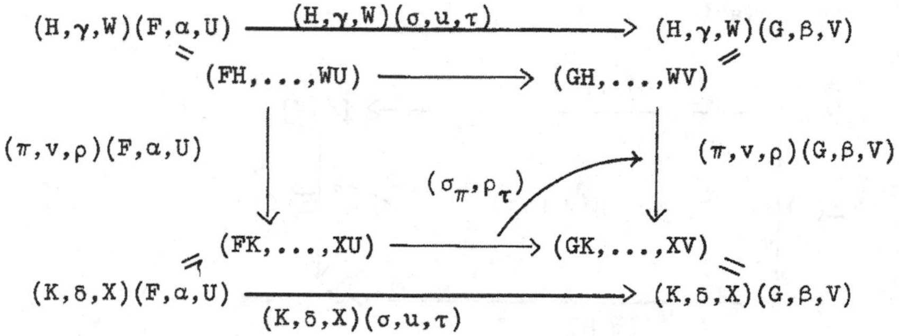

which is the required kind of structure. We omit the lengthy
verification that this satisfies the conditions for a 2-Cat$_\otimes$-category.

I,7.17.5. __Theorem__. There is a quasi-natural transformation between
quasi-functors on 2-Cat$_\otimes$-categories, as indicated:

$$
\begin{array}{ccc}
\mathcal{E} & \xrightarrow{\ \ Pr_1\ \ } & {}^{op}2\text{-Cat}_\otimes{}^{op} \\[2mm]
\scriptstyle Pr_2 \Big\downarrow & \quad \mu \qquad \nearrow & \Big\downarrow \overline{\Phi} \\[2mm]
2\text{-Cat}_\otimes & \xrightarrow[\ \ \Psi\ \ ]{} & \text{Spans (2-Cat)}
\end{array}
$$

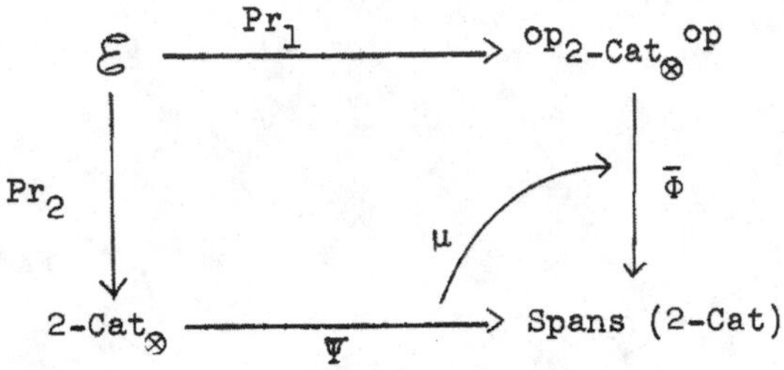

__Proof:__ Here $Pr_1(F,\alpha,U) = F$, etc., and $Pr_2(F,\alpha,U) = U$, etc. The
components of μ are given as follows:

$$\mu_A = 1_A$$

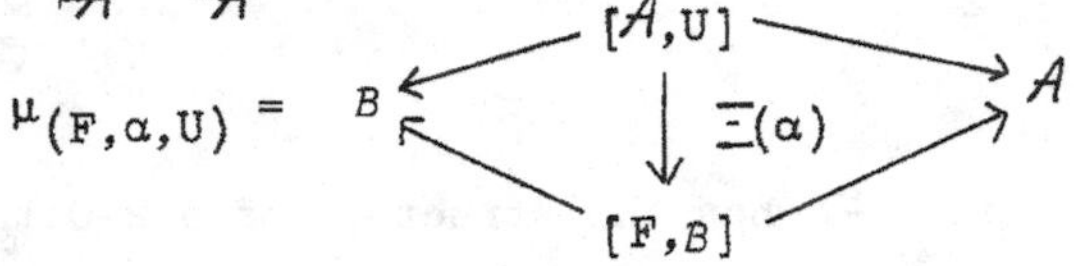

and given $(\sigma,u,\tau): (F,\alpha,U) \longrightarrow (G,\beta,V)$ there is a 2-cell (a Cat-
natural transformation)

$$
\begin{array}{ccc}
[A,U] & \xrightarrow{\ \ [A,\tau]\ \ } & [A,V] \\[2mm]
\scriptstyle \Xi(\alpha)\Big\downarrow & \mu(\sigma,u,\tau) \quad \nearrow & \Big\downarrow \scriptstyle \Xi(\beta) \\[2mm]
[F,B] & \xrightarrow[\ \ [\sigma,B]\ \ ]{} & [G,B]
\end{array}
$$

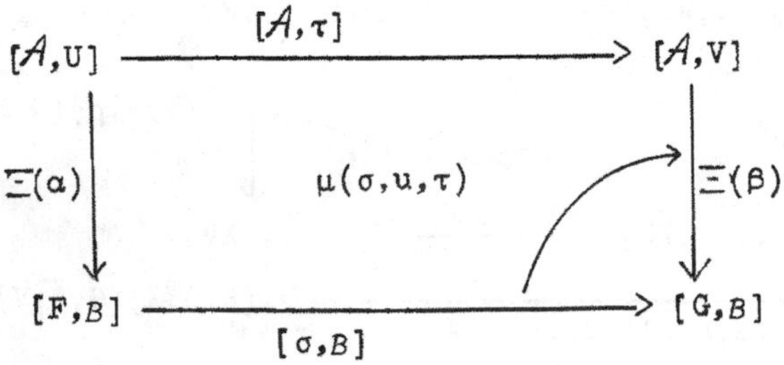

whose component at (f: A ⟶ UB) ∈ [$\mathcal{A}$,U] is the composed morphism in [G,$\mathcal{B}$]

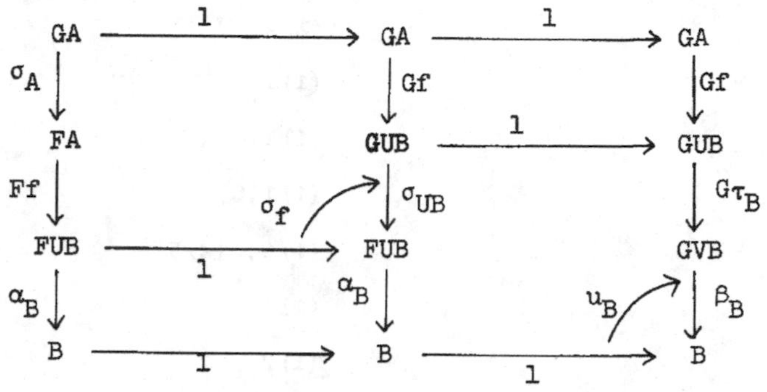

The main step in showing that this is quasi-natural reduces to showing that given (7.77), the cube

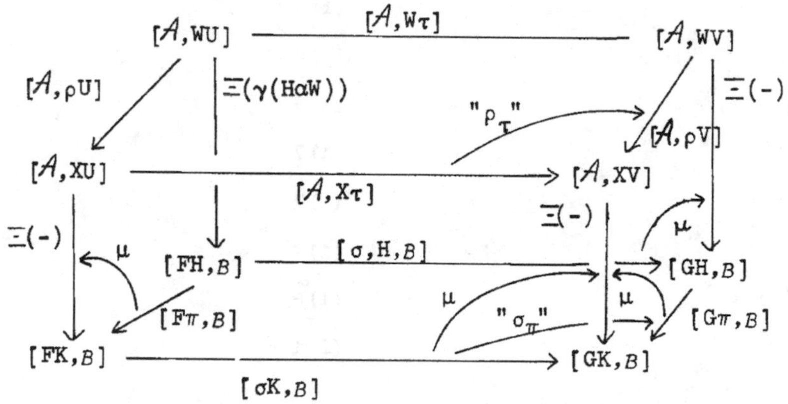

commutes. This follows by an explicite calculation which is too large to include here.

There is a dual treatment of quasi-natural transformations α: $\mathcal{A}$ ⟶ UF. However, since the variances are different there does not seem to be any way to combine the two cases.

Table of symbols

Listed in order of occurrence. (x)y denotes page y in chapter x.

$\longrightarrow\!\!\!\!\!\!\!\!\dashv$	(1)1
$\lvert(-)\rvert$	(1)1
D	(1)1,6
π_o	(1)1,6
Cat	(1)1, (2)5
${}_{\ell}Cat$	(1)1
G	(1)2
$\underline{1}$	(1)3
$\underline{2}$	(1)3
$\underline{3}$	(1)4
$\underline{4}$	(1)4
$\underline{n}$	(1)4
Cat_t	(1)5
$\{\ldots\}$	(1)7
u	(1)7
ℓ	(1)7
$\hat{u}, \check{u}, \hat{\ell}, \check{\ell}$	(1)8
$A(-,-)$	(2)1,6
$\partial_o, \partial_1, \tilde{\partial}_o, \tilde{\partial}_1$	(2)2
$(-)_o$	(2)9
$L\,D$	(2)9
$L\,G$	(2)9
$L\,\pi_o$	(2)9

Index

Note: (x)y denotes page y in chapter x.

References

[BC] Benabou,J. Introduction to Bicategories, Rep. Midw. Cat. Sem. I, Lecture Notes in Mathematics, vol.47, (1967), Springer-Verlag, New York.

[E-K] Eilenberg, S. and Kelly, G.M., Closed Categories, Proc. Conf. on Cat. Alg., La Jolla 1965, Springer-Verlag (1966), p.421-562.

[FCC] Gray, J.W., Fibred and Cofibred Categories, Proc. Conf. on Cat. Alg., La Jolla 1965, Springer-Verlag (1966), p.21-83.

[CCS] Gray, J.W., The Categorical Comprehension Scheme, Category Theory, Homology Theory and their Applications III, Lecture Notes in Mathematics, vol.99 (1969), Springer-Verlag, New York, p.242-312.

[CCFM] Lawvere, F.W., The Category of Categories as a Foundation for Mathematics, Proc. Conf. on Cat. Alg., La Jolla 1965, Springer-Verlag (1966), p.1-20.

Bibliography

[1] Beck, J., Distributive Laws, Seminar on Triples and Categorical Homology Theory, Lecture Notes in Mathematics, vol.80 (1969), Springer-Verlag, New York, p.119-140.

[2] Beck, J., Introduction, Seminar on Triples and Categorical Homology Theory, Lecture Notes in Mathematics, vol.80, (1969), Springer-Verlag, New York, p.1-6bis.

[3] Benabou, J., Catégories avec Multiplication, C.R.Acad. Sci., Paris 256 (1963), p.1887-1890.

[4] Benabou, J., Various unpublished lectures in Oberwolfach, Chicago, Rome, and New Orleans.

[5] Blattner, R.J., Review of Mackey [33], Math. Reviews 29 (1965), # 2325.

[6] Brinkmann, H.-B., and Puppe, D., Abelsche und exakte Kategorien, Korrespondenzen, Lecture Notes in Mathematics, vol.96 (1969), Springer-Verlag, New York.

[7] Bunge, M., Bifibration Induced Adjoint Pairs, Rep. Midw. Cat. Sem. V, Lecture Notes in Mathematics, vol.195 (1971), Springer-Verlag, New York, p.70-122.

[8] Day, B.J., and Kelly, G.M., Enriched Functor Categories, Rep. Midw. Cat. Sem. III, Lecture Notes in Mathematics, vol.106, (1969), Springer-Verlag, New York, p.178-191.

[9] Dubuc, E., Kan Extensions in Enriched Category Theory, Lecture Notes in Mathematics, vol.145, (1970), Springer-Verlag, New York.

[10] Duskin, J., Preliminary Remarks on Groups, preprint (1969).

[11] Duskin, J., Non Abelian Triple Cohomology, preprint (1969).

[12] Ehresmann, C., Gattungen von lokalen Strukturen, Jber. Deutsch. Math. Verein, 60 (1958), p.49-77.

[13] Ehresmann, C., Catégories structurées généralisées, Cahiers Top. et Géom. dif., X, 1 (1968), p.139-168.

[14] Bastiani, A., and Ehresmann, C., Catégories de foncteurs structurés, Cahiers Top. et Géom. dif., XI, 3 (1969), p.329-384.

[15] Ehresmann, C., Catégories et Structures, Dunod, Paris (1965).

[16] Gabriel, P., Construction de Preschemas Quotient, Exp. V, Schémas en Groupes I, Lecture Notes in Mathematics, vol. 151 (1970), Springer-Verlag, New York.

[17] Gabriel, P., and Ulmer, F., Lokal praesentierbare Kategorien, Lecture Notes in Mathematics, vol.221 (1971), Springer-Verlag, New York.

[18] Giraud, J., Méthode de la Descente, Bull. Soc. Math. France, Mémoire 2, VIII + 15 Op., (1964).

[19] Giraud, J., Cohomologie non abélienne, Die Grundlehren der Math. Wissenschaften, Bd. 179, Springer Verlag, New York, 1971.

[20] Godement, R., Topologie Algébrique et Théorie des Faisceaux, Hermann, Paris (1958).

[21] Gray, J.W., The 2-Adjointness of the Fibred Category Construction. Symposia Mathematica IV, Istituto Nazionale di Alta Matematica, Academic Press, London and New York (1970), p.457-492.

[22] Gray, J.W., The Meeting of the Midwest Category Seminar in Zurich, Rep. Midw. Cat. Sem. V, Lecture Notes in Mathematics, vol.195 (1971), Springer-Verlag, New York, p.248-255.

[23] Grothendieck, A., Catégories Fibrées et Descente, Séminaire de Géometrie Algébrique, Institut des Hautes Etudes Scientifiques, Paris (1961).

[24] Kelly, G.M., Adjunction for Enriched Categories, Rep. Midw. Cat. Sem. III, Lecture Notes in Mathematics, vol.106 (1969), Springer-Verlag, New York, p.166-177.

[25] Kleisli, H., Every Standard Construction is Induced by a Pair of Adjoint Functors, Proc. Amer. Math. Soc. 16 (1965), p.544-546.

[26] Lair, C., Construction d'Esquisses-transformations Naturelles Généralisées, Esquisses Mathématiques 2, Dept. de Math., Tours 45-55, 9, Quai Saint Bernard, Paris 5ème.

[27] Lambek, J., Subequalizers, Canad. Math. Bull. 13 (1970), p.337-349.

[28] Lambek, J., Deductive Systems and Categories (II), Category Theory, Homology Theory and their Applications I, Lecture Notes in Mathematics, vol.86 (1969), Springer-Verlag, New York, p.76-122.

[29] Lawvere, F.W., Functorial Semantics of Algebraic Theories, Proc. Nat. Acad. Sci. 50 (1963), p.869-872.

[30] Lawvere, F.W., Ordinal Sums and Equational Doctrines, Seminar on Triples and Categorical Homology Theory, Lecture Notes in Mathematics, vol.80 (1969), Springer-Verlag, New York, p.141-155.

[31] Linton, F.E.J., Autonomous Categories and Duality of Functors, J. Alg. 2 (1965), p.315-349.

[32] Mackey, G.W., Ergodic Theory, Group Theory, and Differential Geometry, Proc. Nat. Acad. Sci. U.S., 50 (1963), p.1184-1191.

[33] Mackey, G.W., Ergodic Theory and Virtual Groups, Math. Ann. (1966), p.187-207.

[34] Mackey, G.W., Virtual Groups, Topological Dynamics, Symposium 1967, Benjamin, New York (1968), p.335-364.

[35] MacLane, S., Coherence and Canonical Maps, Symposia Mathematica IV, Istituto di Alta Matematica, Academic Press, London and New York, (1970), p.231-242.

[36] Maranda, J.M., Formal Categories, Can. J. of Math. 17 (1965), p.758-801.

[37] Palmquist, P.H., The Double Category of Adjoint Squares, Rep. Midw. Cat. Sem. V, Lecture Notes in Mathematics, vol.195 (1971), Springer-Verlag, New York, p.123-153.

[38] Segal, G., Classifying Spaces and Spectral Sequences, Publications Mathématiques No.34, Institut des Hautes Etudes Scientifiques (1969).

[39] Street, R., The Formal Theory of Monoids, preprint, Macquarie University, North Ryde, N.S.W., Australia, (1971).

[40] Wolff, H.E., $\mathfrak{U}$-Localizations and $\mathfrak{U}$-Triples, Thesis, Univ. of Ill., Urbana, Ill. (1970).

Vol. 215: P. Antonelli, D. Burghelea and P. J. Kahn, The Concordance-Homotopy Groups of Geometric Automorphism Groups. X, 140 pages. 1971. DM 16,–

Vol. 216: H. Maaß, Siegel's Modular Forms and Dirichlet Series. VII, 328 pages. 1971. DM 20,–

Vol. 217: T. J. Jech, Lectures in Set Theory with Particular Emphasis on the Method of Forcing. V, 137 pages. 1971. DM 16,–

Vol. 218: C. P. Schnorr, Zufälligkeit und Wahrscheinlichkeit. IV, 212 Seiten. 1971. DM 20,–

Vol. 219: N. L. Alling and N. Greenleaf, Foundations of the Theory of Klein Surfaces. IX, 117 pages. 1971. DM 16,–

Vol. 220: W. A. Coppel, Disconjugacy. V, 148 pages. 1971. DM 16,–

Vol. 221: P. Gabriel und F. Ulmer, Lokal präsentierbare Kategorien. V, 200 Seiten. 1971. DM 18,–

Vol. 222: C. Meghea, Compactification des Espaces Harmoniques. III, 108 pages. 1971. DM 16,–

Vol. 223: U. Felgner, Models of ZF-Set Theory. VI, 173 pages. 1971. DM 16,–

Vol. 224: Revêtements Etales et Groupe Fondamental. (SGA 1). Dirigé par A. Grothendieck XXII, 447 pages. 1971. DM 30,–

Vol. 225: Théorie des Intersections et Théorème de Riemann-Roch. (SGA 6). Dirigé par P. Berthelot, A. Grothendieck et L. Illusie. XII, 700 pages. 1971. DM 40,–

Vol. 226: Seminar on Potential Theory, II. Edited by H. Bauer. IV, 170 pages. 1971. DM 18,–

Vol. 227: H. L. Montgomery, Topics in Multiplicative Number Theory. IX, 178 pages. 1971. DM 18,–

Vol. 228: Conference on Applications of Numerical Analysis. Edited by J. Ll. Morris. X, 358 pages. 1971. DM 26,–

Vol. 229: J. Väisälä, Lectures on n-Dimensional Quasiconformal Mappings. XIV, 144 pages. 1971. DM 16,–

Vol. 230: L. Waelbroeck, Topological Vector Spaces and Algebras. VII, 158 pages. 1971. DM 16,–

Vol. 231: H. Reiter, L^1-Algebras and Segal Algebras. XI, 113 pages. 1971. DM 16,–

Vol. 232: T. H. Ganelius, Tauberian Remainder Theorems. VI, 75 pages. 1971. DM 16,–

Vol. 233: C. P. Tsokos and W. J. Padgett. Random Integral Equations with Applications to stochastic Systems. VII, 174 pages. 1971. DM 18,–

Vol. 234: A. Andreotti and W. Stoll. Analytic and Algebraic Dependence of Meromorphic Functions. III, 390 pages. 1971. DM 26,–

Vol. 235: Global Differentiable Dynamics. Edited by O. Hájek, A. J. Lohwater, and R. McCann. X, 140 pages. 1971. DM 16,–

Vol. 236: M. Barr, P. A. Grillet, and D. H. van Osdol. Exact Categories and Categories of Sheaves. VII, 239 pages. 1971. DM 20,–

Vol. 237: B. Stenström, Rings and Modules of Quotients. VII, 136 pages. 1971. DM 16,–

Vol. 238: Der kanonische Modul eines Cohen-Macaulay-Rings. Herausgegeben von Jürgen Herzog und Ernst Kunz. VI, 103 Seiten. 1971. DM 16,–

Vol. 239: L. Illusie, Complexe Cotangent et Déformations I. XV, 355 pages. 1971. DM 26,–

Vol. 240: A. Kerber, Representations of Permutation Groups I. VII, 192 pages. 1971. DM 18,–

Vol. 241: S. Kaneyuki, Homogeneous Bounded Domains and Siegel Domains. V, 89 pages. 1971. DM 16,–

Vol. 242: R. R. Coifman et G. Weiss, Analyse Harmonique Non-Commutative sur Certains Espaces. V, 160 pages. 1971. DM 16,–

Vol. 243: Japan-United States Seminar on Ordinary Differential and Functional Equations. Edited by M. Urabe. VIII, 332 pages. 1971. DM 26,–

Vol. 244: Séminaire Bourbaki – vol. 1970/71. Exposés 382–399. IV, 356 pages. 1971. DM 26,–

Vol. 245: D. E. Cohen, Groups of Cohomological Dimension One. V, 99 pages. 1972. DM 16,–

Vol. 246: Lectures on Rings and Modules. Tulane University Ring and Operator Theory Year, 1970–1971. Volume I. X, 661 pages. 1972. DM 40,–

Vol. 247: Lectures on Operator Algebras. Tulane University Ring and Operator Theory Year, 1970–1971. Volume II. XI, 786 pages. 1972. DM 40,–

Vol. 248: Lectures on the Applications of Sheaves to Ring Theory. Tulane University Ring and Operator Theory Year, 1970–1971. Volume III. VIII, 315 pages. 1971. DM 26,–

Vol. 249: Symposium on Algebraic Topology. Edited by P. J. Hilton. VII, 111 pages. 1971. DM 16,–

Vol. 250: B. Jónsson, Topics in Universal Algebra. VI, 220 pages. 1972. DM 20,–

Vol. 251: The Theory of Arithmetic Functions. Edited by A. A. Gioia and D. L. Goldsmith VI, 287 pages. 1972. DM 24,–

Vol. 252: D. A. Stone, Stratified Polyhedra. IX, 193 pages. 1972. DM 18,–

Vol. 253: V. Komkov, Optimal Control Theory for the Damping of Vibrations of Simple Elastic Systems. V, 240 pages. 1972. DM 20,–

Vol. 254: C. U. Jensen, Les Foncteurs Dérivés de $\varprojlim$ et leurs Applications en Théorie des Modules. V, 103 pages. 1972. DM 16,–

Vol. 255: Conference in Mathematical Logic – London '70. Edited by W. Hodges. VIII, 351 pages. 1972. DM 26,–

Vol. 256: C. A. Berenstein and M. A. Dostal, Analytically Uniform Spaces and their Applications to Convolution Equations. VII, 130 pages. 1972. DM 16,–

Vol. 257: R. B. Holmes, A Course on Optimization and Best Approximation. VIII, 233 pages. 1972. DM 20,–

Vol. 258: Séminaire de Probabilités VI. Edited by P. A. Meyer. VI, 253 pages. 1972. DM 22,–

Vol. 259: N. Moulis, Structures de Fredholm sur les Variétés Hilbertiennes. V, 123 pages. 1972. DM 16,–

Vol. 260: R. Godement and H. Jacquet, Zeta Functions of Simple Algebras. IX, 188 pages. 1972. DM 18,–

Vol. 261: A. Guichardet, Symmetric Hilbert Spaces and Related Topics. V, 197 pages. 1972. DM 18,–

Vol. 262: H. G. Zimmer, Computational Problems, Methods, and Results in Algebraic Number Theory. V, 103 pages. 1972. DM 16,–

Vol. 263: T. Parthasarathy, Selection Theorems and their Applications. VII, 101 pages. 1972. DM 16,–

Vol. 264: W. Messing, The Crystals Associated to Barsotti-Tate Groups: With Applications to Abelian Schemes. III, 190 pages. 1972. DM 18,–

Vol. 265: N. Saavedra Rivano, Catégories Tannakiennes. II, 418 pages. 1972. DM 26,–

Vol. 266: Conference on Harmonic Analysis. Edited by D. Gulick and R. L. Lipsman. VI, 323 pages. 1972. DM 24,–

Vol. 267: Numerische Lösung nichtlinearer partieller Differential- und Integro-Differentialgleichungen. Herausgegeben von R. Ansorge und W. Törnig, VI, 339 Seiten. 1972. DM 26,–

Vol. 268: C. G. Simader, On Dirichlet's Boundary Value Problem. IV, 238 pages. 1972. DM 20,–

Vol. 269: Théorie des Topos et Cohomologie Etale des Schémas. (SGA 4). Dirigé par M. Artin, A. Grothendieck et J. L. Verdier. XIX, 525 pages. 1972. DM 50,–

Vol. 270: Théorie des Topos et Cohomologie Etale des Schémas. Tome 2. (SGA 4). Dirigé par M. Artin, A. Grothendieck et J. L. Verdier. V, 418 pages. 1972. DM 50,–

Vol. 271: J. P. May, The Geometry of Iterated Loop Spaces. IX, 175 pages. 1972. DM 18,–

Vol. 272: K. R. Parthasarathy and K. Schmidt, Positive Definite Kernels, Continuous Tensor Products, and Central Limit Theorems of Probability Theory. VI, 107 pages. 1972. DM 16,–

Vol. 273: U. Seip, Kompakt erzeugte Vektorräume und Analysis. IX, 119 Seiten. 1972. DM 16,–

Vol. 274: Toposes, Algebraic Geometry and Logic. Edited by. F. W. Lawvere. VI, 189 pages. 1972. DM 18,–

Vol. 275: Séminaire Pierre Lelong (Analyse) Année 1970–1971. VI, 181 pages. 1972. DM 18,–

Vol. 276: A. Borel, Représentations de Groupes Localement Compacts. V, 98 pages. 1972. DM 16,–

Vol. 277: Séminaire Banach. Edité par C. Houzel. VII, 229 pages. 1972. DM 20,–

Vol. 278: H. Jacquet, Automorphic Forms on GL(2). Part II. XIII, 142 pages. 1972. DM 16,–

Vol. 279: R. Bott, S. Gitler and I. M. James, Lectures on Algebraic and Differential Topology. V, 174 pages. 1972. DM 18,–

Vol. 280: Conference on the Theory of Ordinary and Partial Differential Equations. Edited by W. N. Everitt and B. D. Sleeman. XV, 367 pages. 1972. DM 26,–

Vol. 281: Coherence in Categories. Edited by S. Mac Lane. VII, 235 pages. 1972. DM 20,–

Vol. 282: W. Klingenberg und P. Flaschel, Riemannsche Hilbertmannigfaltigkeiten. Periodische Geodätische. VII, 211 Seiten. 1972. DM 20,–

Vol. 283: L. Illusie, Complexe Cotangent et Déformations II. VII, 304 pages. 1972. DM 24,–

Vol. 284: P. A. Meyer, Martingales and Stochastic Integrals I. VI, 89 pages. 1972. DM 16,–

Vol. 285: P. de la Harpe, Classical Banach-Lie Algebras and Banach-Lie Groups of Operators in Hilbert Space. III, 160 pages. 1972. DM 16,–

Vol. 286: S. Murakami, On Automorphisms of Siegel Domains. V, 95 pages. 1972. DM 16,–

Vol. 287: Hyperfunctions and Pseudo-Differential Equations. Edited by H. Komatsu. VII, 529 pages. 1973. DM 36,–

Vol. 288: Groupes de Monodromie en Géométrie Algébrique. (SGA 7 I). Dirigé par A. Grothendieck. IX, 523 pages. 1972. DM 50,–

Vol. 289: B. Fuglede, Finely Harmonic Functions. III, 188. 1972. DM 18,–

Vol. 290: D. B. Zagier, Equivariant Pontrjagin Classes and Applications to Orbit Spaces. IX, 130 pages. 1972. DM 16,–

Vol. 291: P. Orlik, Seifert Manifolds. VIII, 155 pages. 1972. DM 16,–

Vol. 292: W. D. Wallis, A. P. Street and J. S. Wallis, Combinatorics: Room Squares, Sum-Free Sets, Hadamard Matrices. V, 508 pages. 1972. DM 50,–

Vol. 293: R. A. DeVore, The Approximation of Continuous Functions by Positive Linear Operators. VIII, 289 pages. 1972. DM 24,–

Vol. 294: Stability of Stochastic Dynamical Systems. Edited by R. F. Curtain. IX, 332 pages. 1972. DM 26,–

Vol. 295: C. Dellacherie, Ensembles Analytiques, Capacités, Mesures de Hausdorff. XII, 123 pages. 1972. DM 16,–

Vol. 296: Probability and Information Theory II. Edited by M. Behara, K. Krickeberg and J. Wolfowitz. V, 223 pages. 1973. DM 20,–

Vol. 297: J. Garnett, Analytic Capacity and Measure. IV, 138 pages. 1972. DM 16,–

Vol. 298: Proceedings of the Second Conference on Compact Transformation Groups. Part 1. XIII, 453 pages. 1972. DM 32,–

Vol. 299: Proceedings of the Second Conference on Compact Transformation Groups. Part 2. XIV, 327 pages. 1972. DM 26,–

Vol. 300: P. Eymard, Moyennes Invariantes et Représentations Unitaires. II. 113 pages. 1972. DM 16,–

Vol. 301: F. Pittnauer, Vorlesungen über asymptotische Reihen. VI, 186 Seiten. 1972. DM 18,–

Vol. 302: M. Demazure, Lectures on p-Divisible Groups. V, 98 pages. 1972. DM 16,–

Vol. 303: Graph Theory and Applications. Edited by Y. Alavi, D. R. Lick and A. T. White. IX, 329 pages. 1972. DM 26,–

Vol. 304: A. K. Bousfield and D. M. Kan, Homotopy Limits, Completions and Localizations. V, 348 pages. 1972. DM 26,–

Vol. 305: Théorie des Topos et Cohomologie Etale des Schémas. Tome 3. (SGA 4). Dirigé par M. Artin, A. Grothendieck et J. L. Verdier. VI, 640 pages. 1973. DM 50,–

Vol. 306: H. Luckhardt, Extensional Gödel Functional Interpretation. VI, 161 pages. 1973. DM 18,–

Vol. 307: J. L. Bretagnolle, S. D. Chatterji et P.-A. Meyer, Ecole d'été de Probabilités: Processus Stochastiques. VI, 198 pages. 1973. DM 20,–

Vol. 308: D. Knutson, λ-Rings and the Representation Theory of the Symmetric Group. IV, 203 pages. 1973. DM 20,–

Vol. 309: D. H. Sattinger, Topics in Stability and Bifurcation Theory. VI, 190 pages. 1973. DM 18,–

Vol. 310: B. Iversen, Generic Local Structure of the Morphisms in Commutative Algebra. IV, 108 pages. 1973. DM 16,–

Vol. 311: Conference on Commutative Algebra. Edited by J. W. Brewer and E. A. Rutter. VII, 251 pages. 1973. DM 22,–

Vol. 312: Symposium on Ordinary Differential Equations. Edited by W. A. Harris, Jr. and Y. Sibuya. VIII, 204 pages. 1973. DM 22,–

Vol. 313: K. Jörgens and J. Weidmann, Spectral Properties of Hamiltonian Operators. III, 140 pages. 1973. DM 16,–

Vol. 314: M. Deuring, Lectures on the Theory of Algebraic Functions of One Variable. VI, 151 pages. 1973. DM 16,–

Vol. 315: K. Bichteler, Integration Theory (with Special Attention to Vector Measures). VI, 357 pages. 1973. DM 26,–

Vol. 316: Symposium on Non-Well-Posed Problems and Logarithmic Convexity. Edited by R. J. Knops. V, 176 pages. 1973. DM 18,–

Vol. 317: Séminaire Bourbaki – vol. 1971/72. Exposés 400–417. IV, 361 pages. 1973. DM 26,–

Vol. 318: Recent Advances in Topological Dynamics. Edited by A. Beck, VIII, 285 pages. 1973. DM 24,–

Vol. 319: Conference on Group Theory. Edited by R. W. Gatterdam and K. W. Weston. V, 188 pages. 1973. DM 18,–

Vol. 320: Modular Functions of One Variable I. Edited by W. Kuyk. V, 195 pages. 1973. DM 18,–

Vol. 321: Séminaire de Probabilités VII. Edité par P. A. Meyer. VI, 322 pages. 1973. DM 26,–

Vol. 322: Nonlinear Problems in the Physical Sciences and Biology. Edited by I. Stakgold, D. D. Joseph and D. H. Sattinger. VIII, 357 pages. 1973. DM 26,–

Vol. 323: J. L. Lions, Perturbations Singulières dans les Problèmes aux Limites et en Contrôle Optimal. XII, 645 pages. 1973. DM 42,–

Vol. 324: K. Kreith, Oscillation Theory. VI, 109 pages. 1973. DM 16,–

Vol. 325: Ch.-Ch. Chou, La Transformation de Fourier Complexe et L'Equation de Convolution. IX, 137 pages. 1973. DM 16,–

Vol. 326: A. Robert, Elliptic Curves. VIII, 264 pages. 1973. DM 22,–

Vol. 327: E. Matlis, 1-Dimensional Cohen-Macaulay Rings. XII, 157 pages. 1973. DM 18,–

Vol. 328: J. R. Büchi and D. Siefkes, The Monadic Second Orde Theory of All Countable Ordinals. VI, 217 pages. 1973. DM 20,–

Vol. 329: W. Trebels, Multipliers for (C, α)-Bounded Fourier Expan sions in Banach Spaces and Approximation Theory. VII, 103 pages 1973. DM 16,–

Vol. 330: Proceedings of the Second Japan-USSR Symposium o Probability Theory. Edited by G. Maruyama and Yu. V. Prokhorov VI, 550 pages. 1973. DM 36,–

Vol. 331: Summer School on Topological Vector Spaces. Edited b L. Waelbroeck. VI, 226 pages. 1973. DM 20,–

Vol. 332: Séminaire Pierre Lelong (Analyse) Année 1971-1972. V, 13 pages. 1973. DM 16,–

Vol. 333: Numerische, insbesondere approximationstheoretische Be handlung von Funktionalgleichungen. Herausgegeben von R. An sorge und W. Törnig. VI, 296 Seiten. 1973. DM 24,–

Vol. 334: F. Schweiger, The Metrical Theory of Jacobi-Perron Algo rithm. V, 111 pages. 1973. DM 16,–

Vol. 335: H. Huck, R. Roitzsch, U. Simon, W. Vortisch, R. Walden, E Wegner und W. Wendland, Beweismethoden der Differentialgeome trie im Großen. IX, 159 Seiten. 1973. DM 18,–

Vol. 336: L'Analyse Harmonique dans le Domaine Complexe. Edit par E. J. Akutowicz. VIII, 169 pages. 1973. DM 18,–

Vol. 337: Cambridge Summer School in Mathematical Logic. Edite by A. R. D. Mathias and H. Rogers. IX, 660 pages. 1973. DM 42,

Vol: 338: J. Lindenstrauss and L. Tzafriri, Classical Banach Space IX, 243 pages. 1973. DM 22,–

Vol. 339: G. Kempf, F. Knudsen, D. Mumford and B. Saint-Dona Toroidal Embeddings I. VIII, 209 pages. 1973. DM 20,–

Vol. 340: Groupes de Monodromie en Géométrie Algébriqu (SGA 7 II). Par P. Deligne et N. Katz. X, 438 pages. 1973. DM 40,

Vol. 341: Algebraic K-Theory I, Higher K-Theories. Edited b H. Bass. XV, 335 pages. 1973. DM 26,–

Vol. 342: Algebraic K-Theory II, "Classical" Algebraic K-Theor and Connections with Arithmetic. Edited by H. Bass. XV, 52 pages. 1973. DM 36,–